FUNCTIONAL GROUP NAME	FUNCTIONAL GROUP	EXAMPLE	USE OR OCCURRENCE OF EXAMPLE

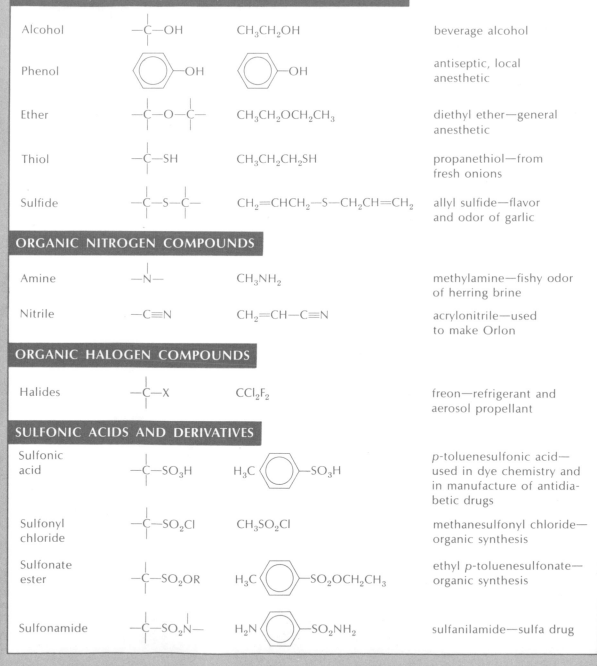

ALCOHOLS, PHENOLS, ETHERS, AND SULFUR ANALOGUES

Alcohol	—C—OH	CH_3CH_2OH	beverage alcohol
Phenol	⬡—OH	⬡—OH	antiseptic, local anesthetic
Ether	—C—O—C—	$CH_3CH_2OCH_2CH_3$	diethyl ether—general anesthetic
Thiol	—C—SH	$CH_3CH_2CH_2SH$	propanethiol—from fresh onions
Sulfide	—C—S—C—	$CH_2{=}CHCH_2—S—CH_2CH{=}CH_2$	allyl sulfide—flavor and odor of garlic

ORGANIC NITROGEN COMPOUNDS

Amine	—N—	CH_3NH_2	methylamine—fishy odor of herring brine
Nitrile	—C≡N	$CH_2{=}CH—C{\equiv}N$	acrylonitrile—used to make Orlon

ORGANIC HALOGEN COMPOUNDS

Halides	—C—X	CCl_2F_2	freon—refrigerant and aerosol propellant

SULFONIC ACIDS AND DERIVATIVES

Sulfonic acid	—C—SO_3H	H_3C—⬡—SO_3H	p-toluenesulfonic acid—used in dye chemistry and in manufacture of antidiabetic drugs
Sulfonyl chloride	—C—SO_2Cl	CH_3SO_2Cl	methanesulfonyl chloride—organic synthesis
Sulfonate ester	—C—SO_2OR	H_3C—⬡—$SO_2OCH_2CH_3$	ethyl p-toluenesulfonate—organic synthesis
Sulfonamide	—C—SO_2N—	H_2N—⬡—SO_2NH_2	sulfanilamide—sulfa drug

ORGANIC CHEMISTRY

Second Edition

ORGANIC CHEMISTRY

A Brief Survey of Concepts and Applications

Philip S. Bailey, Jr., and Christina A. Bailey
California Polytechnic State University
San Luis Obispo

ALLYN AND BACON, INC.
Boston London Sydney Toronto

Production Editor: Mary Hill

Library of Congress Cataloging in Publication Data

Bailey, Philip S
     Organic chemistry.
     Includes index.
     1. Chemistry, Organic.   I. Bailey, Christina A., joint author.
     II. Title. [DNLM: 1. Chemistry, Organic.   QD251.2 B155o]
QD251.2.B34   1981   547   80-27136

ISBN 0-205-07233-X

**Permission for the publication herein of Sadtler Standard
Spectra® has been granted, and all rights are reserved, by
Sadtler Research Laboratories, Division of Bio-Rad Laboratories, Inc. Other spectra are reprinted by permission
of Aldrich Chemical Company, Inc. and of Varian Associates.**

To Karl, Jeni, Kristi, and Michael

Contents

PREFACE xi

INTRODUCTION xv

BONDING IN ORGANIC COMPOUNDS 1

1
1.1 Elements and Compounds; Atoms and Molecules 1
1.2 Electron Configuration and the Periodic Table 2
1.3 Ionic Bonding, Electronegativity, and the Periodic Table 6
1.4 Covalent Bonding 9
1.5 Molecular-Orbital Approach to Covalent Bonding 12
1.6 Polar Covalent Bonds 19
1.7 Lewis Acids and Bases 22

STRUCTURE AND ISOMERISM 26

2
2.1 Chemical Formulas 26
2.2 Skeletal Isomerism 28
2.3 Positional Isomerism 30
2.4 Functional Isomerism 31
2.5 Units of Unsaturation (Multiple Bonds) 33
2.6 Conformational Isomerism 34
2.7 Geometric Isomerism 36
Panel 1: Geometric Isomerism and Vision 37
2.8 Cycloalkanes 38
Panel 2: Diamond and Graphite 43

NONAROMATIC HYDROCARBONS I 48

3
3.1 Classes of Hydrocarbons 48
3.2 IUPAC Nomenclature of Nonaromatic Hydrocarbons 49
3.3 Common Nomenclature of Nonaromatic Hydrocarbons 55
3.4 Hydrocarbons: Relation of Structure and Physical Properties 55
3.5 A Prelude to Organic Reactions 58
3.6 Halogenation of Alkanes: Chlorination and Bromination 60
3.7 Preparation of Alkenes and Alkynes; Elimination Reactions 63
Panel 3: Petroleum and Energy 66

NONAROMATIC HYDROCARBONS II   74

4
4.1 Addition Reactions of Alkenes and Alkynes   74
4.2 Addition Polymers   81
Panel 4: Commercial Applications of Addition Polymers   84
4.3 Natural and Synthetic Rubber—Resonance Structures   87
Panel 5: Two Significant Events in the Development of the Rubber
*   Industry   90*
4.4 Hydroboration   92
4.5 Oxidation of Alkenes   93
4.6 Acidity of Terminal Alkynes   94
Panel 6: Terpenes   95

AROMATIC HYDROCARBONS   104

5
5.1 Introduction to Aromatic Compounds   104
5.2 Benzene: Structure and Bonding   105
5.3 Aromaticity: Structural and Bonding Requirements   107
5.4 Nomenclature of Aromatic Compounds   111
5.5 Uses of Aromatic Compounds   114
5.6 Electrophilic Aromatic Substitution   114
5.7 Oxidation of Alkylbenzenes   123
Panel 7: Gasoline   124
Panel 8: Cancer and Carcinogens   128

OPTICAL ISOMERISM   137

6
6.1 Isomerism   137
6.2 Plane-Polarized Light and the Polarimeter   137
6.3 Structure of Optically Active Compounds   140
6.4 Optical Isomers with Two Chiral Carbons   145
6.5 Optical Isomerism in Cyclic Compounds   148
6.6 Specification of Configuration   148
6.7 Resolution of Enantiomers   153
Panel 9: Optical Isomerism in the Biological World   154

ORGANIC HALOGEN COMPOUNDS   159

7
7.1 Structure, Nomenclature, and Physical Properties   159
7.2 Preparations of Organic Halogen Compounds   161
7.3 Uses and Occurrences of Organic Halogen Compounds   161
7.4 Nucleophilic Substitution   164
7.5 Elimination Reactions of Alkyl Halides   169
Panel 10: Insecticides   171

ALCOHOLS, PHENOLS, AND ETHERS   178

8
8.1 Structure and Nomenclature   178
8.2 Preparations of Alcohols and Ethers   181

8.3 Physical Properties—Hydrogen Bonding 182
8.4 Uses of Alcohols and Ethers 185
Panel 11: Beverage Alcohol 185
8.5 Phenols 188
8.6 Reactions of Alcohols, Phenols, and Ethers 191
8.7 Epoxides 197
8.8 Sulfur Analogues of Alcohols and Ethers 198
Panel 12: Insect Pheromones 199

ALDEHYDES AND KETONES 206

9

9.1 Structure of Aldehydes and Ketones 206
9.2 Nomenclature of Aldehydes and Ketones 207
9.3 Some Preparations of Aldehydes and Ketones 210
9.4 Reactions of Aldehydes and Ketones—Oxidation of Aldehydes 210
9.5 Reactions of Aldehydes and Ketones—Addition 212
9.6 Reactions Involving α-Hydrogens 222
Panel 13: Formaldehyde Polymers 228

CARBOHYDRATES 235

10

10.1 Monosaccharides 235
10.2 Reactions of Monosaccharides 241
10.3 Disaccharides 243
10.4 Polysaccharides 248
10.5 Modified Carbohydrates 250
Panel 14: Metabolic Oversights 252

CARBOXYLIC ACIDS AND THEIR DERIVATIVES 257

11

11.1 Structure and Nomenclature of Carboxylic Acids 257
11.2 Derivatives of Carboxylic Acids 261
11.3 Physical Properties 264
11.4 Some Preparations of Carboxylic Acids 265
11.5 Acidity of Carboxylic Acids 265
11.6 Condensation Reactions of Carboxylic Acids and Their Derivatives 270
11.7 Condensation Polymers 276
11.8 Reactions of Acid Derivatives Involving Carbanions 279
Panel 15: Acid Derivatives and Drugs 281

LIPIDS 292

12

12.1 Lipids 292
12.2 Waxes 292
12.3 Structure of Fats and Oils 293
12.4 Reactions of Fats and Oils 296
12.5 Soaps and Detergents 299
12.6 Biolipids 303
12.7 The Functions of Biolipids 305

AMINES 313

13
 13.1 Structure of Amines 313
 13.2 Nomenclature of Amines 314
 13.3 Physical Properties of Amines 315
 13.4 Basicity of Amines 317
 13.5 Preparations and Reactions of Amines 319
 13.6 Aromatic Diazonium Salts 324
 13.7 Dyes and Dyeing 328
 Panel 16: Alkaloids 334

PROTEINS: AMINO ACID POLYMERS 345

14
 14.1 Sources and Functions 345
 14.2 Structure 345
 14.3 Reactions of Amino Acids: Tools for Study 349
 14.4 Sequence Determination: Primary Structure 351
 14.5 Artificial Polypeptide Synthesis 353
 14.6 The Three-Dimensional Structures of Polypeptides and Proteins 354
 14.7 Conjugated Proteins 361
 14.8 Enzymes 361
 14.9 Pertinent Proteins and Polypeptides of Diverse Function 364
 14.10 Conclusion 370

NUCLEIC ACIDS 373

15
 15.1 Perspective 373
 15.2 Chemical Structure 373
 15.3 The Three-Dimensional Structure of DNA 380
 15.4 The Functions of Nucleic Acids 383
 15.5 The Genetic Code: Its Function and Malfunction 387
 15.6 Viruses 390
 Panel 17: Recombinant DNA—Science Fiction Today? 390

ORGANIC CHEMICAL INSTRUMENTATION 393

16
 16.1 Spectroscopy 393
 16.2 Infrared Spectroscopy 395
 16.3 Ultraviolet and Visible Spectroscopy 397
 16.4 Nuclear Magnetic Resonance 401
 16.5 Mass Spectrometry 408

APPENDIX: SUMMARY OF IUPAC NOMENCLATURE 415
INDEX 419

This book is designed for a short introductory course in organic chemistry of the one-quarter, two-quarter, or one-semester variety. It is written especially for students of agriculture, home economics, biology, health sciences, and others who need a general knowledge of organic chemistry but who do not intend to become professional chemists. The needs and interests of such students have been the prime consideration in determining the level of the text and in selecting and organizing topics.

This Second Edition is a thorough revision. Our main objective is to make the text more concise without sacrificing our original goals: to provide an understandable book that is truly helpful to the student. There are some organizational changes, particularly in the presentation of nomenclature, but the book still follows a functional group approach. We prefer this approach because it gives the student a foundation, the ability to recognize compounds by class and structure, around which to organize organic principles. For each family of organic compounds, we have concentrated on (1) structure, (2) basic nomenclature, (3) a few of the most characteristic, illustrative chemical reactions of the functional group, and (4) occurrences or important applications.

A few comments concerning the organization of this text are in order.

Bonding and Structure: Chapters 1 and 2 present the basic principles of bonding and structure. Chapter 1 concentrates on bonding, molecular orbitals, and electron dot formulas. Chapter 2 covers isomerism and should help a student become proficient in writing organic structures. Optical isomerism is presented in Chapter 6 so that it can be used in subsequent discussions of nucleophilic substitution and biological molecules.

Nomenclature: The IUPAC nomenclature of each functional group is presented in individual chapters as each class of compounds is considered. As each new functional group is introduced, its nomenclature is compared to that of the functional groups already discussed. For those who prefer to present IUPAC nomenclature as one big unit, the Appendix summarizes all aspects of nomenclature covered in the book (the individual sections in the book can act as review and reinforcement throughout the term). For others, the Appendix can act as a review and integrated summary as one begins to develop increased proficiency in naming compounds. Some persistent common nomenclature is briefly presented in individual chapters.

Reactions: The reactions most likely to show, as simply as possible, the characteristic chemical properties of each functional group were selected for coverage. In order to help students organize their study, the reactions are classified as addition, elimination, or substitution. They are arranged within each chapter so that similarities are evident; this acts as a unifying force. Reaction mechanisms have been introduced wherever they can deepen understanding without creating confusion.

Applications: The major applications of organic chemistry are presented in an understandable and retainable manner rather than just in passing. Relationships between organic structure and concepts such as octane number, physiological properties, and color are clearly drawn to allow the student to organize his or her study. Biological examples are included in the chapters whenever they can be presented without confusion. Chapters 10, 12, 14, and 15 are largely devoted to biological chemistry. Interspersed throughout the text are essays that highlight some special applications of organic chemistry. Among the topics presented in this form are petroleum and energy, terpenes, cancer and carcinogens, addition polymers, gasoline production, insect pheromones, insecticides, alkaloids, recombinant DNA, acid derivatives and drugs, and beverage alcohol.

Spectroscopy: Spectroscopy is presented in Chapter 16. None of the preceding chapters depend on it in any way. This chapter could be effectively presented anytime after Chapter 5 if the students have become familiar with the major functional groups.

Study Problems: Exercises can be found in the interior of each chapter as well as at the end. The internal problems are short, quick, usually elementary in nature, and designed to help the student test his or her comprehension of the concepts just presented. The End-of-Chapter Problems are more comprehensive and complete. These are labeled according to subject matter covered. Answers to the problems are available in a supplementary study guide.

Chapter Organization: Most instructors do not cover every topic in a textbook, and usually do some minor editing or rearranging to fit the book better to their course. To assist instructors in coordinating the book with their lectures, each chapter has been thoroughly subdivided by topic for easy reference and study organization. Likewise, exercises have been clearly subdivided for easy assignment.

We wish to acknowledge a number of people for their assistance in making this book possible. Primarily, we wish to thank our students, initially those at Purdue University who endured us as teaching assistants, and for the past dozen years those at Cal Poly, San Luis Obispo. We wish also to thank the reviewers of the Second Edition:

Andrew L. Colb
Bloomsburg State College

Arnold Drucker
University of Connecticut—Stamford

Kevin M. Smith
University of California—Davis

Robert L. Van Etten
Purdue University

In production matters we are grateful to our typist, Jackie Hynes, and particularly to the people at Allyn and Bacon, especially Jim Smith, Jane Dahl, and Mary Hill.

We would be appreciative of any comments or suggestions you the student may have. Please feel free to write us at the following address:

Drs. Philip S. and Christina A. Bailey
Chemistry Department
California Polytechnic State University
San Luis Obispo, California 93407

Since the beginning of recorded time, mechanical, technical, and scientific ingenuity has stood as a testimonial to the human mind. With essentially the same raw materials available during all their time on this planet, human beings have progressed in varying degrees in using these natural resources. The working of gold, silver, lead, copper, iron, and glass were known six thousand years ago. But it is the twentieth century that has witnessed an amazing explosion of information. Any evaluation of the rate of scientific growth throughout history shows that the bulk of the advances in all phases of the physical and biological sciences, medicine, and engineering have taken place in the last few decades. And as civilization faces the possibilities of the twenty-first century, the general populace must become aware of the scientific background that so profoundly affects our lives and also understand it. We must recognize the ultimate advantages of progress as well as fear its excesses.

Consider the pace of the development of chemistry over the past two hundred years. John Dalton presented his ideas on atomic theory during the early 1800s, but it wasn't until 1897 that J. J. Thomson confirmed the existence of the electron. Lord Rutherford's concept of an atom with a small, dense, positive nucleus surrounded by negative electrons came in 1911, followed in 1913 by Moseley's elucidation of the relation of atomic number and nuclear charge. At that same time, the work of Niels Bohr (1922 Nobel Prize winner) gave us the prelude to the modern quantum theory of atomic structure. Even the neutron, theorized about for a long time and now accepted as fact, was not detected until 1932. Yet in the many centuries preceding the advent of an atomic structural theory, scientists were laying the foundations of chemistry by the universal methods of science, observation, and experimentation. Out of the wealth of such accumulated chemical data organic chemistry emerged as the largest branch.

The vital-force theory was the dominant underlying principle for organic chemistry until the nineteenth century. Up to that time living organisms were believed to be the only source of organic chemicals, hence the term *organic*. Only living things were deemed to have the "vital force" necessary to produce organic compounds. In 1828, however, Frederick Wöhler synthesized the organic substance urea from the inorganic compounds lead cyanate and ammonium hydroxide. Urea had until then been found only in nature, in urine. Organic chemistry had to be redefined.

Today organic chemistry is considered the chemistry of carbon compounds. Why carbon? With over a hundred elements known, why should a single element, carbon, be the basis of the largest field of the chemical sciences? And why should all the other elements be relegated to the much smaller field of inorganic chemistry? The answer rests on two unchallenged facts. First, there are several million known organic compounds. Inorganic compounds are less than 10 percent of this amount. So the vast number of organic compounds alone could justify a separate discipline. Second, organic compounds are universal and irreplaceable. Consider the evidence of the following list of items, all related to organic chemistry.

Biochemistry. Living organisms are constructed of organic compounds and also run by them. Biochemistry is a separate field of chemistry, studying the chemicals and processes that sustain life.

Foods. The three main classes of foodstuffs—carbohydrates, fats, and proteins—are organic, and so are many of our current food additives and preservatives.

Fuel. Our civilization depends on coal and petroleum for energy.

Plastics. Polyvinyl chloride (PVC), saran, polyethylene, Styrofoam, Lucite, Melmac, Teflon—these synthetic organics are used to cover, mold, and hold the world we live in.

Natural and synthetic fibers. Besides providing protection as clothing, the materials cotton, wool, silk, Orlon, nylon, Dacron, kodel, and rayon also enhance the aesthetic aspects of society.

Drugs and medicines. Substances such as aspirin, decongestants, sedatives, stimulants, contraceptives, and fertility drugs help to alleviate physical and mental pain.

The list could be continued with such items as soaps and detergents, perfumes, cosmetics, pesticides, herbicides, fertilizers, paints, and dyes.

Like all scientific and technical fields, organic chemistry has advanced at a phenomenal rate. A few familiar examples testify to the extensive change civilization has experienced in about a hundred years. The first well to yield oil was drilled in 1859, but today we have practically depleted this natural resource. Aspirin was introduced in 1899. Teflon, discovered in 1938, took many years to reach the market. The development of synthetic rubber was necessitated by the misfortunes of World War II, and synthetic detergents were not produced until after that conflict. The early 1940s saw the first sales of nylon stockings, the initial tests of penicillin on humans, and the introduction of DDT to rid populations of the insects that caused such often fatal diseases as malaria and typhus. Since 1972, DDT has been almost totally banned in the United States because of environmental and health hazards. Oral contraceptives were readily available in the 1960s and are now undergoing close scrutiny.

As we proceed to our future, we must acknowledge some inescapable truths. Life spans have increased. This coupled with the unchecked birthrate could produce a world population of 6 billion by A.D. 2000. The earth's population was 2 billion in 1930. To accommodate this projected figure by the end of the century would require the construction of a new city every week to house 1.5-2 million people each. The distribution of wealth and with it the benefits of science and technology is uneven. Thousands die daily from factors related to malnutrition, while insects consume one-third of the world's agricultural output. Science cannot exist in a vacuum. It must recognize and embrace the society that allows it to live and flourish. The benefits must be shared efficiently and equitably and any hazards controlled or eliminated. Some individuals will acquire the professional training to face the technical questions relating to scientific advancement. But the most intriguing challenge will fall to all the citizens of the world community—accepting the responsibility for becoming informed consumers of the ideas and products of the scientific society that belongs to everyone.

ORGANIC CHEMISTRY

Chapter 1

Bonding in Organic Compounds

ELEMENTS AND COMPOUNDS; ATOMS AND MOLECULES

1.1   Elements are the fundamental building units of all substances, living and nonliving, in our known universe. They can enter into a seemingly infinite variety of chemical combinations, which are called *compounds*. An *atom* is the smallest particle of an element that retains the properties of that element. The atom itself is made up of electrons, protons, and neutrons, and the characteristic properties of elements are determined by the number and arrangement of these subatomic particles.

Protons and neutrons, which account for essentially the entire mass of the atom, reside in the relatively minute atomic nucleus. The nucleus is surrounded by electrons equal in number to the protons in the nucleus, thereby imparting a neutrality to the atom itself. These electrons occupy a tremendous volume, approximately a billion times the volume of the nucleus, yet they possess an infinitesimal mass. The atom then has a truly amazing substructure consisting of a massive core of negligible volume, the nucleus, surrounded by a tremendous volume of electrons of negligible mass.

To help you in visualizing this unique theory of atomic construction, the following "ifs" may be useful. If an atom were enlarged to the size of the Houston Astrodome, its nucleus would have the size and position of a marble suspended in

TABLE 1.1. COMPARISON OF SUBATOMIC PARTICLES

	Neutron	*Proton*	*Electron*
Charge	0	+1	−1
Mass	1	1	$\frac{1}{1840}$

the center of the open space, with electrons occupying the remaining volume. If all the space occupied by electrons in a nickel could be filled with neutrons and protons, the nickel would increase in weight from 5.5 grams to 100 million tons. If our earth could be compressed so that the nuclei of atoms were touching, the diameter would shrink from approximately 8000 miles to less than a mile.

Chemical compounds are formed by the combination of atoms. Combination is accomplished either by the transfer of electrons from one atom to another to form ions or by the sharing of electron pairs between atoms to form molecules. Both atoms and molecules are extremely small. A single raindrop contains 300 billion times as many molecules of water as there are people on the earth. An individual who takes a couple of aspirin tablets stands to have 2,200,000,000,000,000,000,000 aspirin molecules absorbed into the bloodstream. Despite the minute size of atomic and molecular species, much as been theorized and experimentally substantiated about their composition, behavior, and characteristic properties. To appreciate and understand organic chemistry, one must become intimately familiar with atomic and molecular structure.

ELECTRON CONFIGURATION AND THE PERIODIC TABLE

1.2   ### A. Atomic Orbitals

Although electrons occupy almost all the volume of an atom, they do not exist randomly around the nucleus. Rather, they occupy specific and discrete energy levels that are defined by each individual electron's distance from the nucleus, the shape and geometric orientation of the volume it occupies, and its spin.

The proximity of an electron to the nucleus is described by the principal quantum number, which has values 1, 2, 3, 4, 5, 6, 7. The number 1 stands for the first (lowest) energy level or shell, 2 for the second energy level, and so on. As the principal quantum number increases, the average distance of the electron from the nucleus and the effective volume occupied by the electron also increase.

Within each main energy level are sublevels called atomic orbitals, designated as s, p, d, and f. An *atomic orbital* describes the shape and geometric orientation of the region in space occupied by an electron. An s orbital is spherical with the atom's nucleus located at its center (Figure 1.1). Each main energy level has an s orbital and they differ from one another only in size and average distance from the nucleus; each is spherical. The 2s orbital (an s orbital in the second main energy level) is larger than a 1s orbital and actually encompasses it. Likewise, a 3s orbital is larger than a 2s, and a 4s larger than a 3s (Figure 1.1).

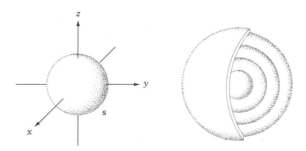

FIGURE 1.1. Spherical s orbitals. Cutaway model on the right indicates the concentric spherical s orbitals of succeeding energy levels.

The p orbitals are dumbbell-shaped with the atomic nucleus located between the two lobes. Each main energy level, other than the first, has three dumbbell-shaped p orbitals that differ only in their geometric orientations in space (Figure 1.2). One, designated p_x, is oriented along the x-axis of a three-dimensional coordinate; one, p_y, along the y-axis; and one, p_z, along the z-axis. The p orbitals in the third shell are larger than those in the second shell, although both have the same shape and orientation.

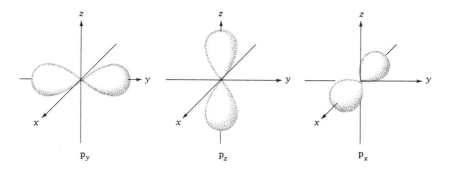

FIGURE 1.2. Representations of p atomic orbitals.

In atomic structure, the first main energy level of electrons contains one s orbital; the second has one s orbital and three p orbitals; the third, one s orbital, three p orbitals, and five d orbitals; and the fourth, one s orbital, three p orbitals, five d orbitals, and seven f orbitals. Each atomic orbital can accommodate a maximum of two electrons of opposite spin. *Electron spin* is simply described as the rotation of an electron on its axis, either clockwise or counterclockwise.

B. Aufbau Principle

Atomic orbitals are occupied by electrons according to the *Aufbau principle*. This states that orbitals of lower energy are filled first, and then orbitals of higher energy. The order can be determined using the guide shown in Figure 1.3. Follow the diagonal arrows, filling the orbitals from lower right (lowest energy) to upper left (highest energy).

Each s orbital can accommodate two electrons. A set of p orbitals within a main energy level can be occupied by a total of six electrons, since there are three orbitals, each of which can hold two spin-paired electrons. The five d orbitals within a main shell can hold ten electrons, and the seven f orbitals a total of fourteen electrons. Electrons with similar spin enter a set of orbitals of a given type within a principal quantum shell singly. Each orbital will accept a single electron before any pairing of electrons occurs (see Table 1.2 for examples).

PROBLEM

1.1. Write electron configurations for the following elements: **(a)** $_{14}$Si; **(b)** $_{15}$P; **(c)** $_{16}$S; **(d)** $_{17}$Cl; **(e)** $_{18}$Ar; **(f)** $_{36}$Kr.

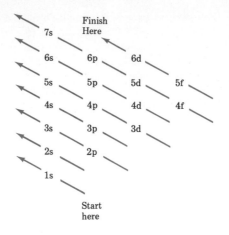

FIGURE 1.3. Aufbau principle. Atomic orbitals are filled beginning with the lowest-energy orbitals and proceeding to ones of higher energy. Follow the diagonal arrows to determine the order of priority for filling atomic orbitals. The s orbitals within a principal quantum shell can accommodate 2 electrons; p orbitals, 6; d orbitals, 10; and f orbitals, 14.

TABLE 1.2. ELECTRON CONFIGURATIONS

	Atomic Number	*Number of Electrons*	*Electron Configuration*
Li	3	3	$1s^2 2s^1$
Be	4	4	$1s^2 2s^2$
B	5	5	$1s^2 2s^2 2p_x{}^1$
C	6	6	$1s^2 2s^2 2p_x{}^1 2p_y{}^1$
N	7	7	$1s^2 2s^2 2p_x{}^1 2p_y{}^1 2p_z{}^1$
O	8	8	$1s^2 2s^2 2p_x{}^2 2p_y{}^1 2p_z{}^1$
F	9	9	$1s^2 2s^2 2p_x{}^2 2p_y{}^2 2p_z{}^1$
Ne	10	10	$1s^2 2s^2 2p_x{}^2 2p_y{}^2 2p_z{}^2$

C. Electron Configuration and the Periodic Table

The organization of all known elements into a concise form is called the *periodic table*. This table is constructed in accordance with the Aufbau principle, that is, according to electron configuration.

Using the Aufbau principle (Figure 1.3), let us consider the logical construction of the periodic table (Figure 1.4). The first period contains two elements, $_1$H ($1s^1$) and $_2$He ($1s^2$). These represent the orderly completion of the first energy level and its one s orbital. According to Figure 1.3, the next element, $_3$Li, should have an electron configuration of $1s^2\ 2s^1$ and begin period 2. Succeeding elements should finish filling the 2s orbital and proceed to fill the three 2p orbitals—which Be, B, C, N, O, F, and Ne do accomplish. Period 3, in which the 3s and 3p orbitals are filled, begins with $_{11}$Na ($1s^2\ 2s^2\ 2p^6\ 3s^1$) and ends with $_{18}$Ar ($1s^2\ 2s^2\ 2p^6\ 3s^2\ 3p^6$). Notice that only the outer-shell electron configuration of each element is shown in Figure 1.4.

FIGURE 1.4. The periodic table.

Periodic table with groups I–VIII (GROUPS) across the top and periods 1–7 (PERIODS) down the sides.

Group I
- 1 H $1s^1$
- 3 Li $2s^1$
- 11 Na $3s^1$
- 19 K $4s^1$
- 37 Rb $5s^1$
- 55 Cs $6s^1$
- 87 Fr $7s^1$

Group II
- 4 Be $2s^2$
- 12 Mg $3s^2$
- 20 Ca $4s^2$
- 38 Sr $5s^2$
- 56 Ba $6s^2$
- 88 Ra $7s^2$

Transition series (III–VIII, d orbitals)
- 21 Sc $4s^2 3d^1$ · 22 Ti $4s^2 3d^2$ · 23 V $4s^2 3d^3$ · 24 Cr $4s^1 3d^5$ · 25 Mn $4s^2 3d^5$ · 26 Fe $4s^2 3d^6$ · 27 Co $4s^2 3d^7$ · 28 Ni $4s^2 3d^8$ · 29 Cu $4s^1 3d^{10}$ · 30 Zn $4s^2 3d^{10}$
- 39 Y $5s^2 4d^1$ · 40 Zr $5s^2 4d^2$ · 41 Nb $5s^1 4d^4$ · 42 Mo $5s^1 4d^5$ · 43 Tc $5s^2 4d^5$ · 44 Ru $5s^1 4d^7$ · 45 Rh $5s^1 4d^8$ · 46 Pd $4d^{10}$ · 47 Ag $5s^1 4d^{10}$ · 48 Cd $5s^2 4d^{10}$
- 57 La $6s^2 5d^1$ · 72 Hf $6s^2 5d^2$ · 73 Ta $6s^2 5d^3$ · 74 W $6s^2 5d^4$ · 75 Re $6s^2 5d^5$ · 76 Os $6s^2 5d^6$ · 77 Ir $6s^0 5d^9$ · 78 Pt $6s^1 5d^9$ · 79 Au $6s^1 5d^{10}$ · 80 Hg $6s^2 5d^{10}$
- 89 Ac $7s^2 6d^1$ · 104 Ku $7s^2 6d^2$ · 105 Ha $7s^2 6d^3$

Groups III–VIII (p orbitals)
- 5 B $2s^2 2p^1$ · 6 C $2s^2 2p^2$ · 7 N $2s^2 2p^3$ · 8 O $2s^2 2p^4$ · 9 F $2s^2 2p^5$ · 10 Ne $2s^2 2p^6$ · 2 He $1s^2$
- 13 Al $3s^2 3p^1$ · 14 Si $3s^2 3p^2$ · 15 P $3s^2 3p^3$ · 16 S $3s^2 3p^4$ · 17 Cl $3s^2 3p^5$ · 18 Ar $3s^2 3p^6$
- 31 Ga $4s^2 4p^1$ · 32 Ge $4s^2 4p^2$ · 33 As $4s^2 4p^3$ · 34 Se $4s^2 4p^4$ · 35 Br $4s^2 4p^5$ · 36 Kr $4s^2 4p^6$
- 49 In $5s^2 5p^1$ · 50 Sn $5s^2 5p^2$ · 51 Sb $5s^2 5p^3$ · 52 Te $5s^2 5p^4$ · 53 I $5s^2 5p^5$ · 54 Xe $5s^2 5p^6$
- 81 Tl $6s^2 6p^1$ · 82 Pb $6s^2 6p^2$ · 83 Bi $6s^2 6p^3$ · 84 Po $6s^2 6p^4$ · 85 At $6s^2 6p^5$ · 86 Rn $6s^2 6p^6$

At this point, the 4f orbitals are filled forming the lanthanide series.

Lanthanide series
- 58 Ce $6s^2 4f^2$ · 59 Pr $6s^2 4f^3$ · 60 Nd $6s^2 4f^4$ · 61 Pm $6s^2 4f^5$ · 62 Sm $6s^2 4f^6$ · 63 Eu $6s^2 4f^7$ · 64 Gd $6s^2 5d^1 4f^7$ · 65 Tb $6s^2 4f^9$ · 66 Dy $6s^2 4f^{10}$ · 67 Ho $6s^2 4f^{11}$ · 68 Er $6s^2 4f^{12}$ · 69 Tm $6s^2 4f^{13}$ · 70 Yb $6s^2 4f^{14}$ · 71 Lu $6s^2 5d^1 4f^{14}$

At about this point, the 5f orbitals are filled forming the actinide series.

Actinide series
- 90 Th $7s^2 6d^2$ · 91 Pa $7s^2 6d^1 5f^2$ · 92 U $7s^2 6d^1 5f^3$ · 93 Np $7s^2 6d^1 5f^4$ · 94 Pu $7s^2 5f^6$ · 95 Am $7s^2 5f^7$ · 96 Cm $7s^2 6d^1 5f^7$ · 97 Bk $7s^2 5f^9$ · 98 Cf $7s^2 5f^{10}$ · 99 Es $7s^2 5f^{11}$ · 100 Fm $7s^2 5f^{12}$ · 101 Md $7s^2 5f^{13}$ · 102 No $7s^2 5f^{14}$ · 103 Lw $7s^2 6d^1 5f^{14}$

FIGURE 1.4. The periodic table. Outer-shell electron configurations and also the orbitals being filled are shown below each element. Note that in groups I and II, s orbitals are being filled and in groups III–VIII, p orbitals are being occupied. The transition series is formed by the filling of d orbitals. In the lanthanide and actinide series, f orbitals are being filled. Note the similarity of outer-shell electron configurations within each group. Although most of the elements have electron configurations in accordance with the Aufbau principle (Figure 1.3), there are some exceptions.

Since the 4s orbital has less energy than the five 3d orbitals, period 4 starts with $_{19}K$ ($1s^2\ 2s^2\ 2p^6\ 3s^2\ 3p^6\ 4s^1$). The placement in the table of the ten elements from $_{21}Sc$ through $_{30}Zn$ results from the occupation of the five 3d orbitals (ten electrons), these elements having a section of the table to themselves. Most of these d-orbital elements have the same *outer*-shell configuration, that of calcium, $_{20}Ca$ ($4s^2$), and they are called the *transition metals*. The filling of the remaining 4p orbitals accounts for the rest of the elements in the fourth period.

Period 5, again following the Aufbau principle, involves the 5s orbital (two elements), 4d orbitals (ten elements), and 5p orbitals (six elements).

Period 6 begins with the 6s orbital and period 7 with the 7s orbital, and each also introduces seven f orbitals. Lanthanum, $_{57}La$, begins a series of fourteen elements having a new subshell structure, the lanthanide series, in which the 4f orbitals are being occupied. The placement of the actinide series is a result of the filling of the 5f orbitals.

Notice that each vertical group (I–VIII), or family, has the same outer-shell electron configuration (in different quantum levels) and the elements within a particular group show similar chemical properties (Table 1.3).

The elements in group VIII, the noble gases, are the most stable and least reactive in the periodic table, entering into chemical combination infrequently and with difficulty. All these elements except helium have the same outer-shell configuration, s^2p^6, which is known as a *stable octet*. Helium has a $1s^2$ configuration. We shall find that elements, in forming chemical compounds, tend to form a stable configuration, that is, the same electron configuration possessed by an inert gas of group VIII.

PROBLEM

1.2. Write the outer-shell electron configuration (including principal quantum number) of every element in group VIII.

IONIC BONDING, ELECTRONEGATIVITY, AND THE PERIODIC TABLE

1.3   *Ionic bonding* involves the complete transfer of electrons between two atoms of widely different electronegativities to form ions. The atom losing electrons becomes positive, a *cation,* and the one gaining electrons becomes negative, an *anion.* The ionic bond results from the electrostatic attraction between these two oppositely charged species.

What elements form ionic bonds with one another? For a complete transfer of one or more electrons from one atom to another to occur, one atom must have a very strong attraction for electrons and the other a very weak attraction. *Electronegativity* is defined as the attraction of an atom for its outer-shell electrons. Electronegativity increases left to right within a period since the number of protons per nucleus increases and the electrons are entering the same main energy level (outer shell). The attraction between the nucleus and electrons thus becomes stronger. Electronegativity decreases from top to bottom within a group, even though the nuclear charge increases, since the outer shell gets farther and farther from the nucleus and is shielded from the nucleus by inner-shell electrons. As a result of their electronegativities, elements on the far left of the periodic table (low electronegativities) tend to lose electrons, and elements on the far right (high electronegativities) tend to gain these electrons.

TABLE 1.3. OUTER-SHELL CONFIGURATION WITHIN GROUPS

| | *GROUPS* | | | | | | | |
	I	II	III	IV	V	VI	VII	VIII
Number of electrons in outer shell	1	2	3	4	5	6	7	8
Outer-shell electron configuration	s^1	s^2	s^2p^1	s^2p^2	s^2p^3	s^2p^4	s^2p^5	s^2p^6

Elements gain and lose electrons to obtain a stable outer-shell configuration, in most instances a stable octet ($1s^2$ or s^2p^6). Consider as an example the reaction between sodium and chlorine atoms to form sodium chloride (table salt). Sodium can obtain a complete outer shell of eight electrons by gaining seven electrons, or by losing one and leaving the underlying complete inner shell to become the outer shell. Similarly, chlorine can achieve an octet of electrons by gaining one electron, thus completing its outer shell, or by losing seven electrons. The simpler of these two possibilities occurs if one electron is transferred from the sodium to the chlorine, resulting in a noble gas outer shell for each (NaCl).

$$_{11}\text{Na } 2)\text{e } 8)\text{e } 1)\text{e } + {}_{17}\text{Cl } 2)\text{e } 8)\text{e } 7)\text{e } \longrightarrow$$
$$\quad 1s^2 \ 2s^2 \ 2p^6 \ 3s^1 \qquad 1s^2 \ 2s^2 \ 2p^6 \ 3s^2 \ 3p^5$$

$$\qquad\qquad\qquad _{11}\text{Na}^+ 2)\text{e } 8)\text{e} \qquad + {}_{17}\text{Cl}^- 2)\text{e } 8)\text{e } 8)\text{e}$$
$$\qquad\qquad\qquad 1s^2 \ 2s^2 \ 2p^6 \qquad\quad 1s^2 \ 2s^2 \ 2p^6 \ 3s^2 \ 3p^6$$
$$\qquad\qquad (_{10}\text{Neon configuration}) \quad (_{18}\text{Argon configuration})$$

Elements on the far left of the periodic table obtain a stable octet by releasing the small number of electrons in their outer shells, thereby leaving the underlying complete inner shell. The acquisition of these electrons by elements on the far right of the periodic table completes their outer shells, resulting in a noble gas configuration. Elements on the left of the periodic table have low electronegativities and can lose electrons easily; those on the right have high electronegativities and easily attract electrons. The intrinsic logic of the periodic table is evident. You should review the valences of the ions commonly found in ionic compounds as presented in Table 1.4.

Ionic compounds do not exist as molecules. In the crystalline state of sodium chloride, each sodium ion is surrounded by six chloride ions and each chloride ion by six sodium ions (Figure 1.5). The number of sodium ions equals the number of chloride ions, but no one sodium can be identified as belonging to a particular chloride and vice versa.

PROBLEM

1.3. Using electron configurations like those in section 1.3, show the reactions between the elements of the following pairs that form ionically bonded compounds: **(a)** Li and F; **(b)** Mg and O; **(c)** Al and Br.

TABLE 1.4.   IONIC VALENCES

Group Valences					
			Other Elements		
Group I	1+	H^+	Zn^{2+}	Cu^+, Cu^{2+}	Fe^{2+}, Fe^{3+} Pb^{2+}, Pb^{4+}
Group II	2+	Ag^+	Cd^{2+}	Hg^+, Hg^{2+}	Sn^{2+}, Sn^{4+}
Group III	3+		Ni^{2+}		
Group VI	2−		Co^{2+}		
Group VII	1−				

Polyatomic Ions

NH_4^+	Ammonium	HCO_3^-	Bicarbonate	SO_4^{2-}	Sulfate	PO_4^{3-}	Phosphate
OH^-	Hydroxide	ClO_4^-	Perchlorate	CO_3^{2-}	Carbonate		
NO_2^-	Nitrite	MnO_4^-	Permanganate	CrO_4^{2-}	Chromate		
NO_3^-	Nitrate	CN^-	Cyanide	$Cr_2O_7^{2-}$	Dichromate		

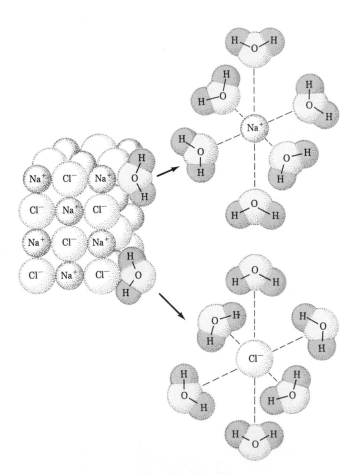

FIGURE 1.5. Crystalline structure of sodium chloride. Dark circles represent Na^+ and light circles Cl^-. When the ions are dissolved in water, they become independent and solvated (surrounded) by water molecules.

COVALENT BONDING

1.4

A. Covalent Bonding, Electron Configuration, and the Periodic Table

Covalent bonds involve the sharing of electron pairs between atoms of similar electronegativity. Because of the similarity in electronegativity, neither atom can relieve the other of its outer-shell electrons as is the case in ionic bonding.

In the simplest kind of covalent bond formation, each atom involved provides one electron to the bond. A shared pair of electrons results, becoming part of the outer shell of each atom. Consider, for example, hydrogen, chlorine, and hydrogen chloride, which are represented below by electron dot formulas showing both bonding and nonbonding outer-shell electrons.

Atoms   H ·   · H   : Cl ·   · Cl :   H ·   · Cl :

Molecules   H : H   : Cl : Cl :   H : Cl :

Usually, electron sharing occurs in a way that provides one or both atoms in the bond with the outer-shell configuration of a noble gas (in these cases two outer-shell electrons for hydrogen and eight for chlorine).

In Table 1.5, bond formation is related to groups on the periodic table. The number of covalent bonds an element may commonly form is explained by bonding each group to hydrogen. Hydrogen forms only one covalent bond since it has only one electron ($H^\times$).

B. Covalent Bonding in Organic Compounds

In organic compounds, carbon forms four covalent bonds since it has four electrons in its outer shell (group IV) and needs four more to attain a stable octet. More than any other element, carbon tends to share electrons with atoms of its own kind. The chemical results are an infinite variety of compounds, from chains of a few carbon atoms to gigantic chains of hundreds or thousands of carbon atoms.

 *Electrons are still available for bonding with carbon or other elements.*

Other elements are also found in organic compounds in addition to carbon. A listing of the elements most commonly found in organic compounds and the number of covalent bonds each usually forms follows:

C	4
N	3
O, S	2
H	1
F, Cl, Br, I	1

Achieving the required covalence of an atom does not necessitate using only single covalent bonds. Any combination of single bonds (one electron pair shared), double bonds (two electron pairs shared), or triple bonds (three electron pairs shared)

TABLE 1.5. COVALENT BOND FORMATION AND THE PERIODIC TABLE

Group	Outer-Shell Dot Structure	Covalent Compound with Hydrogen	Number of Covalent Bonds (Covalence)
I	A·	A⤫H	1

Since there is only one electron in the outer shell, only one covalent bond forms even though a noble gas configuration is not achieved by A. Elements in group I usually form ionic bonds.

II	·A·	H⤫A⤫H	2

Only two electrons occupy the outer shell and are available for sharing.

III	·A̤·	H⤫A̤⤫H (with H above)	3

Only three covalent bonds are possible since only three electrons are available for bonding. As in groups I and II, a noble gas outer shell is not achieved for A.

IV	·A̤·	H⤫A⤫H (with H above and H below)	4

Four bonds can form since there are four electrons in the outer shell. In forming these bonds, A gains four electrons and achieves the stable octet outer shell of a noble gas.

V	·A̤·	H⤫A⤫H (with H below)	3

With five electrons in the outer shell, only three more are needed to achieve a stable octet. In the compound there are three bonding pairs and one nonbonding pair of electrons.

VI	·Ä·	H⤫Ä⤫H	2

Two electrons are needed to complete the outer shell and two covalent bonds are formed.

VII	:Ä·	:Ä⤫H	1

Only one electron is needed to complete the octet and thus only one covalent bond forms. There are one bonding pair and three nonbonding pairs of electrons.

can be used so long as the total adds up to the required covalence. The four covalent bonds necessary for carbon, for example, can be achieved in the following ways:

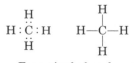

Four single bonds

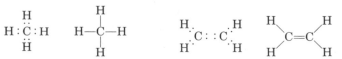

One double and two single bonds

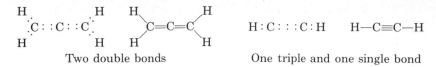

Two double bonds   One triple and one single bond

To write the structural formula for an organic molecule, one must arrange the atoms so that each fulfills its required valence. Consider, for example, the biological preservative formaldehyde, CH_2O:

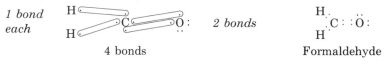

4 bonds   Formaldehyde

Such formulas showing all bonding and nonbonding electron pairs are called *electron dot formulas*.

PROBLEM

1.4. Draw electron dot formulas for the following compounds showing all bonding and nonbonding electron pairs: **(a)** $CHCl_3$ (chloroform); **(b)** CO_2 (carbon dioxide).

C. The Structural Nature of Compounds

Ionic compounds are composed of positive and negative ions in a formula ratio providing for an electrically neutral compound. If an ionic compound is melted or dissolved in water, the positive and negative ions go into independent motion. For example, crystalline or liquid salt or salt in solution contains no molecules of sodium chloride, merely a conglomeration of positive and negative ions, as in Figure 1.5.

Covalent compounds are composed of molecules. The atoms in a molecule belong exclusively to that particular molecule and travel together as a fixed unit. In the solid state of sugar, $C_{12}H_{22}O_{11}$, each unit of the crystal lattice is a complete sugar molecule. When the sugar crystal is dissolved in water, it disperses into sugar molecules with each molecule containing 12 carbons, 22 hydrogens, and 11 oxygens (Figure 1.6). Not only are the atoms of a covalent molecule exclusive to that particular molecule but they are arranged in a specific pattern. The molecule has a definite shape and three-dimensional geometry, with definite bond lengths and angles.

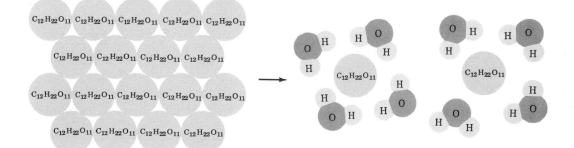

FIGURE 1.6. Sugar molecules dissolving in water.

MOLECULAR-ORBITAL APPROACH TO COVALENT BONDING

1.5   A. Molecular Orbitals

A covalent bond is formed by the sharing of electron pairs. This is accomplished by the overlap of atomic orbitals (each with one electron) to form a *molecular orbital* consisting of two spin-paired electrons. There are two important types of molecular orbitals—sigma bonds and pi bonds.

A *sigma bond* (σ bond) is formed by the overlap of atomic orbitals in one position. As shown in Figure 1.7, one could be formed by the overlap of s orbitals, as in hydrogen, end-to-end overlap of p orbitals, as in chlorine, or s-p overlap, as in hydrogen chloride.

A *pi bond* (π bond) is formed by the overlap of atomic orbitals in two positions, as in the overlap of two p atomic orbitals, each with one electron (Figure 1.8).

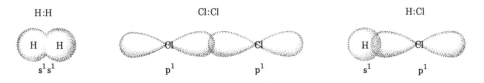

FIGURE 1.7. Sigma (σ) bonds. Atomic orbitals overlap in one position to form σ molecular orbitals.

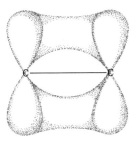

FIGURE 1.8. Pi (π) bonds. Each lobe of each p orbital overlaps with its counterpart in the other orbital to form a π molecular orbital.

B. Electron Configuration of Carbon

The chemical properties of an element depend on the electronic configuration of the outer shell. Carbon has four electrons in its outer shell, two in the 2s orbital, and one each in the $2p_x$ and $2p_y$ orbitals. One would expect carbon, with this configuration, to be divalent, since the 2s orbital is filled and only the $2p_x$ and $2p_y$ orbitals have an unpaired electron to share. Carbon's tetravalence is explained by promoting one 2s electron to a 2p orbital, creating four unpaired electrons during bonding (Figure 1.9). Since bond formation is an energy-releasing process and since the formation of four bonds creates a stable octet in carbon's outer shell, the energy required for promotion is more than returned. Figure 1.10 shows the shapes and geometric orientations of the four half-filled orbitals in carbon's outer shell.

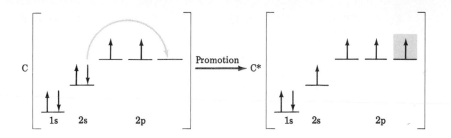

FIGURE 1.9. Promotion of an electron in carbon, allowing formation of four covalent bonds. It will be seen later that this does not describe the whole story; orbital hybridization must occur simultaneously.

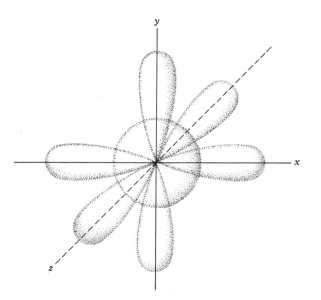

FIGURE 1.10. Outer-shell atomic orbitals of carbon. The carbon nucleus is at the origin. It is surrounded by one spherical s orbital and three dumbbell-shaped p orbitals (p_x, p_y, and p_z), which are identical except in geometric orientation. Each orbital possesses one electron.

C. Shapes of Organic Molecules

Let us consider a carbon with two atoms bonded to it by a triple bond and a single bond. How would these two atoms orient themselves around the carbon? There are two extreme possibilities: one in which the two atoms are as close to one another as possible, and one in which they are as far apart as possible. Common sense would lead us to choose the latter case in which the two atoms are a maximum distance from each other. A linear arrangement in which the two atoms are on opposite sides of the central carbon would allow this (for example, hydrogen cyanide, H—C≡N).

In most cases, atoms are oriented in a molecule so that repulsion between electron pairs (either bonding or nonbonding) around an atom is minimized. The following simple principle is useful for predicting the shape of a molecule or the geometry of a portion of a molecule: *atoms (and nonbonding electron pairs) bonded to a common central atom are arranged as far apart as possible in space.* Depending on the types of bonds involved, a carbon will have four, three, or two atoms bonded to it (see section 1.4B). If there are four bonded groups, the geometric orientation that will position these atoms around the central carbon as far from one another as possible is a tetrahedron (four-cornered pyramid). The geometric orientation will be a triangle if there are three bonded groups, and a straight line if there are two.

D. Carbon Bonded to Four Atoms

The simplest example of an organic compound with a carbon bonded to four atoms is methane, or natural gas, CH_4. To satisfy the valence of all five atoms, each of the hydrogens must be bonded to the carbon by a single bond. The most stable molecular geometry calls for the four hydrogens to be a maximum distance from one another. Placing the hydrogens at the four corners of a tetrahedron with the carbon in the center accomplishes this (Figure 1.11). The bond angles between hydrogens is 109.5°, and all the carbon-hydrogen bonds are equivalent.

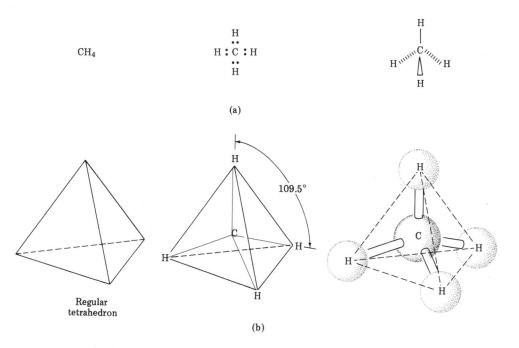

FIGURE 1.11. Methane. (a) Methods of representing the chemical formula. (b) Tetrahedral geometry. Bond angles are 109.5°.

If the hydrogens, with their 1s orbitals, were to bond to carbon's outer-shell atomic orbitals, as pictured in Figure 1.10, this stable tetrahedral molecule could not be formed. Recall that the angles between the p orbitals are 90°, not 109.5°. Furthermore, the carbon-hydrogen bonds would not be equivalent since there are two types of atomic orbitals, s and p, in carbon's outer shell. To establish the more stable tetrahedral geometry, the outer-shell orbitals (2s, $2p_x$, $2p_y$, and $2p_z$) hybridize, or blend, to form four new orbitals that are equivalent and at the ideal 109.5° angle from one another. The four new orbitals, called sp^3 hybrid orbitals because they were formed from one s and three p orbitals, are directed toward the corners of a tetrahedron. Four σ bonds will form by the overlap of the four raindrop-shaped sp^3 orbitals of carbon and the spherical s orbitals of hydrogen. Methane, CH_4, is sp^3-hybridized and tetrahedral, with four equivalent σ bonds, that is, four equivalent hydrogens all having bond angles equal to 109.5° (Figure 1.12).

PROBLEM

1.5. Draw a bonding picture for chloroform, $CHCl_3$, showing all molecular orbitals. Indicate the shape, bond angles, and hybridization of the molecule.

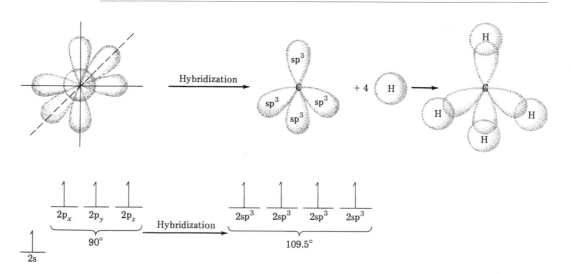

FIGURE 1.12. Hybridization and bonding in methane. Carbon's four outer-shell atomic orbitals (s, p_x, p_y, p_z) are converted into four new sp^3 hybrid orbitals with angles of 109.5° separating them. Each sp^3 orbital overlaps the s orbital of a hydrogen atom to form a σ bond.

E. Carbon Bonded to Three Atoms

Ethene, $CH_2\!=\!CH_2$ (from which the plastic polyethylene is made), has three atoms bonded to each carbon: two hydrogens and the other carbon. The geometric arrangement that allows three atoms bonded to a central carbon atom to be as far apart in space as possible is triangular, or trigonal. In ethene, each carbon is at the center of a triangle with the two hydrogens and the other carbon occupying the three corners. The bond angles are each approximately 120° (Figure 1.13).

Once again, the electron configuration of carbon as shown in Figure 1.10 would not allow a trigonal arrangement because the three p orbitals are perpendicular to one another. The outer-shell orbitals must be hybridized to create an orbital geometry consistent with the preferred triangular shape. In this case, only three of the four orbitals have to be hybridized since only three bonded atoms

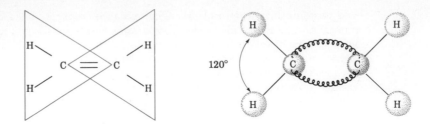

FIGURE 1.13.   Triangular structure of ethene, with 120° bond angles.

must be arranged in space.   The s and two of the p orbitals are combined to form three new sp² hybrid orbitals.   These sp² orbitals are directed toward the corners of an equilateral triangle (Figure 1.14).   An unhybridized p orbital remains unchanged on each carbon, perpendicular to the hybridized orbitals.

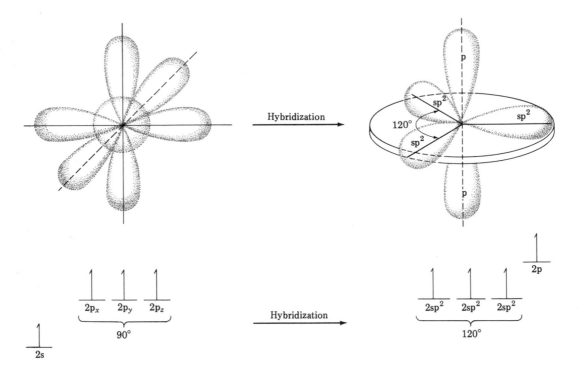

FIGURE 1.14.   sp² Hybridization.   The s and two p orbitals hybridize to form three new sp² orbitals that are directed to the corners of a triangle.   A p orbital remains unhybridized.

If the two hybridized carbons are now brought together, a σ bond can form between them by the overlap of two sp² hybrid orbitals.   Both carbons also have unhybridized p orbitals, which can be oriented parallel to one another and thereby overlap.   Both lobes of the p orbitals merge above and below the σ bond, forming a

π molecular orbital (Figure 1.15). Thus a double bond is composed of a σ bond and a π bond. The molecule is completed when σ bonds are formed between the overlapping sp^2 hybrid orbitals of the carbons and the spherical s orbitals of the hydrogens.

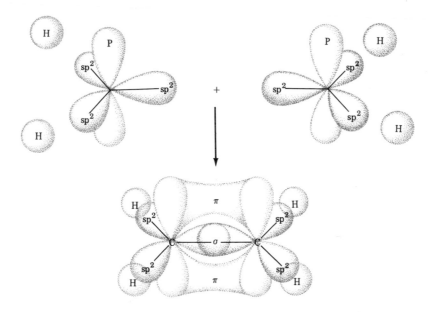

FIGURE 1.15. Orbital overlap of two sp^2-hybridized carbons and four hydrogens with spherical s orbitals to form ethene. The carbon-hydrogen bonds are s-sp^2 σ bonds and the double bond is composed of one sp^2-sp^2 σ bond and one p-p π bond.

PROBLEM

1.6. Draw a bonding picture for formaldehyde, H—$\overset{\overset{\text{O}}{\|}}{\text{C}}$—H (a biological preservative), showing all σ and π bonds. Indicate the shape, bond angles, and hybridization of the molecule.

F. Carbon Bonded to Two Atoms

The carbons in acetylene (used in oxyacetylene welding torches) are bonded to only two other atoms, a hydrogen and another carbon. These two atoms are positioned as follows:

Acetylene   H—C≡C—H

180°

180°

The two bonded atoms are on opposite sides of the central carbon at a maximum distance from one another. The molecule is linear, with 180° bond angles. To produce two equivalent orbitals directed 180° to one another, the 2s orbital and a 2p orbital on each carbon hybridize, forming two sp hybrid orbitals.

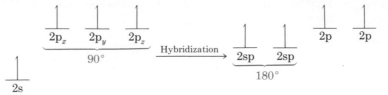

The overlapping of an sp orbital on each carbon joins the two carbons by a σ bond. The hydrogens are connected on the other sides of the carbons by σ bond formation between the remaining sp orbitals and the s orbitals of hydrogen. As in methane and ethene, the geometry of the molecule is determined by σ bond formation (Figure 1.16).

FIGURE 1.16. The linear geometry of acetylene is determined by the orientation of the two sp hybrid orbitals that engage in σ bond formation.

Two unhybridized, perpendicular p orbitals still remain on each carbon. The carbons are oriented so that the p orbitals on one carbon are parallel to the corresponding ones on the other carbon. These orbitals overlap to form two π bonds, one above and below and the other in front of and behind the σ bond. A triple bond then is composed of a σ bond and two π bonds. A virtual cylinder of electrons surrounds the two carbons of the triple bond (Figure 1.17).

PROBLEM

1.7. Draw a bonding picture for the following molecule, showing all molecular orbitals. Indicate the shape, bond angles, and hybridization at each carbon.

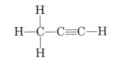

G. Bonding in Organic Compounds—A Summary

1. Geometry and hybridization: A carbon with
   a. four bonded groups is tetrahedral, sp^3 hybridized, and has 109.5° bond angles.
   b. three bonded groups is triangular, sp^2 hybridized, and has 120° bond angles.
   c. two bonded groups is linear, sp hybridized, and has a 180° bond angle.
2. Types of bonds:
   a. All single bonds are σ bonds.
   b. A double bond is made up of a σ and a π bond.
   c. A triple bond is a σ and two π bonds (see Table 1.6).
3. Bond strength:

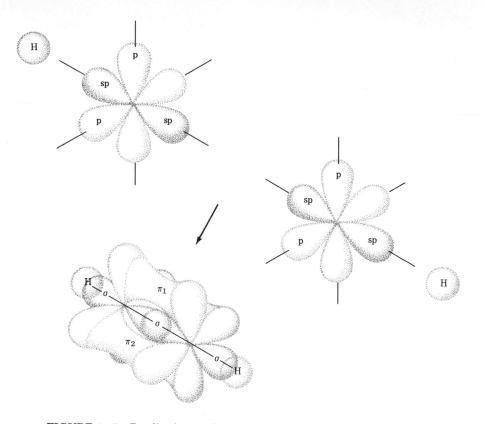

FIGURE 1.17. Bonding in acetylene. The carbon-hydrogen bonds are s-sp σ bonds and the carbon-carbon triple bond is composed of one sp-sp σ bond and two p-p π bonds. The π bonds are perpendicular, one in the plane of the paper and one extending in front of the plane of the paper and behind.

$$C\equiv C > C=C > C-C$$

4. Bond length:

$$C-C > C=C > C\equiv C$$

Table 1.6 presents bond lengths and bond energies (the energy necessary to break a bond).

POLAR COVALENT BONDS

1.6   If the atoms in a covalent bond have similar attractions for electrons, that is, similar electronegativities, the bonding electrons are shared equally. Consider, for example, the hydrogen and chlorine molecules, in which each of the hydrogen-hydrogen and chlorine-chlorine bonds involves the sharing of one pair of electrons (Figure 1.18). Since two hydrogen atoms or two chlorine atoms obviously have identical electronegativities, the electron pair is shared equally and the molecular orbital (σ bond) is symmetrically distributed. However, in the covalent bond between hydrogen and chlorine, the electron pair is not shared equally because chlo-

TABLE 1.6. BONDING IN ORGANIC COMPOUNDS

Number of Atoms Bonded to Central Carbon	Hybridization	Geometry	Bond Angles, deg	Example	Types of Bonds around Each Carbon	Molecular Orbitals	Carbon-Carbon Bond Length, Å	Carbon-Carbon Bond Energy, kcal/mole
4	sp^3	Tetrahedral	109.5	$\begin{array}{c} \text{H} \quad \text{H} \\ \text{H--C--C--H} \\ \text{H} \quad \text{H} \end{array}$	4 Single	$4\,\sigma$	1.54	83
3	sp^2	Triangular	120	$\begin{array}{c} \text{H} \qquad \text{H} \\ \text{C=C} \\ \text{H} \qquad \text{H} \end{array}$	2 Single 1 Double	$2\,\sigma$ $1\,\sigma, 1\,\pi$	1.34	146
2	sp	Linear	180	$\text{H--C}\equiv\text{C--H}$	1 Single 1 Triple	$1\,\sigma$ $1\,\sigma, 2\,\pi$	1.20	200

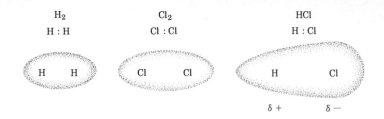

FIGURE 1.18. Hydrogen and chlorine are nonpolar molecules and the σ molecular orbital is symmetrically distributed. In HCl, the σ bond is distorted toward the more electronegative chlorine, forming a polar bond.

rine has a greater electronegativity than hydrogen and therefore a stronger attraction for the shared pair of electrons. Instead, the electrons are distorted toward the chlorine, giving it a partial negative charge ($\delta-$) and the hydrogen a partial positive charge ($\delta+$). This type of bond, which involves the unequal sharing of electron pairs between atoms of different electronegativities, is called a *polar covalent bond*. It is intermediate between an ionic bond and a covalent bond in that although one atom has a greater share of the electrons than the other, the electrons are not actually transferred, a contrast to the ionic bond. The symbol δ (*delta*) is used to signify that only partial charges are formed and not full ones, in contrast to ionic compounds.

A polar covalent bond can be represented as

$$\overset{\delta+ \quad \delta-}{A - B}$$

where atom B is greater than atom A in electronegativity. An oversimplified but useful way to use electronegativity for predicting polarity is to consider that carbon and hydrogen have almost identical electronegativities. Of the atoms commonly found in organic compounds, those to the right of carbon and hydrogen in the periodic table are more electronegative and those to the left, less electronegative.

Most carbon-carbon and carbon-hydrogen bonds are nonpolar. The accompanying examples show how polarity is predicted.

$$\overset{\delta+ \quad \delta-}{H - O} \qquad \overset{\delta+ \quad \delta-}{C - Cl} \qquad \overset{\delta+ \quad \delta-}{N = O} \qquad \overset{\delta- \quad \delta+}{C - B} \qquad \overset{\delta+ \quad \delta-}{C \equiv N}$$

The concept of polar bonds will be used frequently to predict and explain reactions of organic compounds.

PROBLEM

1.8. Show the distortion of the molecular orbitals and the charge separation in each of the following polar bonds: **(a)** C—Br; **(b)** C=O; **(c)** N—H; **(d)** C=N; **(e)** C—O; **(f)** C—S.

LEWIS ACIDS AND BASES

1.7   A *Lewis base* is a substance that has an unused pair of electrons (nonbonding pair) in an outer shell and is willing to share them in a chemical reaction. A *Lewis acid* is a substance that can accept this lone pair of electrons.

Ammonia, NH_3, is a good example of a Lewis base. Elemental nitrogen has five electrons in its outer shell and needs to form only three covalent bonds to attain a stable octet. Notice that there is a pair of electrons on the nitrogen not used in bonding. This pair of electrons can be shared with substances that may be deficient in electrons. An example is the hydrogen ion, H^+, with which ammonia forms the ammonium ion.

$$\underset{\text{Lewis base}}{H\overset{\displaystyle H}{\underset{\displaystyle H}{:N:}}} \qquad \underset{\text{Lewis acid}}{H^+} \longrightarrow H\overset{\displaystyle H}{\underset{\displaystyle H}{:\overset{+}{N}:}}H$$

The nitrogen in the ammonium ion is positive by the following reasoning. In order to be neutral the nitrogen must have complete ownership of five of its outer-shell electrons (since it is in group V of the periodic table). In ammonia, the nitrogen owns the nonbonding pair and half of the three bonding pairs, a total of five electrons, and is neutral. In the ammonium ion, it owns half of each of the four bonding pairs of electrons and consequently has a $+1$ charge. This Lewis acid-Lewis base reaction is used to make the fertilizers ammonium nitrate and ammonium sulfate.

$$NH_3 + HNO_3 \longrightarrow NH_4NO_3 \qquad \text{Ammonium nitrate}$$

$$2NH_3 + H_2SO_4 \longrightarrow (NH_4)_2SO_4 \qquad \text{Ammonium sulfate}$$

The formation of the hydronium ion, H_3O^+, in solutions of protonic acids, and the neutralization reaction of a hydrogen ion and hydroxide ion can be visualized using the Lewis concept.

$$\underset{\text{Lewis acid}}{H^+} \quad + \quad \underset{\text{Lewis base}}{H:\overset{\cdot\cdot}{O}:H} \longrightarrow H:\overset{\displaystyle H}{\underset{\cdot\cdot}{O}:}\overset{+}{H}$$

$$\underset{\text{Lewis acid}}{H^+} \quad + \quad \underset{\text{Lewis base}}{:\overset{\cdot\cdot}{O}H^-} \longrightarrow H:\overset{\cdot\cdot}{O}:H$$

Some compounds of boron and aluminum, such as BF_3 and $AlCl_3$, are common examples of Lewis acids. Both elements are in group III, have three electrons in the outer shell, and normally form only three covalent bonds. Neither

$$:\overset{\cdot\cdot}{\underset{\cdot\cdot}{F}}:\overset{\cdot\cdot}{\underset{\cdot\cdot}{B}}:\overset{\cdot\cdot}{\underset{\cdot\cdot}{F}}: \qquad :\overset{\cdot\cdot}{\underset{\cdot\cdot}{Cl}}:\overset{\cdot\cdot}{\underset{\cdot\cdot}{Al}}:\overset{\cdot\cdot}{\underset{\cdot\cdot}{Cl}}:$$

the boron nor the aluminum atom has a stable octet in its outer shell in these compounds. In fact, each has an empty atomic orbital. If this empty atomic orbital were to overlap with the atomic orbital of a Lewis base containing two electrons, the resulting bond would still have only two electrons and the outer shell of the group III atom would be complete. Consider, for example, the reaction between boron trifluoride and ammonia:

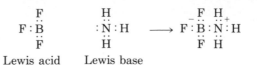

Now both boron and nitrogen show eight electrons in the outer shell but have complete ownership of only four. Therefore, the boron is negative since it is in group III and requires only three electrons to be neutral. In contrast, the nitrogen is positive since it needs five electrons to be neutral and has only four.

PROBLEM

1.9. **(a)** Ammonium phosphate (also a fertilizer) can be prepared from ammonia and phosphoric acid (H_3PO_4). Write an equation for this reaction. Emphasize the Lewis acid-base aspect of the reaction. **(b)** Hydrogen sulfide, found in volcanic gases, sulfur springs, and in rotten eggs (responsible for the smell) can be formed by combining NaSH and HCl. Using electron dot formulas, illustrate the Lewis acid-base reaction between the hydrogen ion and hydrogen sulfide ion.

END-OF-CHAPTER PROBLEMS

1.10 Electron Configurations: Write complete electron configurations for the following elements: **(a)** Na; **(b)** Ca; **(c)** Al; **(d)** Pb; **(e)** Br; **(f)** Rn; **(g)** Hg.

1.11 Electron Configurations: Write specific outer-shell electron configurations for each element in group **(a)** I; **(b)** II; **(c)** III; **(d)** IV; **(e)** V; **(f)** VI; **(g)** VII. Try this without using Figure 1.4; then check your answers with the figure.

1.12 Electron Configurations: Write the outer-shell electron configurations for the following elements from their positions on the periodic table. Do this initially without using Figure 1.4: **(a)** Na; **(b)** Mg; **(c)** Al; **(d)** Si; **(e)** P; **(f)** S; **(g)** Cl; **(h)** Ar; **(i)** Ga; **(j)** I; **(k)** Ca; **(l)** Kr.

1.13 Electron Configurations: Identify the elements that have the following outer-shell electron configurations. Use a periodic table without electron configurations: **(a)** $7s^1$; **(b)** $5s^25p^2$; **(c)** $3s^23p^5$; **(d)** $3s^2$; **(e)** $2s^22p^1$; **(f)** $4s^24p^4$; **(g)** $1s^2$; **(h)** $5s^25p^6$; **(i)** $4s^24p^3$; **(j)** $2s^22p^2$.

1.14 Ionic Compounds: Write chemical formulas for the following ionic compounds: **(a)** sodium fluoride (in some fluoride toothpastes); **(b)** magnesium hydroxide (milk of magnesia, antacid); **(c)** calcium carbonate (limestone, stomach antacid); **(d)** sodium nitrite (food preservative); **(e)** potassium chlorate (component of match heads); **(f)** lead II bromide (in auto exhaust if leaded gasoline is used); **(g)** lithium carbonate (used in the treatment of manic psychosis); **(h)** calcium oxide (lime); **(i)** sodium hydrogen carbonate (bicarbonate, baking soda); **(j)** calcium sulfate (gypsum).

1.15 Electron Dot Formulas: Write electron dot formulas for the following covalent molecules showing all bonding and nonbonding electron pairs: **(a)** CF_2Cl_2 (a freon); **(b)** CH_4O (wood alcohol); **(c)** CH_5N (odor of fish); **(d)** H_2S (hydrogen sulfide, odor of rotten eggs); **(e)** C_3H_8 (propane, fuel gas for rural areas); **(f)** C_2Cl_4

(a dry cleaning agent); **(g)** CO_2 (carbon dioxide, dry ice); **(h)** $COCl_2$ (the nerve gas phosgene); **(i)** HCN (hydrogen cyanide, a poisonous gas); **(j)** BH_3O_3 (boric acid, eyewash, veterinary antiseptic; no O—O bonds in formula).

1.16 Electron Dot Formulas: Write electron dot formulas for the following molecules, showing all bonding and nonbonding electron pairs:
(a) CH_3CH_2OH (beverage alcohol)   **(b)** $CH_3CH=CH_2$ (precursor of polypropy-

lene)   **(c)** $CH_3\overset{\overset{\displaystyle O}{\|}}{C}OH$ (acetic acid, sour taste of vinegar)   **(d)** $CH_2=CH—C\equiv N$

(acrylonitrile, from which orlon is made)   **(e)** $H_2N\overset{\overset{\displaystyle O}{\|}}{C}NH_2$ (urea, means by which nitrogen is excreted in urine)   **(f)** $CH_3CH=CHCH_2SH$ (a constituent of skunk

scent)   **(g)** $H\overset{\overset{\displaystyle O}{\|}}{C}OCH_2CH_3$ (artificial rum flavor)

1.17 Electron Dot Formulas: There are two possible covalent compounds for each of the following molecular formulas. Write electron dot formulas for each, showing all bonding and nonbonding electron pairs: **(a)** C_4H_{10}; **(b)** C_2H_6O; **(c)** C_3H_7Br; **(d)** $C_2H_4Cl_2$; **(e)** C_2H_7N; **(f)** C_3H_6.

1.18 Electron Dot Formulas: Write electron dot formulas showing all bonding and nonbonding electron pairs for carbonic acid, H_2CO_3, the bicarbonate ion, HCO_3^-, and the carbonate ion, CO_3^{2-}. All are components of the blood buffer (none have any O—O bonds).

1.19 Bonding, Molecular Orbitals, Geometry, Hybridization: Construct a bonding picture for each of the following molecules showing all σ and π bonds. Indicate for each carbon the hybridization, geometric shape, and bond angles:
(a) CH_3CH_3 **(b)** $CH_3CH=CHCH_3$ **(c)** $ClC\equiv CCl$ **(d)** $CH_2=C=CH_2$
(e) $CH_3C\equiv N$ **(f)** $CH_2=CHCH_2C\equiv CH$

1.20 Polar Covalent Bonds: Identify and show the polarity of the polar covalent bonds in the following molecules:

(a) $CH_3\underset{\underset{\displaystyle OH}{|}}{C}HCH_3$ **(b)** $H\overset{\overset{\displaystyle O}{\|}}{C}OH$ **(c)** $H\underset{\displaystyle N}{}CH_2\overset{\overset{\displaystyle O}{\|}}{C}OH$

Rubbing   Formic acid   Amino acid
alcohol   ("ant sting")   glycine

(d) $ClCH_2CH_2SCH_2CH_2Cl$ Mustard gas once used in chemical warfare

1.21 Lewis Acids and Bases: Determine whether the following are Lewis acids or Lewis bases: **(a)** $AlBr_3$; **(b)** CH_3NHCH_3; **(c)** CH_3OH; **(d)** BH_3.

1.22 Lewis Acids and Bases: Write, using electron dot formulas, Lewis acid-Lewis base reactions between the following species: **(a)** H^+, CH_3OH; **(b)** $AlCl_3$, H_2O; **(c)** CH_3NH_2, HCl; **(d)** BF_3, $(CH_3)_3N$.

1.23 Shapes of Molecules: Ammonia, NH_3, and water, H_2O, have bond angles similar to the tetrahedral bond angles. Explain the deviation from the orientation of the unhybridized atomic orbitals.   Explain also why ammonia is not triangular and water is not linear.

1.24 Silicon: Life on earth is based on the element carbon.   One of the episodes of the television series "Star Trek" centered around a life form on another planet based on the element silicon.   Do you think the author picked this element randomly or was there logic involved in the choice?   Explain.

Chapter 2

Structure and Isomerism

The chemistry of carbon compounds merits a separate branch of chemistry. This is partly due to the theoretically infinite number of organic compounds that are possible. In this chapter you will learn how to write organic chemical formulas and begin to gain an appreciation for the depth and variety of this fascinating branch of chemistry.

CHEMICAL FORMULAS

2.1 Chemical compounds are described by formulas, of which there are three types: empirical, molecular, and structural. An *empirical formula* tells us what kinds of atoms there are in a molecule and the simplest relative number ratio of the atoms. For example, the empirical formula C_2H_5 indicates that the carbon-to-hydrogen atomic ratio in the compound is $2:5$ but does not give the exact number of each atom.

A *molecular formula* describes the kinds of atoms in a molecule and the exact number of each. If the molecular weight of the compound with the empirical formula C_2H_5 is 58, the molecular formula must be C_4H_{10}, since the molecular weight, 58, is exactly double the empirical formula weight, 29: (C_2H_5; $2 \times 12 + 5 \times 1 = 29$). The molecular formula is always a whole-number multiple (1, 2, 3, . . .) of the empirical formula.

A *structural formula* not only informs us of the kinds of atoms and the exact number of each in the molecule but also describes the actual arrangement of these atoms. In other words, it states exactly how the atoms are bonded. For the molecular formula C_4H_{10}, two structural formulas are possible (Figure 2.1). Note that in each structure each carbon has four bonds and each hydrogen one bond.

Compounds that have identical molecular formulas (the same kinds and numbers of atoms) but different structural formulas (different atom arrange-

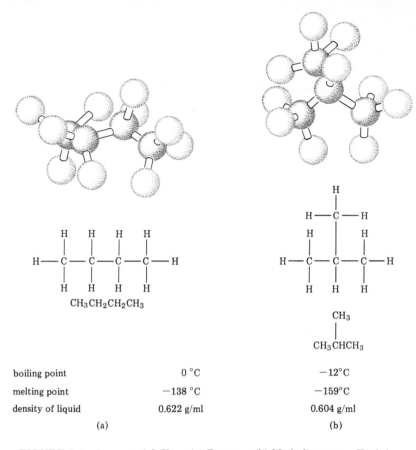

boiling point	0 °C	−12°C
melting point	−138 °C	−159°C
density of liquid	0.622 g/ml	0.604 g/ml
	(a)	(b)

FIGURE 2.1. Isomers of C_4H_{10}. (a) Butane. (b) Methylpropane. Each has the same molecular formula as the other but a unique structure.

ments) are called *isomers*. The compounds in Figure 2.1 are isomers because although each contains four carbons and ten hydrogens, the structural arrangements of those atoms differ. They are therefore different compounds, with different physical and chemical properties. No amount of rotating, bending, or twisting—short of actually breaking the bonds—can cause these two structures to be identical.

There are five types of isomerism of immediate interest: skeletal, positional, functional, conformational, and geometric. A sixth type, optical isomerism, will be considered separately. In drawing structural isomers, one must obey two rules: (1) Every atom in the molecular formula must be used—no more, no less. (2) The covalence of each atom must be satisfied. The covalences of the elements commonly found in organic compounds are

C	4
N	3
O, S	2
H, F, Cl, Br, I	1

SKELETAL ISOMERISM

2.2   The class of organic compounds called *saturated hydrocarbons,* or *alkanes,* demonstrates skeletal isomerism. These compounds contain only the elements carbon and hydrogen (hydrocarbon) and have all single bonds (saturated). The simplest member of this class is methane, or natural gas, CH_4 (Figure 2.2). As discussed in section 1.5D, methane is a tetrahedral molecule with bond angles of 109.5°.

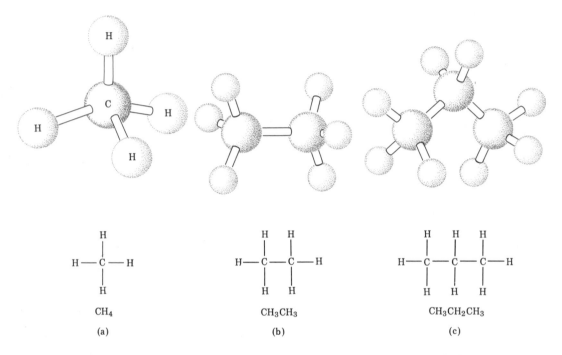

FIGURE 2.2. Representations of (a) methane (b) ethane and (c) propane.

The next two members of this series are ethane, C_2H_6, and propane, C_3H_8 (Figure 2.2). These two compounds, and the butanes (C_4H_{10}), make up the lique-fied petroleum gas (LPG) used for heating and cooking in rural areas. Notice that in ethane and propane each carbon has four bonds and each hydrogen one bond. Furthermore, each carbon is tetrahedral (sp^3-hybridized), like any carbon bonded to four atoms. For each of the molecular formulas CH_4, C_2H_6, and C_3H_8, there is only one arrangement of the atoms, as shown in Figure 2.2. There are no iso-mers. For the next member of the alkane series, C_4H_{10}, however, there are two arrangements of the four carbons and ten hydrogens (Figure 2.1). The two ar-rangements are skeletal isomers of one another, since the major difference be-tween them is in the carbon skeleton. In the first structure (butane), the carbon skeleton is a continuous four-carbon chain, whereas in methylpropane the skele-ton is a three-carbon chain with a one-carbon branch in the middle. To a person just learning organic chemistry, it may seem as though there should be several isomers of C_4H_{10}. For example, why aren't the accompanying structures different from butane and considered isomers?

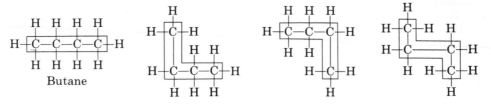

Butane

Close examination of these structures reveals that each has a continuous chain of four carbons and is therefore butane. All that has been done is to twist the molecule around into various contortions like a snake. Each structure retains a four-carbon continuous chain and all structures are identical.

The next molecular formula in this series is C_5H_{12}, for which there are three skeletal isomers (Figure 2.3). The first has a continuous five-carbon chain (only

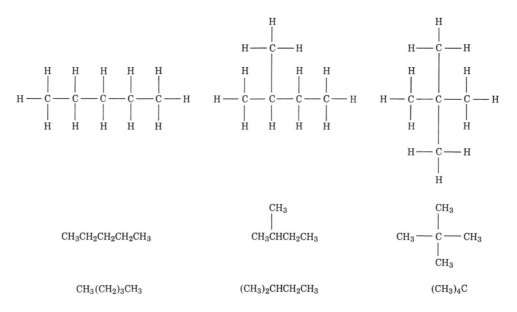

$CH_3CH_2CH_2CH_2CH_3$ $CH_3CHCH_2CH_3$ $CH_3 \!-\! C \!-\! CH_3$

$CH_3(CH_2)_3CH_3$ $(CH_3)_2CHCH_2CH_3$ $(CH_3)_4C$

FIGURE 2.3. Structural-formula representations for the isomers of C_5H_{12}.

one isomer of C_5H_{12} can have such a chain). The second has a four-carbon chain with a one-carbon branch on the second carbon. Note that it makes no difference on which of the two interior carbons the one-carbon branch is placed. In either case, it would be on the second carbon from the end of a four-carbon chain. The third isomer has two one-carbon branches on the middle carbon of a three-carbon chain. Note that if the two one-carbon branches had been put on the first carbon, or if one two-carbon chain had been put on the second carbon of the three-carbon chain, we would have repeated the second isomer, since the longest chain would be extended to four carbons.

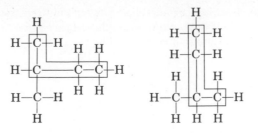

Figures 2.1–2.3 show condensed structural formulas for the compounds under discussion. Since this type of formula is the most frequently used, you should compare the expanded and condensed formulas to make certain of their meaning.

For the molecular formula C_6H_{14}, there are five skeletal isomers. A systematic approach should be followed in drawing isomers of molecular formulas to avoid repetition or omission of an isomer. Begin by arranging the carbons in one continuous chain:

$$CH_3CH_2CH_2CH_2CH_2CH_3$$

Now reduce the length of the longest chain by one carbon and place the remaining one-carbon chain in as many different places as possible. After forming as many

$$\overset{\displaystyle CH_3}{\underset{\displaystyle |}{CH_3CHCH_2CH_2CH_3}} \qquad \overset{\displaystyle CH_3}{\underset{\displaystyle |}{CH_3CH_2CHCH_2CH_3}}$$

isomers as we can with a five-carbon chain, we now reduce the longest chain length to four carbons and consider the remaining carbons as one two-carbon branch ($-CH_2CH_3$) and as two one-carbon branches ($-CH_3, -CH_3$). There is no place a two-carbon branch can be placed without extending the length of the longest chain. If it were attached to the end, the longest chain would become six carbons. If it were placed on an interior carbon, the chain would be extended to five. There are, however, two arrangements of two one-carbon branches on a four-carbon chain.

$$\overset{\displaystyle CH_3}{\underset{\displaystyle \underset{\displaystyle CH_3}{|}}{\underset{|}{CH_3CCH_2CH_3}}} \qquad \overset{\displaystyle CH_3\ \ CH_3}{\underset{\displaystyle |\ \ \ \ |}{CH_3CH-CHCH_3}}$$

The molecular formula C_7H_{16} has 9 possible isomers; C_8H_{18} has 18; C_9H_{20} has 35; $C_{10}H_{22}$, 75; $C_{20}H_{42}$, 366,319; and $C_{40}H_{82}$, 62,491,178,805,831. Very few of the isomers of $C_{20}H_{42}$ or $C_{40}H_{82}$ have been synthesized, isolated from natural sources, or characterized. Yet the possibility of their existence aptly illustrates the enormous scope of organic chemistry.

PROBLEM

2.1. Draw the nine skeletal isomers with the formula C_7H_{16}.

POSITIONAL ISOMERISM

2.3   Skeletal isomers differ in the position of carbon atoms, that is, the arrangement of the carbon skeleton. *Positional isomers* differ in the position of a noncarbon group; there is no change in the carbon skeleton. Let us consider, for example, the four isomers of C_4H_9Br.

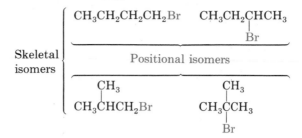

The members of the first pair of compounds are positional isomers because each has a four-carbon continuous chain, and they differ only in the position of the bromine. Likewise, the members of the second pair have identical carbon skeletons and differ only in the position of the bromine. The first pair of compounds and the second pair differ, however, in the carbon skeleton, and the pairs are therefore related as skeletal isomers. The following examples illustrate other types of positional isomers.

$$CH_3CCH_2CH_2CH_3 \quad CH_3CH_2CCH_2CH_3$$
$$O \qquad\qquad O$$

$$CH_3CH_2CH{=}CH_2 \quad CH_3CH{=}CHCH_3$$

PROBLEM

2.2. Draw positional isomers of the following compounds:

$$CH_3$$

(a) $CH_3CH_2CH_2CH_2OH$ **(b)** $CH_3CHCH_2CH_2Br$ **(c)** $HC{\equiv}CCH_2CH_2CH_2CH_3$

FUNCTIONAL ISOMERISM

2.4   Compounds that have the same molecular formula but belong to different classes of organic compounds are called *functional isomers*. The functional group is usually the site of the characteristic reactions of a particular class of compounds. Consider, for example, the two isomers with the molecular formula C_2H_6O.

$$CH_3CH_2OH \qquad CH_3OCH_3$$
An alcohol   An ether
(beverage alcohol)

Each member of the class of compounds called alcohols possesses a saturated

carbon with a bonded hydroxy group ($-\overset{|}{\underset{|}{C}}-OH$), whereas ethers possess a unit of

two saturated carbons separated by an oxygen ($-\overset{|}{\underset{|}{C}}-O-\overset{|}{\underset{|}{C}}-$). These character-

istic structural units are called functional groups. They illustrate an especially important type of isomerism. Different compounds possessing the same func-

TABLE 2.1.   MAJOR CLASSES OF ORGANIC COMPOUNDS

Functional Group Name	Functional Group Structure	Example	Application of Example
Alkane	$-\overset{\mid}{\underset{\mid}{C}}-\overset{\mid}{\underset{\mid}{C}}-$	$CH_3CH_2CH_3$	propane (rural or camping gas)
Alkene	$\underset{/}{\overset{\backslash}{C}}=\underset{\backslash}{\overset{/}{C}}$	$CH_2{=}CH_2$	ethene (precursor of polyethylene)
Alkyne	$-C{\equiv}C-$	$HC{\equiv}CH$	acetylene (used in oxyacetylene torches)
Aromatic	⬡	⬡$-CH_3$	benzene, toluene (high-octane gasoline components)
Carboxylic acid	$-\overset{O}{\overset{\|}{C}}OH$	$CH_3\overset{O}{\overset{\|}{C}}OH$	acetic acid (vinegar acid)
Aldehyde	$-\overset{O}{\overset{\|}{C}}H$	$H\overset{O}{\overset{\|}{C}}H$	formaldehyde (biological preservative)
Ketone	$-\overset{\mid}{\underset{\mid}{C}}-\overset{O}{\overset{\|}{C}}-\overset{\mid}{\underset{\mid}{C}}-$	$CH_3\overset{O}{\overset{\|}{C}}CH_3$	acetone (fingernail polish remover)
Alcohol	$-\overset{\mid}{\underset{\mid}{C}}-OH$	CH_3CH_2OH	beverage alcohol
Ether	$-\overset{\mid}{\underset{\mid}{C}}-O-\overset{\mid}{\underset{\mid}{C}}-$	$CH_3CH_2OCH_2CH_3$	diethyl ether (general anesthetic)
Amine	$-\overset{\mid}{N}-$	CH_3NH_2	methylamine (fishy odor of herring brine)

tional group have similar chemical properties; those with different functional groups often undergo distinctively different chemical reactions. The functional group is often the basis for the naming of an organic compound. Table 2.1 summarizes some of the major classes of organic compounds. A more complete list appears inside the front cover.

Following are some examples of functional isomers with the formula C_4H_8O (not all isomers are shown):

$$CH_3CH_2CH_2\overset{O}{\overset{\|}{C}}H \qquad CH_3CH_2\overset{O}{\overset{\|}{C}}CH_3 \qquad CH_2{=}CHCH_2CH_2OH$$

Aldehyde   Ketone   Alkene–Alcohol

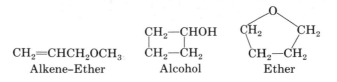

$$CH_2\!\!=\!\!CHCH_2OCH_3$$
Alkene–Ether   Alcohol   Ether

Alcohol   Ether

PROBLEM

2.3. Draw a specific example containing three carbons of the following types of compounds: **(a)** alkene; **(b)** alkyne; **(c)** carboxylic acid; **(d)** aldehyde; **(e)** ketone; **(f)** alcohol; **(g)** amine; **(h)** ether.

UNITS OF UNSATURATION (MULTIPLE BONDS)

2.5   Sections 2.3 and 2.4 presented examples of compounds with double bonds. How can one determine whether or not the isomers of a molecular formula have multiple (double or triple) bonds and how many such bonds there are? To answer this question, consider the formulas C_3H_8 and C_3H_6. Begin drawing these compounds by putting the carbons in a continuous chain connected by single bonds and fill in the available hydrogens.

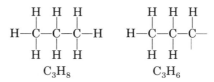

As this is done, we see that all atoms of C_3H_8 have satisfied covalences but the third carbon of C_3H_6 needs to complete two more bonds. To remedy this situation, try placing a double bond into the carbon skeleton.

$$
\begin{array}{ccc}
 & H & H & H \\
 & | & & | \\
H\!-\!C\!-\!C\!=\!C\!-\!H \\
 & | \\
 & H
\end{array}
$$

This time all atoms have satisfied valences. If they did not, it would be necessary to put in another double bond or to convert the double bond to a triple bond. Another solution is to divide the hydrogens evenly among the three carbons. Now each of the end carbons needs an additional bond, a problem solved by connecting them.

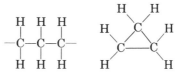

Any compound with three or more carbons that has a double bond must have at least one isomer with a ring; that is, it must form a cyclic compound.

For molecular formulas with other elements in addition to carbon and hydrogen, arrange all the elements with a covalence greater than 1 in a continuous chain and insert multiple bonds between them until all covalences can be satisfied by bonding with the monovalent elements (valence = 1). Doing this for the molecular formula C_2H_4O, we find one unit of unsaturation, which can be expressed as either a double bond or a ring.

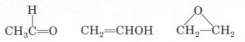

Now, let us consider the molecular formula C_4H_6. Following the procedure used, one finds that two units of unsaturation are required.

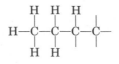

The two units of unsaturation can be expressed as

1. one triple bond.

$$CH_3CH_2C\equiv CH \qquad CH_3C\equiv CCH_3$$

2. two double bonds.

$$CH_3CH=C=CH_2 \qquad CH_2=CH-CH=CH_2$$

3. one double bond and one ring.

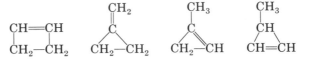

4. two rings.

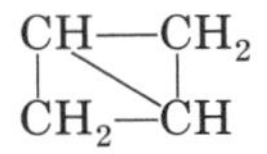

$$\begin{array}{c} CH-CH_2 \\ | \quad\quad | \\ CH_2-CH \end{array}$$

In summary, a unit of unsaturation can be expressed as a multiple bond (a double bond is one unit, a triple bond is two) or as a ring.

PROBLEM

2.4. Determine the number of units of unsaturation in each of the following molecular formulas: **(a)** C_5H_{10}; **(b)** $C_8H_{12}Br_2$; **(c)** C_6H_6; **(d)** $C_7H_7NO_3$.

CONFORMATIONAL ISOMERISM

2.6   A single bond between two atoms, such as in ethane (CH_3CH_3), Figure 2.2b, is free to rotate more or less unrestrictedly. As a result, an infinite number of atomic arrangements, called *conformations,* of ethane are possible. There are two extreme forms: the staggered conformation, which is the most stable and therefore the most abundant; and the eclipsed, which is the least stable and least abundant. They can be represented by sawhorse diagrams or Newman projections, as illustrated in Figure 2.4.

In the Newman projection, one is viewing the carbon-carbon bond end-on along the axis of connection. The bonds emanating from the center of the circle are on the front carbon and those from the perimeter are on the hidden rear carbon. This eclipsed conformation is least stable because the hydrogen atoms and bonding pairs of electrons on adjacent carbons are as near one another as possible. This causes maximum repulsion. If the carbons are rotated 60° with

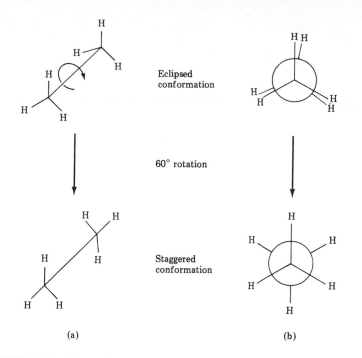

FIGURE 2.4. The conformations of ethane, CH_3—CH_3. (a) Sawhorse diagram. (b) Newman projection.

respect to each other, the most stable conformation, the staggered, is effected, in which the hydrogens and bonding pairs are a maximum distance apart and repulsion is minimized. Although the conformers of ethane are of different stabilities, the energy differences are not great and the carbon-carbon bond rotation occurs with almost no restriction. For this reason it is impossible to isolate the different conformers of ethane—they are constantly interconverting.

In 1,2-dibromoethane ($BrCH_2$—CH_2Br), two staggered and two eclipsed conformations are possible, with all four having different stabilities (Figure 2.5). The

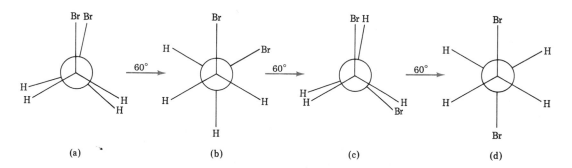

FIGURE 2.5. Newman projections of the conformations of 1,2-dibromoethane (CH_2Br—CH_2Br). (a) eclipsed (least stable); (b) staggered; (c) eclipsed; (d) staggered (most stable).

conformation in which the two bromines are eclipsed is the least stable, owing to the large size and high electron density of the bromine atoms. For like reasons, the staggered conformer in which the two bromines are maximally separated is most stable. Still, all samples of 1,2-dibromoethane are identical and are composed of these four and other intermediate conformers in concentrations approximately related to their stability. Although conformational isomers are different from one another, unlike the other types of isomers we have so far studied, they are not separable or isolable. The sample of a compound as we know it is merely a dynamic, uncontrollable mixture of these conformers.

PROBLEM

2.5. Draw the four conformational isomers of butane, $CH_3CH_2CH_2CH_3$, formed by 60° rotations about the single bond between carbons two and three. Use Newman projections. Identify the most and least stable.

GEOMETRIC ISOMERISM

2.7   An example of a substance in which there is free rotation around carbon-carbon single bonds is 1,2-dibromoethane ($BrCH_2CH_2Br$). Although this substance has several conformers of different energies and different abundances (Figure 2.5), they cannot be separated or isolated, because of constant interconversion. Rotation is unrestricted because it is not affecting the degree of orbital overlap in the σ bond between the two singly bonded carbons and the energy barriers between the conformers are small.

        Such is not the case with 1,2-dibromoethene (BrCH=CHBr). In general, free rotation about carbon-carbon double bonds is not possible, since this would involve a decrease in orbital overlap of the parallel p orbitals composing the π bond (Figure 2.6). This is not a problem with σ bonds because they involve overlap in only one position. For rotation, the π bond would have to be broken, a process that is not energetically favorable. Owing to the restricted rotation in 1,2-dibromoethene, two isomers are possible—one in which the two bromines are on the same side (*cis,* Latin, "same side"), and another in which they are on opposite sides of the double bond (*trans,* Latin, "across").

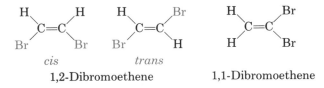

cis   *trans*
        1,2-Dibromoethene   1,1-Dibromoethene

Unlike conformational isomers, these are distinct and isolable compounds. They are called *geometric isomers* because they differ in the geometric orientation of atoms, not in the structural (atom-to-atom) arrangement. For geometric (cis-trans) isomerism to be possible, each carbon involved in a carbon-carbon double bond must have two different groups attached. For example, 1,2-dibromoethene exhibits cis-trans isomerism, as illustrated, but 1,1-dibromoethene (CH_2=CBr_2) does not. Rotation around the double bond is still restricted but it is evident that rotation would make no difference in the compound.

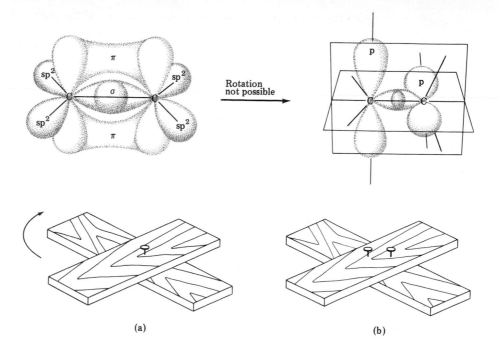

FIGURE 2.6. Rotation around a double bond is not usually possible, since the π bond, which has two positions of overlap, must be broken. The difference between rotation about a carbon-carbon single bond and a carbon-carbon double bond is analogous to the difference in ability to rotate (a) two pieces of wood connected by one nail, versus (b) two pieces connected by two nails.

PROBLEM

2.6. (a) Draw the geometric isomers of 2-butene, CH_3CH=$CHCH_3$. Label them *cis* and *trans*. **(b)** Does 1-butene, CH_3CH_2CH=CH_2, have geometrical isomers? Explain.

Panel 1

GEOMETRIC ISOMERISM AND VISION

A significant example of the importance of cis-trans isomerism is evidenced by the reaction of vitamin A in the visual cycle. In the rods of the eye vitamin A, in the form of the compound retinal, combines with a large protein molecule called opsin to produce rhodopsin, or visual purple. Retinal, drawn below, has four carbon-carbon double bonds outside the ring. While three of these double bonds remain in the stable trans configuration, one double bond can be converted from cis to trans upon contact with light.

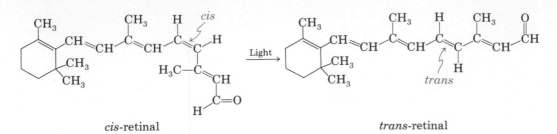

cis-retinal   trans-retinal

[PANEL 1] This interconversion of geometric isomers effects a series of nerve impulses and we "see" light. Other proteins called enzymes aid in the regeneration of *cis*-retinal from *trans*-retinal. Bright light will temporarily deplete the store of rhodopsin, causing impairment of vision until the enzymatic trans-to-cis conversion has time to occur.

A deficiency of vitamin A causes night blindness in adults and can result in total blindness in children as well as serious impairments in growth and development. Consumption of excessive amounts of vitamin A can also prove to be toxic.

CYCLOALKANES

2.8   A. Structure and Stability

Saturated hydrocarbons possessing one or more rings are called *cycloalkanes*. They are often described by regular polygons, each corner of which represents a carbon with enough hydrogens to satisfy the valence. The smallest member of this class, cyclopropane, has three carbons. Cyclopropane and cyclobutane (four-carbon ring) have been shown to be less stable than larger-ring cycloalkanes.

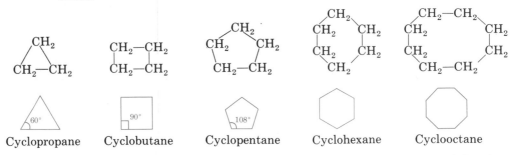

Cyclopropane   Cyclobutane   Cyclopentane   Cyclohexane   Cyclooctane

The source of this relative instability is within the internal angles of the ring. Each carbon in the ring has four bonded atoms, is sp³-hybridized, and should be tetrahedral, with 109.5° bond angles. Since cyclopropane has a three-membered ring, and three points define a plane, the molecule as a whole must have the geometry of an equilateral triangle with internal angles of 60°. This angle differs significantly from the preferred tetrahedral angle, causing decreased orbital overlap in the σ bond and internal angle strain. Although cyclobutane need not be planar, it still geometrically approximates a square, with internal bond angles of 90°, and it thereby suffers from ring strain. The internal angle of a pentagon is very close to the 109.5° tetrahedral angle, and as a result, cyclopentane is energetically very stable.

In larger-ring cycloalkanes, such as cyclohexane (six-membered ring) and cyclooctane (eight-membered ring), the rings are large enough and have sufficient flexibility through bond rotation to bend, twist, and pucker out of the plane until each carbon has the stable tetrahedral angle. Figure 2.7 illustrates a stable conformation of cholesterol.

Organic chemists have long been fascinated by the variety of strange and unusual shapes possible for cyclic compounds. Figure 2.8 shows some of these interesting compounds. Note that all are polycyclic compounds; that is, each has more than one ring. In one case the physical connection between rings is not a chemical bond; rather it is analogous to the connections of a magician's unconnected yet interlocking metal rings.

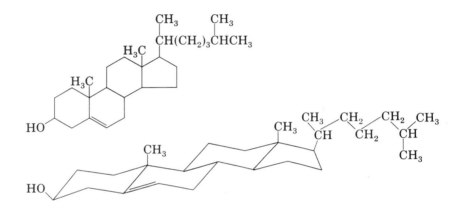

FIGURE 2.7. Cholesterol shown in a simple formula and also in its actual three-dimensional conformation. Note that none of the rings are planar but are puckered to provide more stable bond angles. Cholesterol occurs widely in the body and can be isolated from nearly all animal tissues. It is the primary constituent of human gallstones and has been implicated in hardening of the arteries.

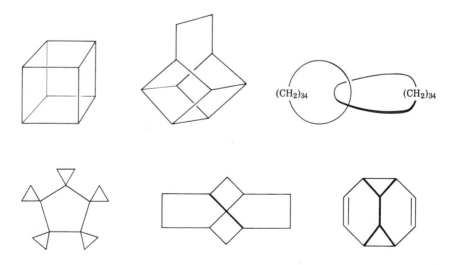

FIGURE 2.8. Some unusual polycyclic hydrocarbons.

B. Conformational Isomerism in Cyclohexane

To visualize the puckering in cyclohexane that provides for 109.5° bond angles, let us begin with cyclohexane as a regular hexagon. The simplest way to make this molecule nonplanar, minus internal strain, is to bend the two end carbons out of the plane of the ring. Both carbons can be "puckered" in the same direction to form the "boat" conformation (Figure 2.9a). Or one carbon can be pulled above the plane of the ring and the other below the plane, producing the "chair" conformation (Figure 2.9b). In both conformations, each carbon is tetrahedral and all bond angles are 109.5°. The two conformations are not of equal stability, however. The chair form is more stable and by far the predominant conformer of cyclohexane. The difference in stability is evident by comparing the structures of the boat and chair forms.

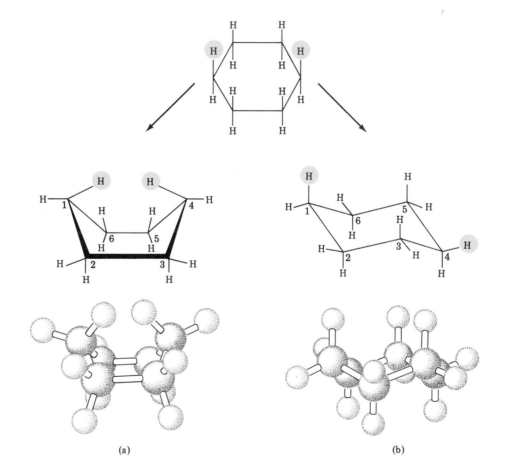

FIGURE 2.9. Boat and chair conformations of cyclohexane. (a) Boat form. (b) Chair form.

In the boat form, the carbons on opposite ends (carbons 1 and 4) are pulled toward one another, causing steric interactions between the "flagpole" hydrogens because of their proximity (see Figures 2.9 and 2.10). In the chair form, these same two carbons are bent away from one another—one up and one down—and thus they are not subject to mutual repulsion. A second destabilizing factor can be found by viewing the C_2—C_3 and C_5—C_6 carbon-carbon bonds end-on, using Newman projections (Figure 2.10). In the boat form, the bonded atoms are in the

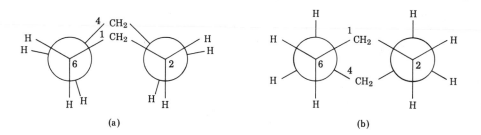

(a)   (b)

FIGURE 2.10. Newman projections of (a) the boat and (b) the chair conformation of cyclohexane. Compare the numbered carbons to those in the boat and chair forms in Figure 2.9.

less stable eclipsed conformation, whereas in the chair form they are staggered (see section 2.6). For these two reasons, the chair form is the more stable, predominating conformation.

Close examination of the chair form of cyclohexane reveals that there are two basic orientations of the hydrogens (Figure 2.11). Six of the hydrogens are approximately perpendicular to the ring and are called *axial hydrogens*. There are three above and three below the ring on alternating carbons (Figures 2.9 and 2.11a). The other six hydrogens, one on each carbon, protrude from the perimeter of the ring and are called *equatorial hydrogens*. The axial hydrogens are nearer

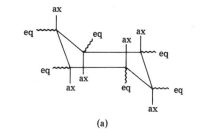

(a)

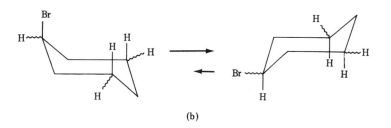

(b)

FIGURE 2.11. (a) Chair form of cyclohexane, showing axial and equatorial positions. (b) Chair forms of bromocyclohexane. Since bromine is a large atom, it is usually found in the roomier equatorial position.

to one another than the equatorial hydrogens are. Substituted cyclohexanes will then exist predominantly in conformations in which the group or groups that have replaced hydrogens are in the roomier equatorial positions (for example, bromocyclohexane, Figure 2.11b).

2.7. To develop a feel for the chair conformation of cyclohexane and axial and equatorial positions, draw chair forms of cyclohexane with the following substituents: **(a)** axial CH_3; **(b)** equatorial CH_3; **(c)** Br's on carbons 1 and 2, both axial; **(d)** Br's on carbons 1 and 4, one axial and one equatorial; **(e)** Br's on carbons 1 and 3, both equatorial.

C. Geometric Isomerism in Cyclic Compounds

Relatively free rotation around a single bond is possible because it does not cause a decrease in orbital overlap in the σ molecular orbital. Now consider the case for which rotation about one single bond could cause decreased overlap, or even cleavage, of another bond within the molecule. This is the situation in small-ring compounds such as cyclopropane and cyclobutane.

$$CH_2 \diagdown \atop CH_2 \!-\! CH_2 \qquad \begin{matrix} CH_2 \!-\! CH_2 \\ | \qquad | \\ CH_2 \!-\! CH_2 \end{matrix}$$

Cyclopropane   Cyclobutane

If any one of the carbon-carbon bonds were to rotate, the rest of the molecule would experience a severe strain. In cyclopropane, for example, if any two carbons began to rotate in opposite directions, the third carbon would be forced to break its attachments because it is bonded to both and obviously cannot follow the opposing rotations. Even in more flexible large-ring molecules, the rotation around one carbon-carbon bond would be transmitted and compensated for throughout the molecule to relieve any resulting strain.

Because of this restricted rotation around single bonds, cyclic compounds are capable of displaying geometric isomerism in a manner similar to the alkenes. For example, compare 1,2-dibromocyclopropane with 1,2-dibromoethene (discussed in Section 2.7). In both cases, the relation of the bromines to one another is maintained because of the impossibility of carbon-carbon bond rotation. Thus 1,2-dibromocyclopropane has a cis and a trans isomer, just like 1,2-dibromoethene.

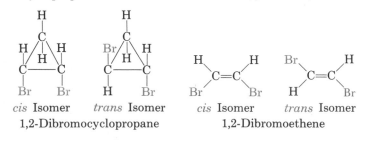

<table>
<tr><td align="center">*cis* Isomer</td><td align="center">*trans* Isomer</td><td align="center">*cis* Isomer</td><td align="center">*trans* Isomer</td></tr>
<tr><td align="center" colspan="2">1,2-Dibromocyclopropane</td><td align="center" colspan="2">1,2-Dibromoethene</td></tr>
</table>

2.8. Draw the geometric isomers of the following compounds. Label them *cis* and *trans*.

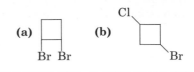

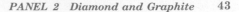

Panel 2

DIAMOND AND GRAPHITE

Diamond and graphite are two of the crystalline forms of elemental carbon. Both are composed only of carbon atoms, yet their characteristic differences are due solely to a variation in the geometric arrangement and crystalline bonding of the carbon atoms. The dissimilarities in crystalline structure give rise to a striking difference in properties. Diamond is the hardest, most abrasive mineral known; graphite is soft, slippery, and often used as a lubricant—for example, to relieve stiff locks. Diamonds can be colorless and transparent, whereas graphite is a black, opaque material (as found in pencils). Graphite can conduct an electric current; diamond cannot. And the difference in price is tremendous, even though each is merely a collection of carbon atoms.

To explain these diverse properties, let us delve into the crystalline structure on an atomic scale. Both substances are covalently bonded. In diamond, each carbon is singly bonded to four other carbons in a tetrahedral arrangement (sp^3-hybridized) with 109.5° bond angles. This pattern extends continuously throughout a vast network, as shown in Figure 2.12. Note the recurring chair conformations of cyclohexane throughout the network. The strength and hardness of diamond are a result of this crystalline structure. To break a diamond involves not merely cleavage of the substance between molecules but the actual cracking of the molecule along innumerable strong covalent bonds. In reality, a diamond is a single gigantic molecule. Therefore, to break it requires breaking the molecule itself.

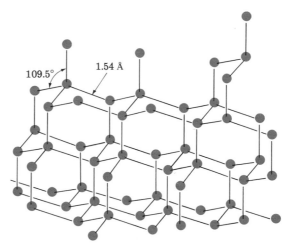

FIGURE 2.12. The covalent network in diamond.

[PANEL 2]   In graphite, each carbon is bonded to three other carbon atoms. The geometry around each carbon is that of a planar equilateral triangle with 120° bond angles (sp^2-hybridized). Like all sp^2-hybridized carbons, those in graphite have an unhybridized p orbital, in this case possessing one electron. Because of this geometry, all carbons in a molecule of graphite are necessarily in the same plane. The p orbitals can thereby overlap continuously, creating a mobile π cloud of electrons above and below each large graphite molecule. These large hexagonal sheets are layered on one another, cushioned by the π electron cloud (Figure 2.13). The loosely held electron mass is responsible for the electrical conductivity of graphite. Furthermore, gas molecules can be absorbed into the π cloud, where they act as ball bearings. This allows the carbon sheets to slide by one another—hence the lubricating properties of graphite.

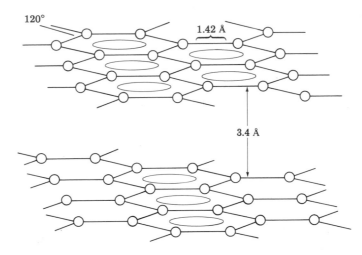

FIGURE 2.13. Crystalline structure of graphite.

END-OF-CHAPTER PROBLEMS

2.9 Skeletal Isomers: Draw the isomers described.
(a) nine isomers with the formula C_7H_{16}
(b) eighteen isomers with the formula C_8H_{18}
(c) eight isomers of C_9H_{20}, with five carbons in the longest chain
(d) twenty-five isomers of $C_{10}H_{22}$, with seven carbons in the longest chain
(e) five cyclic compounds of C_5H_{10} ⎤ disregard
(f) eleven cyclic compounds of C_6H_{12} ⎦ geometrical isomers

2.10 Positional Isomers: Suppose you have a method by which you could remove a single hydrogen from a molecule and replace it with a chlorine. For each of the following molecules, determine how many different isomers, each possessing one chlorine, could be made by replacing a single hydrogen with a chlorine.

(a) $CH_3CH_2CH_2CH_2CH_2CH_3$

(b)
$$CH_3\overset{\displaystyle CH_3}{\underset{\displaystyle CH_3}{C}}CH_2CH_2\overset{\displaystyle CH_3}{C}HCH_3$$

$$\underset{\text{(c)}}{} \quad CH_3\overset{\overset{\displaystyle CH_3}{|}}{C}HCH_2\overset{\overset{\displaystyle CH_3}{|}}{C}HCH_2\overset{\overset{\displaystyle CH_3}{|}}{C}HCH_3 \qquad \text{(d)}$$

2.11 Skeletal and Positional Isomers: Draw the isomers described.

(a) five isomers of C_3H_6BrCl (b) four isomers of C_4H_9Br

(c) eight isomers of $C_5H_{11}F$ (d) twelve isomers of C_4H_8BrF

(e) four isomers of C_3H_9N (f) eight isomers of $C_4H_{11}N$

2.12 Functional Isomers: Using the formula $C_4H_8O_2$, draw a (an)

(a) carboxylic acid   (b) alcohol-aldehyde   (c) alcohol-ketone

(d) ether-aldehyde   (e) ether-ketone   (f) alkene-dialcohol

(g) alkene-diether   (h) alcohol-ether   (i) dialcohol

(j) diether

2.13 Skeletal, Positional and Functional Isomers: Draw the isomers described.

(a) aldehydes with the formula C_4H_8O

(b) ketones with the formula $C_6H_{12}O$

(c) aldehydes and ketones with the formula $C_5H_{10}O$ (7 total)

(d) carboxylic acids with the formula $C_6H_{12}O_2$ (7 total)

(e) alcohols and ethers with the formula $C_4H_{10}O$ (7 total)

(f) alcohols and ethers with the formula $C_5H_{12}O$ (12 total)

(g) six functional isomers of $C_5H_{10}O$

(h) alkynes and alkenes with the formula C_5H_8 (9 total). Disregard geometric isomers.

2.14 Expressing Units of Unsaturation; Functional Isomers: Determine the units of unsaturation in the following molecular formulas and then draw the isomers described.

(a) Using the formula C_8H_{10} draw three compounds, one with as many triple bonds as possible, one with as many double bonds as possible, and one with as many rings as possible.

(b) Draw six isomers of C_6H_8 so that each differs from the others in the number of triple bonds or double bonds or rings.

(c) Describe in words how four units of unsaturation can be expressed in terms of triple bonds, double bonds, and rings.

2.15 Conformational Isomers: Using Newman projections or sawhorse diagrams, draw the conformational isomers of

(a) propane, $CH_3CH_2CH_3$— (cylinder gas for rural areas).

(b) ethylene glycol, $HOCH_2CH_2OH$— (antifreeze; production of dacron).

2.16 Geometric Isomers: Draw the two geometric isomers of each of the following compounds:

(a) $BrCH=CHCl$ (b) $BrFC=CHCl$ (c) $CH_3CH_2CH=CHCH_3$

(d) $CH_3\overset{\overset{\displaystyle CH_3}{|}}{C}HCH=C\overset{\overset{\displaystyle |}{CH_2CH_3}}{} CH_2CH_2CH_2OH$ (e) $Br\text{—}\bigwedge\text{—}Br$

2.17 Geometric Isomers: Each of the following compounds has three or four geometric isomers. Draw them.

(a) $CH_3CH{=}CH{-}CH{=}CHCH_3$ **(b)** $BrCH{=}CH{-}CH{=}CHF$

(c)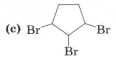

2.18 Isomers: Draw the isomers of the following molecular formulas: **(a)** C_4H_8 (6 isomers); **(b)** C_3H_5Br (5 isomers); **(c)** C_5H_{10} (12 isomers); **(d)** C_3H_4BrF (16 isomers).

2.19 Ring Stability: Which of the following is most likely to be stable? Explain.

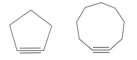

2.20 Conformational Isomerism in Cyclohexane: Draw the most stable and least stable chair conformations of the following molecules:

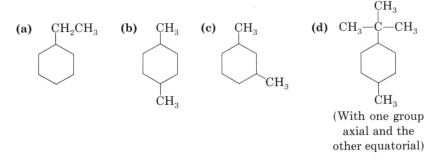

(With one group axial and the other equatorial)

2.21 Conformational Isomerism in Cyclohexane: Camphor, a compound long respected for its medicinal qualities (actually it has been found to have little or no value), exists in the boat conformation. Draw the molecule in this conformation and explain why it cannot exist in the chair form.

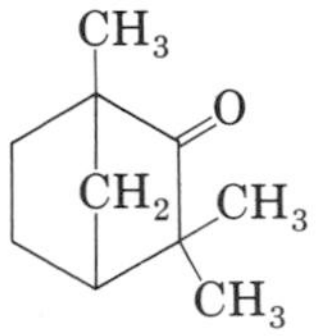

2.22 Types of Isomerism: Indicate the type of isomerism displayed in each of the following cases:

(a) $CH_3CH_2CH_2CH_2CH_3$ $CH_3{-}\overset{\underset{\displaystyle CH_3}{|}}{\overset{\displaystyle CH_3}{C}}{-}CH_3$ **(b)** 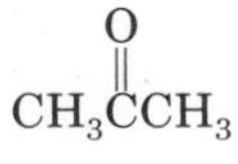 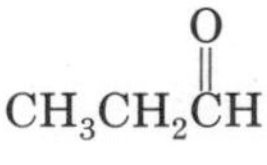 $CH_3\overset{\displaystyle O}{\overset{||}{C}}CH_3$ $CH_3CH_2\overset{\displaystyle O}{\overset{||}{C}}H$

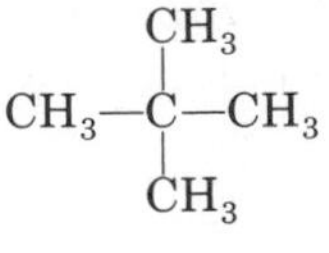

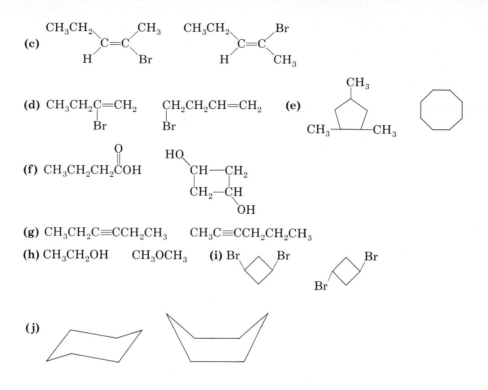

(c) 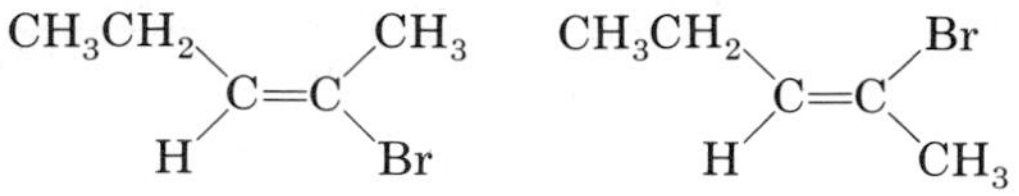

(d) 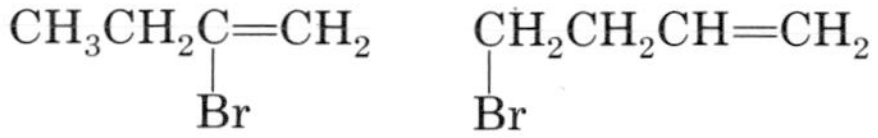

(e) 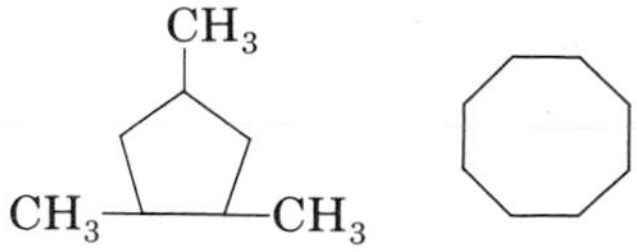

(f) $CH_3CH_2CH_2\overset{\overset{O}{\|}}{C}OH$ 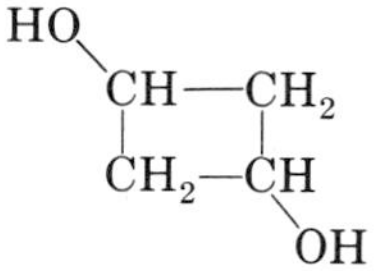

(g) $CH_3CH_2C{\equiv}CCH_2CH_3$ $CH_3C{\equiv}CCH_2CH_2CH_3$

(h) CH_3CH_2OH CH_3OCH_3 **(i)** 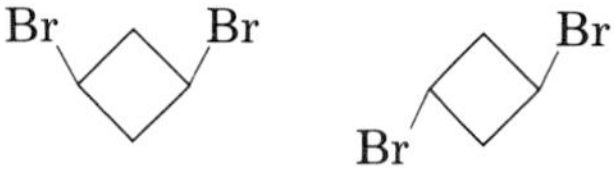

(j) 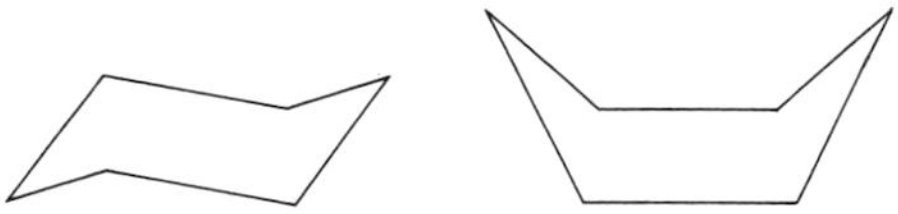

Chapter 3

Nonaromatic Hydrocarbons I

CLASSES OF HYDROCARBONS

3.1   *Hydrocarbons* are organic compounds composed only of carbon and hydrogen. Currently, the major sources of hydrocarbons are petroleum and, to a lesser extent, coal. Petroleum is a complex mixture of literally thousands of compounds, most of them hydrocarbons, formed by the decay and degradation of marine plants and animals (see Panel 3 at the end of this chapter). The ultimate source of carbon for these organisms was carbon dioxide. This, in the presence of water and sunlight, was converted by green plants into biological materials through photosynthesis. Modern civilization uses hydrocarbons (natural gas, gasoline, diesel fuel, fuel oil) to produce energy and through combustion it is converting this treasure of organic compounds back to the materials from whence it came—carbon dioxide and water. A small portion of petroleum (5%–15%) is converted into petrochemicals—plastics, dyes, medicines, fibers, detergents, insecticides, and other items. The bulk is burned as fuel.

Hydrocarbons fall into two major classes: saturated hydrocarbons in which all bonds, C—C and C—H, are single bonds; and unsaturated hydrocarbons in which the molecule has at least one carbon-carbon double bond or triple bond. The terms *saturated* and *unsaturated* refer to the number of atoms bonded to a carbon. Carbon must have four bonds. If all the bonds are single then the carbon will have four attached atoms. This is the maximum number of attachments and if all carbons in the molecule have four atoms attached to them, the molecule is saturated. Since a carbon in a double bond has only three attached atoms and that in a triple bond has but two, molecules with multiple bonds are said to be unsaturated.

The saturated hydrocarbons, or alkanes, have the general formula C_nH_{2n+2}. The first four members of the series—methane, CH_4; ethane, C_2H_6; propane, C_3H_8; and butane, C_4H_{10}—are the main components of natural gas. There are three major types of unsaturated hydrocarbons—alkenes, alkynes, and aromatics. Alkenes have a double bond and are of the general formula C_nH_{2n}. Alkynes, with a triple bond, take the form C_nH_{2n-2}. Aromatics do not assume a general formula. Alkenes or alkynes with more than one multiple bond can be further classified as to the relative locations of the bonds. Consider the following dienes, for example. If the double bonds are consecutive, the diene is cumulative, whereas if they alternate with single bonds the compound is conjugated. If two or more single bonds separate the double bonds, they are isolated. Table 3.1 summa-

$$CH_2=C=CH_2 \qquad CH_2=CH-CH=CH_2 \qquad CH_2=CH-CH_2-CH=CH_2$$

Propadiene	1,3-Butadiene	1,4-Pentadiene
cumulative	conjugated	isolated

rizes the classes of hydrocarbons. Bonding in the hydrocarbons was covered in section 1.5. Remember, all single bonds are σ bonds, a double bond is a σ and a π bond, and a triple bond is constructed of one σ bond and two π bonds.

TABLE 3.1. CLASSES OF HYDROCARBONS

Class	General Formula	Examples	IUPAC Suffix
Saturated			
Alkanes	C_nH_{2n+2}	CH_4, CH_3CH_3, $CH_3CH_2CH_3$	*-ane*
Unsaturated			
Alkenes	C_nH_{2n}	$CH_2=CH_2$, $CH_3CH=CH_2$	*-ene*
Alkynes	C_nH_{2n-2}	$HC\equiv CH$, $CH_3C\equiv CH$	*-yne*
Aromatics	. . .		. . .

PROBLEMS

3.1. Draw a bonding picture for the following molecule showing all σ and π bonds. Describe the geometric arrangement of the groups bonded to each carbon, the bond angles, and hybridization.

$$CH_2=CH-CH_2-C\equiv CH$$

3.2. What are the general molecular formulas (such as C_nH_{2n+2} for alkanes) for cycloalkanes, cycloalkenes, and dienes?

IUPAC NOMENCLATURE OF NONAROMATIC HYDROCARBONS

3.2   **A. Introduction to IUPAC Nomenclature**

There are several million organic compounds. Each has its individual structural characteristics just as every person does. Yet five "John Smith" compounds would hamper the rapid pace of a research chemist and be intolerably confusing to the student. The early history of chemistry shows that a system of classification developed in a rather random way. A name could indicate a compound's source—like *pinene,* from pine trees; *cocaine,* from Peruvian cocoa leaves; or *bu-*

tyric acid, from rancid butter. Smell was also considered a rational distinction, with *putrescine* and *cadaverine* named after decaying flesh, and *caproic, caprylic,* and *capric* acids after the Latin for *goat.* Color (*Congo red, malachite green*), geometry (*basketane, cubane*), and public adulation (*dieldrin* and *aldrin,* insecticides named for Nobel Prize winners Otto Diels and Kurt Alder) added to the long list of popular jargon. These common, or trivial, names were frequently assigned by alchemists who may have been attempting to conceal a substance's source, identity, or importance. A more legitimate reason for common names was the ignorance of early organic chemists about the structures of the compounds they were investigating.

As structural theory developed and organic chemistry began to grow, however, the need for structurally descriptive and systematic names became imperative. Such an approach to nomenclature had to be able to generate terms that described the kinds of atoms, the number of each kind, and the arrangement of the atoms in the molecule. Systematic nomenclature had its genesis in the early 1800s but did not really become organized until the turn of the century, when an international commission met in Geneva. Today, compounds are named systematically by the rules of nomenclature of the International Union of Pure and Applied Chemists (IUPAC).

The IUPAC nomenclature for nonaromatic hydrocarbons (alkanes, alkenes, and alkynes) will be presented in this chapter. Nomenclature for other classes of compounds will be presented in individual chapters as the chemistry of the functional groups is covered. (The appendix summarizes IUPAC nomenclature for all organic compounds.) The common, nonsystematic names of some compounds have persisted. A few of the most important of these will be covered in individual chapters.

B. Alkanes, Saturated Hydrocarbons

1. Continuous-Chain Unbranched Alkanes. These are compounds containing only carbon and hydrogen in a continuous, unbranched carbon chain. All carbon-carbon bonds are single bonds. The names of these compounds constitute the basis of the nomenclature of hydrocarbon derivatives and organic compounds in general.

Nomenclature Rule: The names of the continuous-chain (unbranched) saturated hydrocarbons (alkanes) are derived from the Greek names for the *numbers of carbon atoms* present, followed by the suffix *-ane.* A cyclic (ring) hydrocarbon is designated by the prefix *cyclo-.*

Table 3.2 presents structural formulas of enough saturated hydrocarbons to represent their nomenclature. The first four alkanes have trivial names that because of their continuing use were made a part of the IUPAC system. As a minimum, it helps to learn the names of the first ten hydrocarbons.

Ring, or cyclic, compounds are designated by the prefix *cyclo-.*

Cyclopropane   Cyclobutane   Cyclopentane   Cyclohexane   Cyclooctane

TABLE 3.2.   CONTINUOUS-CHAIN HYDROCARBONS

First Ten Hydrocarbons

CH_4	Methane	$CH_3(CH_2)_4CH_3$	Hexane
CH_3CH_3	Ethane	$CH_3(CH_2)_5CH_3$	Heptane
$CH_3CH_2CH_3$	Propane	$CH_3(CH_2)_6CH_3$	Octane
$CH_3(CH_2)_2CH_3$	Butane	$CH_3(CH_2)_7CH_3$	Nonane
$CH_3(CH_2)_3CH_3$	Pentane	$CH_3(CH_2)_8CH_3$	Decane

2. Branched-Chain Alkanes.   These are compounds in which short carbon chains (alkyl groups) are attached to a longer carbon skeleton. To name these compounds, first find the longest continuous chain of carbon atoms. Name it according to Table 3.2 by using the Greek word for the number of carbons present, followed by the suffix *-ane*. In cyclic compounds the ring is usually used as the base of the name regardless of the longest continuous carbon chain. The shorter chains are then named by prefixes derived by changing the *-ane* ending of the corresponding alkane to *-yl* (see Table 3.3). The following rule summarizes the nomenclature of these compounds:

Nomenclature Rule: The names of branched-chain hydrocarbons are based on the name of the longest *continuous* carbon chain in the molecule. The names of attached hydrocarbon substituents (alkyl groups) are derived by changing the ending of the hydrocarbon from *-ane* to *-yl*. To locate the position of a substituent, one should number the carbon chain consecutively from one end to the other, starting at the end that will give the lowest number to the substituent.

TABLE 3.3.   ALKYL GROUPS

CH_3-	Methyl	$CH_3(CH_2)_4CH_2-$	Hexyl
CH_3CH_2-	Ethyl	$CH_3(CH_2)_5CH_2-$	Heptyl
$CH_3CH_2CH_2-$	Propyl	$CH_3(CH_2)_6CH_2-$	Octyl
$CH_3(CH_2)_2CH_2-$	Butyl	$CH_3(CH_2)_7CH_2-$	Nonyl
$CH_3(CH_2)_3CH_2-$	Pentyl	$CH_3(CH_2)_8CH_2-$	Decyl

Branched alkyl groups

		CH_3	CH_3
		$\mid$	$\mid$
CH_3CHCH_3	$CH_3CHCH_2CH_3$	CH_3CHCH_2-	CH_3CCH_3
$\mid$	$\mid$		$\mid$
Isopropyl	Secondary butyl	Isobutyl	Tertiary butyl
	(*sec-*)		(*tert-*, or *t-*)

The alkyl groups with three and four carbon chains deserve special mention. A three-carbon-chain alkyl group could be attached to a longer chain at either the outside carbons or the middle carbon; the groups are called propyl and isopropyl respectively.

Propyl $CH_3CH_2CH_2-$ Isopropyl CH_3CHCH_3
$\mid$

There are two structural isomers of a four-carbon alkyl group, each of which has two different points of connection.

$$CH_3CH_2CH_2CH_2-$$

Butyl

$$CH_3\overset{|}{\underset{|}{C}}HCH_2CH_3$$

Secondary butyl

$$CH_3\overset{\overset{\displaystyle CH_3}{|}}{\underset{|}{C}}HCH_2-$$

Isobutyl

$$CH_3\overset{\overset{\displaystyle CH_3}{|}}{\underset{|}{C}}CH_3$$

Tertiary butyl

Now, carefully correlate these principles of nomenclature to the examples which follow.

Example 3.1. Name

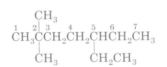

$$\overset{5}{C}H_3\overset{4}{C}H_2\overset{3}{C}H_2\overset{2}{C}H\overset{1}{C}H_3$$
$$\overset{|}{C}H_3$$

1. The longest continuous chain has five carbons and thus the base of the name is *pentane*.
2. A one-carbon substituent, *methyl-,* is attached to the longest chain. The compound is so far a methylpentane.
3. Numbering left to right locates the methyl group on carbon 4. Conversely, numbering right to left puts it on carbon 2. The second alternative gives the lowest number to the substituent. The compound name is 2-methylpentane.

Example 3.2. Name

$$\overset{1}{C}H_3\overset{\overset{\displaystyle CH_3}{\overset{2|}{}}}{\underset{\underset{\displaystyle CH_3}{|}}{C}}\overset{3}{C}H_2\overset{4}{C}H_2\overset{5}{C}H\underset{\underset{\displaystyle CH_2CH_3}{|}}{\overset{6}{C}}H_2\overset{7}{C}H_3$$

1. The longest continuous chain has seven carbons: a heptane.
2. There are three shorter chain branches on the heptane chain: 2 one-carbon chains and 1 two-carbon chain. One-carbon substituents are called methyl groups. Since there are two methyl groups, they are referred to as dimethyl (three, trimethyl; four, tetramethyl; five, pentamethyl; etc.). The two-carbon chain is called an ethyl group. The parent compound is an ethyl dimethyl heptane.
3. The substituents must now be placed on the longest chain. Numbering from left to right allows the lowest designations. The two methyl groups are on carbon 2 (2,2-dimethyl) and the ethyl is on carbon 5 (5-ethyl). Each substituent gets its own number. The complete name is 5-ethyl-2,2-dimethylheptane.

Example 3.3. Name

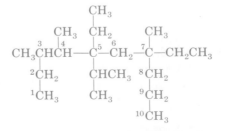

1. Finding the longest continuous chain may require careful inspection. We are not looking for the longest straight chain; rather, the longest continuous carbon chain. In this case, it has ten carbons: a decane.
2. The chain is numbered from left to right to obtain the lowest numbering for the location of the substituents.
3. Identify the substituents on the decane chain. On carbons 3, 4, and 7 there are methyl groups (3,4,7-trimethyl). Attached to carbons 5 and 7 are ethyl groups (5,7-diethyl). Carbon 5 also has an isopropyl group (5-isopropyl). Arranging the substituents in alphabetical order, we get the name 5,7-diethyl-5-isopropyl-3,4,7-trimethyldecane.

Example 3.4. Name

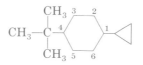

1. The cyclohexane ring is the base of the name.
2. Numbering in this case is alphabetical. The substituent on carbon 1 should be designated cyclopropyl- and the one on carbon 4, *tert*-butyl-.
3. The name is 1-cyclopropyl-4-*tert*-butylcyclohexane.

PROBLEM

3.3. Name by the IUPAC system of nomenclature

$$\text{CH}_2\text{CH}_2\text{CH}_3$$

(a) $\text{CH}_3\text{CH}_2\text{CH}_2\text{CH}_2\text{CHCH}_2\text{CH}_2\text{CH}_3$

$$\text{CH}_3 \quad \text{CH}_3$$

(b) $\text{CH}_3\text{CCH}_2\text{CHCHCH}_3$

$$\text{CH}_3 \qquad \text{CH}_2\text{CH}_3$$

$$\text{CH}_3$$

(c) CH_2CHCH_3

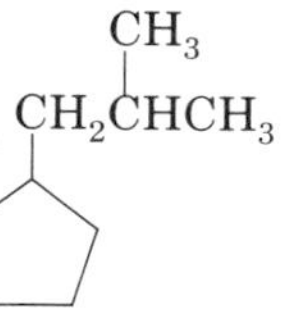

3. Halogenated Hydrocarbons. Halogens attached to a hydrocarbon chain are named by the prefixes *F* (*fluoro*-), *Cl* (*chloro*-), *Br* (*bromo*-), and *I* (*iodo*-).

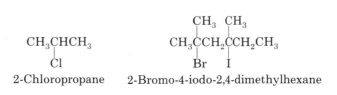

2-Chloropropane   2-Bromo-4-iodo-2,4-dimethylhexane

PROBLEM

3.4. Name by the IUPAC system of nomenclature

$$\text{CH}_3 \quad \text{Br}$$

(a) CCl_2F_2 **(b)** $\text{CH}_3\text{CHCH}_2\text{CCH}_2\text{CH}_3$ **(c)** CH_3CH_2—⬦—I

$$\text{Br}$$

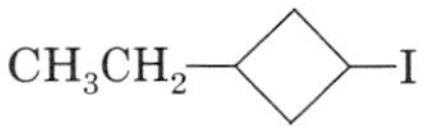

C. Alkenes and Alkynes

The names of hydrocarbons are based on the Greek word indicating the number of carbons in the longest continuous chain of carbon atoms. This is followed by a suffix that indicates whether the compound is saturated or unsaturated. For alkanes in which all carbon-carbon bonds are single bonds, the suffix is *-ane*. Carbon-carbon double bonds are designated by the suffix *-ene* (alkenes) and triple bonds by *-yne* (alkynes).

C—C	C=C	C≡C
Alkanes	Alkenes	Alkynes
(*-ane*)	(*-ene*)	(*-yne*)

Double and triple bonds must be specifically identified and their position determined in order to name a compound. All other carbon-carbon attachments are assumed to be single bonds.

Nomenclature Rule: Double bonds in hydrocarbons are indicated by changing the suffix *-ane* to *-ene;* and triple bonds are indicated by the suffix *-yne*. The position of the multiple bond is indicated by the number of the first, or lowest-numbered, carbon atom involved in the multiple bond.

Example 3.5. Name

$$\overset{4}{C}H_3\overset{3}{C}H_2\overset{2}{C}H=\overset{1}{C}H_2 \qquad \overset{1}{C}H_3\overset{2}{C}H=\overset{3}{C}H\overset{4}{C}H_3$$
$$A \qquad\qquad\qquad B$$

1. Both A and B have a four-carbon continuous chain; therefore the prefix *but-*.
2. Both have a double bond designated by the suffix *-ene:* butene.
3. Number the carbon chain from the end that will give the lowest number to the double bond. The double bond is located by the lowest-numbered carbon involved. A is 1-butene; B is 2-butene.
4. Compound B is capable of geometric isomerism (section 2.7) because each carbon involved in the double bond has two different groups attached. To designate the geometric isomers, we use the terms *cis* and *trans*.

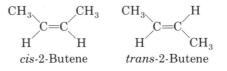

cis-2-Butene   *trans*-2-Butene

Example 3.6. Name

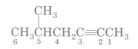

1. There are six carbons in the longest continuous chain: *hex-*.
2. The triple bond is designated by the suffix *-yne:* hexyne.
3. Number so as to give the multiple bond the lowest number, right to left: 2-hexyne.
4. Name the branched methyl group with a prefix. The complete name is 5-methyl-2-hexyne.

Example 3.7. Name

1. There are five carbons in the ring. This is designated by *cyclopent-*.
2. The two double bonds are indicated by *-diene* (*di,* meaning "two"), cyclopenta-diene (the syllable *a* is added for smoother pronounciation).
3. The ring is numbered, giving the lowest possible numbers to the carbons involved in the double bonds. The complete name is 1,3-cyclopentadiene.

PROBLEM

3.5. Name the following by the IUPAC system of nomenclature:

(a) $CH_3CH_2CH_2CH_2CH_2CH{=}CH_2$ **(b)** $CH_3C{\equiv}CCH_2CH_2CH_2CH_3$

(c) [hexene ring]—CH_3 **(d)** $CH_3C{\equiv}C-C{\equiv}C-C{\equiv}C-C{\equiv}C-CH_3$

COMMON NOMENCLATURE OF
NONAROMATIC HYDROCARBONS

3.3   A few common names have been so persistent that it is necessary to learn them. Some alkenes are referred to by an alkylene type nomenclature, such as ethylene ($CH_2{=}CH_2$), from which polyethylene is made, and propylene ($CH_3CH{=}CH_2$), the precursor of polypropylene. Occasionally, ethylene and propylene appear as the prefixes *vinyl-* and *allyl-,* respectively (the plastic PVC, polyvinyl chloride, is made from vinyl chloride).

Vinyl $CH_2{=}CH-$ Allyl $CH_2{=}CHCH_2-$

Alkynes are sometimes named as derivatives of acetylene ($HC{\equiv}CH$, used in oxyacetylene welding torches), the simplest alkyne. For example, $HC{\equiv}CCH_3$ is methyl acetylene.

PROBLEM

3.6. Draw structures for the following molecules: **(a)** butylene; **(b)** vinyl bromide; **(c)** allyl alcohol; **(d)** ethylmethylacetylene.

HYDROCARBONS: RELATION OF STRUCTURE
AND PHYSICAL PROPERTIES

3.4   The solid, liquid, and gaseous states of a compound do not represent differences in the structure of the individual molecules. They represent, rather, variations in the arrangement of the molecules. In a solid, the molecules are arranged very compactly and are relatively immobilized in an orderly crystal lattice. Attractive forces between molecules are at a maximum (Figure 3.1). In the liquid state, the intermolecular attractions still exist, but the molecules are mobile; they have greater kinetic energy. Molecular mobility in the vapor phase has increased to the point that all intermolecular attractions are practically nonexistent and each molecule is theoretically independent of the others. Energy, in the form of heat, is required to provide molecules with the impetus and mobility to break out of a crystal lattice and form a liquid or to sever all attractive forces and become a vapor. Here we will consider the factors that influence the melting points and boiling points of nonaromatic hydrocarbons (see Tables 3.4–3.6).

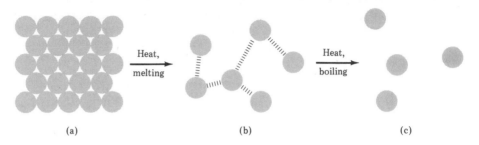

FIGURE 3.1. Physical states of matter. (a) Solid. (b) Liquid. (c) Gas.

TABLE 3.4. MELTING POINTS AND BOILING POINTS OF ALKANES

Name	Formula	Molecular Weight	Melting Point, °C	Boiling Point, °C
Methane	CH_4	16	−182	−164
Ethane	CH_3CH_3	30	−183	− 89
Propane	$CH_3CH_2CH_3$	44	−190	− 42
Butane	$CH_3(CH_2)_2CH_3$	58	−138	− 1
Pentane	$CH_3(CH_2)_3CH_3$	72	−130	36

TABLE 3.5. MELTING POINTS AND BOILING POINTS OF ALKENES

Name	Formula	Molecular Weight	Melting Point, °C	Boiling Point, °C
Ethene	$CH_2{=}CH_2$	28	−169	−104
Propene	$CH_3CH{=}CH_2$	42	−185	− 47
1-Butene	$CH_3CH_2CH{=}CH_2$	56	−185	− 6.3
1-Pentene	$CH_3(CH_2)_2CH{=}CH_2$	70	−138	30
1-Decene	$CH_3(CH_2)_7CH{=}CH_2$	140	− 66	171

TABLE 3.6. MELTING POINTS AND BOILING POINTS OF ALKYNES

Name	Formula	Molecular Weight	Melting Point, °C	Boiling Point, °C
Acetylene	$HC{\equiv}CH$	26	− 81	− 84
Propyne	$CH_3C{\equiv}CH$	40	−102	− 23
1-Butyne	$CH_3CH_2C{\equiv}CH$	54	−126	8
1-Pentyne	$CH_3(CH_2)_2C{\equiv}CH$	68	− 90	40
1-Decyne	$CH_3(CH_2)_7C{\equiv}CH$	138	− 36	174

A. Melting Points, Boiling Points, and Molecular Weight

Within a homologous series,* melting points and boiling points increase with increasing molecular weight for most classes of organic compounds. Consider your own experience with the following hydrocarbon fractions of increasing molecular weight: natural gas is, of course, a gas, gasoline is a volatile liquid, motor oil is a thick nonvolatile liquid, and paraffin wax (candles) is a solid. These trends in melting point and boiling point are understandable on two bases. First, the larger the molecule, the more numerous the sites that exist for intermolecular attractions. These attractions must be either weakened or broken in any transition from the solid to the liquid or the liquid to the gaseous states. Second, the heavier the substance, the greater the energy needed to give the molecules sufficient impetus to break these intermolecular forces. Melting point trends are less regular than the trends for boiling points since melting also depends on the correct fit of a molecule into its crystal lattice. We can see in Tables 3.4–3.6 that these generalizations hold for the hydrocarbons. We shall see similar trends with other functional groups.

B. Melting Point, Boiling Point, and Molecular Structure

Boiling points decrease with chain branching in hydrocarbons. Branching makes a molecule more compact and decreases the surface area. The smaller the surface area, the fewer the opportunities for intermolecular attraction; consequently, the branched molecules have lower boiling points. On the other hand, compactness or molecular symmetry usually increases the melting point of a compound. Such molecules fit more easily into a crystal lattice. A more stable crystal lattice requires a larger energy to disrupt. Consequently, the melting points are higher for highly branched compounds than for compounds with a longer, straighter chain.

Consider, for example, the isomers of C_5H_{12} (all with molecular weight 72).

	$CH_3CH_2CH_2CH_2CH_3$	$CH_3\overset{\displaystyle CH_3}{\underset{\displaystyle |}{C}}HCH_2CH_3$	$CH_3\overset{\displaystyle CH_3}{\underset{\displaystyle \underset{\displaystyle CH_3}{|}}{\overset{|}{C}}}CH_3$
MP, °C	−130	−160	−17
BP, °C	36	28	9.5

Note the progressive decrease in boiling point with branching. Also observe the large difference in melting point between pentane and the highly compact, symmetrical dimethylpropane.

A dramatic difference in melting point occurs between the isomers octane and 2,2,3,3-tetramethylbutane.

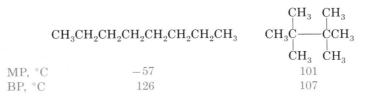

	$CH_3CH_2CH_2CH_2CH_2CH_2CH_2CH_3$	$CH_3\overset{CH_3}{\underset{CH_3}{C}}{-}{-}\overset{CH_3}{\underset{CH_3}{C}}CH_3$
MP, °C	−57	101
BP, °C	126	107

---

* A *homologous series* of compounds is a series in which each compound differs from the one preceding it by a constant amount. For example, each of the members of the homologous series methane, ethane, propane, butane, pentane, and so on differs from the preceding by a —CH_2— group.

Fitting octane into an orderly crystal lattice would be like trying to stack wet spaghetti, whereas arranging the 2,2,4,4-tetramethylbutane is analogous to stacking wooden blocks. However, octane, with more surface area, has a higher boiling point.

C. Solubility

Hydrocarbons are nonpolar molecules and as a result are not soluble in water, which dissolves polar substances. Since hydrocarbons are less dense than water, they will float on its surface (oil spills stay on the ocean's surface). Hydrocarbons are miscible with many organic liquids such as alcohols, ethers, ketones, and carboxylic acids.

A PRELUDE TO ORGANIC REACTIONS

3.5   Part of the attraction of organic chemistry lies in the variety of chemical changes that occur among organic compounds. Equally fascinating is the fact that most of these reactions can be rationalized and unified around a few basic principles. In this text, only the most characteristic reactions of each class of organic compounds will be presented. As you study the reactions of each functional group, you should try to organize them in the context of the ideas presented in this section.

A. Reaction Types

Most of the reactions we shall encounter can be classified as substitution, elimination, or addition.

Substitution. In a substitution reaction, an atom or group of atoms is replaced by another species.

$$-\overset{|}{\underset{|}{C}}-A + B \longrightarrow -\overset{|}{\underset{|}{C}}-B + A$$

Elimination. An elimination reaction involves the removal of a pair of atoms or groups from adjacent carbon atoms. This necessarily results in the formation of a multiple bond.

$$-\overset{|}{\underset{|}{C}}-\overset{|}{\underset{|}{C}}- \longrightarrow -\overset{|}{C}=\overset{|}{C}- + AB$$
$$\quad B \quad A$$

Addition. In an addition reaction, atoms or groups add to the adjacent carbons of a multiple bond. To maintain the proper valence the multiplicity of the bond decreases.

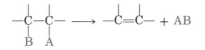

$$-\overset{|}{C}=\overset{|}{C}- + A-B \longrightarrow -\overset{|}{\underset{|}{C}}-\overset{|}{\underset{|}{C}}-$$
$$\qquad\qquad\qquad\qquad A \quad B$$

PROBLEM

3.7. Classify the following reactions, which involve either a preparation or a reaction of ethyl alcohol (beverage alcohol), as substitution, elimination, or addition reactions:

(a) $CH_3CH_2Br + NaOH \longrightarrow CH_3CH_2OH + NaBr$

(b) $CH_3CH_2OH \xrightarrow[\text{catalyst}]{H_2SO_4} CH_2{=}CH_2 + H_2O$

(c) $CH_3\overset{\overset{\displaystyle O}{\|}}{C}H + H_2 \xrightarrow[\text{catalyst}]{Ni} CH_3CH_2OH$

(d) $CH_2{=}CH_2 + H_2O \xrightarrow[\text{catalyst}]{H_2SO_4} CH_3CH_2OH$

(e) $CH_3CH_2OH \xrightarrow[\text{catalyst}]{\text{Heat, Cu}} CH_3\overset{\overset{\displaystyle O}{\|}}{C}H + H_2$

(f) $CH_3CH_2OH + HCl \xrightarrow{ZnCl_2} CH_3CH_2Cl + H_2O$

B. Reaction Mechanisms

To understand organic reactions completely and to interrelate them effectively, we must delve into the step-by-step process by which a chemical change occurs, in other words, the reaction mechanism. A chemical equation describes what happens, whereas a reaction mechanism describes how it happens. Consider, for example, the possible ways, or reaction mechanisms, by which HCl could add to ethene:

$$CH_2{=}CH_2 + HCl \longrightarrow \underset{\underset{Cl \quad\; H}{|\qquad |}}{CH_2{-}CH_2}$$

Do the hydrogen and chlorine add simultaneously? Does the hydrogen bond first, followed by the chlorine? Or is it possible that the chlorine bonds first, followed by the hydrogen? Do the hydrogen and chlorine add as charged or neutral species? Are any short-lived intermediate species formed during the steps of the reaction? A reaction mechanism answers these questions in describing the reaction. It is postulated and then supported by experiment.

C. Reaction Intermediates

Multistep reaction mechanisms usually involve the formation of unstable, short-lived species called *reaction intermediates*. There are three major types: carbocations (also called carbonium ions), carbanions, and free radicals.

Carbocation   Free radical   Carbanion

Each of these species is unstable for one or both of the following reasons: (1) the particle is charged (carbocation, carbanion); (2) the particle does not have an octet of electrons in the outer shell (carbocation, free radical). In the carbocation, carbon "owns" half of the bonding pairs of electrons, a total of three electrons. Since it is in group IV of the periodic table, to be neutral it must formally own four outer-shell electrons. Consequently, carbocations are positive. Free radicals have one additional electron and are neutral, and carbanions have formal ownership of five outer-shell electrons and are negative.

D. Polar Bonds

Polar bonds (section 1.6) involve unequal sharing of electron pairs between atoms of different electronegativity. Because of this uneven electron distribution, one atom in the bond becomes partially positive and the other partially negative. Opposite charges attract. One might expect therefore that bringing together two molecules, each with polar bonds, could result in attraction between opposite charges and consequent reaction.

Consider the reaction between methanal (formaldehyde) and hydrogen cyanide, for example. Both molecules are polar. It can be predicted that the positive hydrogen of HCN will be attracted to the negative oxygen of methanal and the negative CN^- to the positive carbon.

$$
\underset{\delta+}{\overset{\overset{\displaystyle O}{\underset{\scriptstyle \delta-}{\|}}}{H-\underset{\delta+}{C}-H}} + \overset{\delta+}{H}-\overset{\delta-}{CN} \longrightarrow \overset{\displaystyle O-H}{\underset{\displaystyle CN}{H-C-H}}
$$

In reality this does occur. An addition reaction is the result.

HALOGENATION OF ALKANES: CHLORINATION AND BROMINATION

3.6   When alkanes are treated with either chlorine or bromine in the presence of light or heat, haloalkanes are formed. This is a substitution reaction and involves the replacement of a hydrogen atom with a halogen. The result is the simultaneous formation of a haloalkane and a hydrogen halide (HCl or HBr). The following is an expression of the reaction in a general form, with only the participating bonds shown:

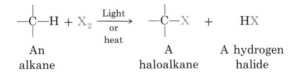

$$
-\overset{|}{\underset{|}{C}}-H + X_2 \xrightarrow[\text{or heat}]{\text{Light}} -\overset{|}{\underset{|}{C}}-X \quad + \quad HX
$$

| An | A | A hydrogen |
| alkane | haloalkane | halide |

where X = Cl, Br.

Fluorine is too reactive with most alkanes, often rupturing the carbon chain. In contrast, iodine is so unreactive that the reaction proceeds at a negligible rate.

A. Chlorination of Methane

Although halogenation of alkanes is used industrially, it has limited synthetic utility in the laboratory because of the multiplicity of products formed.

For example, methane has four carbon-hydrogen bonds and even when mixed with chlorine in a 1:1 molar ratio forms products ranging from the replacement of one to all four hydrogens.

$$CH_4 + Cl_2 \longrightarrow CH_3Cl + CH_2Cl_2 + CHCl_3 + CCl_4 + HCl$$

This occurs because once chloromethane is formed it competes with methane for the remaining chlorine because both methane and chloromethane have carbon-hydrogen bonds. The result is that some of the chloromethane is converted to dichloromethane. The dichloromethane in turn will react with some of the chlorine to form trichloromethane (chloroform), which can react further to produce tetrachloromethane (carbon tetrachloride). At the conclusion of the reaction, when all the chlorine has been consumed, the reaction vessel contains four chlorination products (CH_3Cl, CH_2Cl_2, $CHCl_3$, and CCl_4), 1 mole of HCl, and some unreacted CH_4. It is this polyhalogenation that is a serious detriment to the synthetic use of alkane halogenation.

PROBLEM

3.8. Show all the possible products formed from the chlorination of ethane in light.

B. Control of the Halogenation Reaction

As with many potentially useful reactions, the challenge of halogenation lies in the control, so that an acceptable yield of desired product is obtained. How could the halogenation reaction conditions be adjusted so as to obtain, say, predominantly carbon tetrachloride (CCl_4)? The formation of carbon tetrachloride from methane involves replacement of all four hydrogens. This requires 4 moles of chlorine for each mole of methane. Thus, if the reactants CH_4 and Cl_2 are combined in a 1:4 ratio, CCl_4 can be obtained almost exclusively.

$$CH_4 + 4Cl_2 \xrightarrow{\text{Light}} CCl_4 + 4HCl$$

For polyhalogenation to be effected in the extreme, an alkane should be exposed to at least as many moles of halogen as there are hydrogen atoms in the molecule.

On the other hand, how could the predominance of chloromethane (CH_3Cl) be assured? For such an outcome, conditions have to be such that chlorine is more likely to encounter a methane molecule than it is any of the chloromethane that forms. This can be accomplished by running the reaction in a large excess of methane (for example, $10CH_4:1Cl_2$):

$$\underset{\text{Large excess}}{CH_4} + Cl_2 \xrightarrow{\text{Light}} CH_3Cl + HCl$$

To favor monohalogenation, the alkane is introduced in excess so that the halogen molecules will always have a higher probability of reacting with the alkane rather than monohalogenated product. Even when monohalogenation is predominant, however, the less symmetrical the alkane the greater the number of products possible. Compare the possible monochlorination products of pentane and its symmetrical skeletal isomer, dimethylpropane.

$$CH_3CH_2CH_2CH_2CH_3 + Cl_2 \xrightarrow{\text{Light}}$$

Excess

$$\underset{\overset{|}{Cl}}{CH_2CH_2CH_2CH_2CH_3} + \underset{\overset{|}{Cl}}{CH_3CHCH_2CH_2CH_3} + \underset{\overset{|}{Cl}}{CH_3CH_2CHCH_2CH_3}$$

Three possible monochlorination products

$$\underset{\overset{|}{CH_3}}{\overset{CH_3}{CH_3-C-CH_3}} + Cl_2 \xrightarrow{\text{Light}} \underset{\overset{|}{CH_3}}{\overset{CH_3}{CH_3-C-CH_2Cl}} + HCl$$

Excess   One monochlorination product

PROBLEM

3.9. Draw all the possible products formed by the monochlorination of 2-methylbutane, an isomer of pentane and dimethylpropane, discussed above.

C. Mechanism of Halogenation

To discuss the mechanism of halogenation, let us consider the monochlorination of methane.

$$CH_4 + Cl_2 \xrightarrow{\text{Light}} CH_3Cl + HCl$$

For the reaction to occur a C—H and Cl—Cl bond must be broken and a C—Cl and H—Cl bond must be formed. The Cl—Cl bond (bond energy = 58 kcal/mol) is weaker than the C—H bond (bond energy = 102 kcal/mol) and is cleaved by heat or light to form two chlorine atoms (free radicals).

$$:\overset{..}{\underset{..}{Cl}}:\overset{..}{\underset{..}{Cl}}: \xrightarrow[\text{heat}]{\text{Light or}} 2:\overset{..}{\underset{..}{Cl}}\cdot \qquad \text{Chlorine atom (free radical)}$$

The chlorine atoms will immediately seek a method for completing their octets. This can be accomplished by abstracting a hydrogen atom from methane thus cleaving a C—H bond and forming one of the reaction products.

$$\underset{\overset{|}{H}}{\overset{H}{H:\overset{..}{C}:H}} + \cdot\overset{..}{\underset{..}{Cl}}: \longrightarrow \underset{\overset{|}{H}}{\overset{H}{H:\overset{..}{C}\cdot}} + H:\overset{..}{\underset{..}{Cl}}:$$

Free radical

Now we have a carbon lacking an octet of electrons—a methyl free radical. It can alleviate the situation by abstracting a chlorine from an undissociated chlorine molecule to form the other reaction product.

$$\underset{\overset{|}{H}}{\overset{H}{H:\overset{..}{C}\cdot}} + :\overset{..}{\underset{..}{Cl}}:\overset{..}{\underset{..}{Cl}}: \longrightarrow \underset{\overset{|}{H}}{\overset{H}{H:\overset{..}{C}:\overset{..}{\underset{..}{Cl}}:}} + :\overset{..}{\underset{..}{Cl}}\cdot$$

Another chlorine free radical is formed in this step that can attack yet another methane molecule thus continuing the process.

The chlorination of methane (and halogenation of alkanes in general) occurs by a free radical chain reaction. It is initiated by the light- or heat-induced cleavage of a chlorine molecule. Once initiated, the reaction will proceed in the absence of light or heat. This is due to the alternate formation of methyl and chlorine free radicals in the two propagation steps just described. Each step produces a product and a reactive intermediate, the second of which participates in the other step. The reaction will not proceed indefinitely, however, since chain termination steps, although not as statistically probable as propagation, do occur. They result in the consumption of free radicals without producing new ones to continue the chain. The entire mechanism is summarized below.

$$\textit{Chain initiation} \qquad Cl_2 \xrightarrow[\text{light}]{\text{Heat or}} 2Cl\cdot$$

$$\textit{Chain propagation} \begin{cases} CH_4 + Cl\cdot \longrightarrow CH_3\cdot + HCl \\ CH_3\cdot + Cl_2 \longrightarrow CH_3Cl + Cl\cdot \end{cases}$$

$$\textit{Chain termination} \begin{cases} Cl\cdot + Cl\cdot \longrightarrow Cl_2 \\ CH_3\cdot + Cl\cdot \longrightarrow CH_3Cl \\ CH_3\cdot + CH_3\cdot \longrightarrow CH_3CH_3 \end{cases}$$

PROBLEM

3.10. Write a step-by-step free radical chain reaction for the light-induced monobromination of ethane.

PREPARATION OF ALKENES AND ALKYNES; ELIMINATION REACTIONS

3.7 To produce a carbon-carbon double bond between two carbons initially united by a single bond requires eliminating an atom or a group from each of two adjacent carbons.

$$-\overset{|}{\underset{\underset{B}{|}}{C}}-\overset{|}{\underset{\underset{A}{|}}{C}}- \longrightarrow -\overset{|}{C}=\overset{|}{C}- + A{-}B$$

For a triple bond to be generated, the process must be carried out twice. This general type of reaction is termed *elimination*.

$$-\overset{\overset{B}{|}}{\underset{\underset{B}{|}}{C}}-\overset{\overset{A}{|}}{\underset{\underset{A}{|}}{C}}- \longrightarrow -C{\equiv}C- + 2AB$$

A. Dehydrohalogenation

Alkenes and alkynes can be prepared by the elimination of hydrogen and halogen from adjacent carbons. Elimination of HX once yields a double bond, and elimination of two HX units produces a triple bond. Since the elements of a hydrohalic acid are being removed, a base—such as KOH in alcoholic solution or the stronger base $NaNH_2$ for preparing alkynes—is used to effect the conversion. A mixture of alcohol and water is used as the solvent in alkene preparation because it dissolves both the KOH and the alkyl halide. The mechanism of dehydrohalogenation will be discussed in section 8.6B.

Alkenes $\begin{array}{c}|\ |\\ -C-C-\\ |\ |\\ H\ X\end{array}$ + KOH $\longrightarrow$ $\begin{array}{c}|\ |\\ -C=C-\\ \end{array}$ + KX + H_2O

Alkynes $\begin{array}{c}H\ X\\ |\ |\\ -C-C-\\ |\ |\\ H\ X\end{array}$ or $\begin{array}{c}X\ H\\ |\ |\\ -C-C-\\ |\ |\\ H\ X\end{array}$ + $\begin{array}{c}2KOH\\ \text{(or 2NaNH}_2)\end{array}$ $\longrightarrow$ $-C\equiv C-$ + $\begin{array}{c}2KX\\ \text{(2NaX)}\end{array}$ + $\begin{array}{c}2H_2O\\ \text{(2NH}_3)\end{array}$

X = Cl, Br, I

B. Dehydration

1. Preparation of Alkenes. Elimination of the elements of water from two adjacent carbons of an alcohol results in the formation of a double bond. Acids such as sulfuric and phosphoric acids are effective dehydrating agents. They act as catalysts and are not consumed in the reaction. Dehydration is not an effective method for making alkynes.

Alkenes $\begin{array}{c}|\ \ |\\ -C-C-\\ |\ \ |\\ HO\ H\end{array}$ $\xrightarrow[H_3PO_4]{H_2SO_4,}$ $\begin{array}{c}|\ \ |\\ -C=C-\\ \end{array}$ + H_2O

2. Mechanism of Dehydration. Let us consider the conversion of ethanol to ethene to illustrate the mechanism of dehydration.

$$CH_3CH_2OH \xrightarrow{H_2SO_4} CH_2{=}CH_2 + H_2O$$

For a reaction to occur the two reagents, ethanol and sulfuric acid, must get together. Ethanol is a Lewis base because of the unshared electron pairs on the oxygen. Consequently, an acid-base reaction occurs in which the sulfuric acid donates a hydrogen ion to the alcohol.

$$CH_3CH_2\overset{..}{\underset{..}{O}}H + \underset{\text{(from }H_2SO_4)}{H^+} \rightleftharpoons CH_3CH_2\overset{\overset{\displaystyle H}{..+}}{\underset{..}{O}}H$$

The reaction is analogous to the ionization of a strong acid in water to form the hydronium ion ($H_2O + H_2SO_4 \longrightarrow H_3O^+ + HSO_4^-$). In the second step of the mechanism, the protonated alcohol loses a molecule of water producing a carbocation; the —OH is thereby eliminated.

$$CH_3CH_2 : \overset{\overset{\displaystyle H}{..+}}{O} : H \rightleftharpoons CH_3CH_2^+ + H_2\overset{..}{\underset{..}{O}} :$$

Note that the oxygen retained the pair of electrons in the carbon-oxygen bond, creating a neutral water molecule and leaving a positive charge on the carbon. In the final step of the mechanism, the carbocation eliminates a hydrogen ion to form a neutral ethene molecule.

$$\begin{array}{c}H\ H\\ |\ \ |\\ H-C{\underset{\curvearrowright}{}}C-H\\ |\\ \overset{..}{H}\ \ +\end{array} \rightleftharpoons \begin{array}{c}H\ H\\ |\ \ |\\ H-C=C-H\end{array} + H^+$$

Note that the hydrogen ion provided by H_2SO_4 in the first step is returned in the last; the acid is truly a catalyst. The reaction is entirely reversible and each step of the mechanism is an equilibrium.

PROBLEM

3.11. Write equations for the reaction of **(a)** 1-bromopropane with KOH; **(b)** 2-propanol ($CH_3CHOHCH_3$) with H_2SO_4; and **(c)** 1,2-dichloroethane with KOH or $NaNH_2$ (2 moles).

C. Orientation of Elimination

Consider the dehydration of 1-butanol and 2-butanol. Only one product is possible from 1-butanol.

$$CH_3CH_2CH_2CH_2OH \xrightarrow{H_2SO_4} CH_3CH_2CH\!=\!CH_2 + H_2O$$

Two products are formed from 2-butanol since the hydrogen that leaves after the hydroxy can come from either carbon-1 or carbon-3. The alkenes are not formed in equal amounts, however; 2-butene occurs in the greater amount.

$$\underset{\underset{\displaystyle OH}{|}}{CH_3CH_2CHCH_3} \xrightarrow{H_2SO_4} CH_3CH_2CH\!=\!CH_2 + \underset{\substack{\text{Predominant}\\ \text{alkene}}}{CH_3CH\!=\!CHCH_3} + H_2O$$

The following rule is useful in predicting the predominant product of an elimination reaction:

Rule: In dehydration and dehydrohalogenation, if elimination can result in the formation of more than one alkene, the most stable alkene is formed predominantly. The most stable alkene is the one most highly substituted with alkyl groups.

Stability and Ease of Formation of Alkenes

$$CH_2\!=\!CH_2 < RCH\!=\!CH_2 < \begin{matrix} R_2C\!=\!CH_2 \\ RCH\!=\!CHR \end{matrix} < R_2C\!=\!CHR < R_2C\!=\!CR_2$$

Least substituted,   Most substituted,
least stable   most stable

To determine the degree of substitution of an alkene, count the number of carbons directly attached to the two carbons involved in the double bond. For example, there are three possible alkenes formed from the dehydrobromination of 2,3-dimethyl-3-bromopentane. By the rule above, we should predict that the tetrasubstituted one will be the major product.

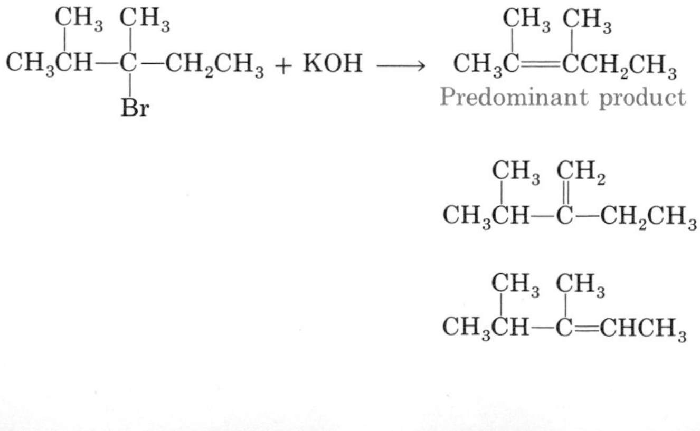

$$\underset{\underset{\displaystyle Br}{|}}{\underset{|}{CH_3CH\!-\!\underset{|}{C}\!-\!CH_2CH_3}} + KOH \longrightarrow \underset{\text{Predominant product}}{CH_3\underset{|}{C}\!=\!\underset{|}{C}CH_2CH_3} \qquad \text{Tetrasubstituted}$$

with CH_3 CH_3 groups above carbons.

$$CH_3\underset{|}{C}H\!-\!\underset{\|}{C}\!-\!CH_2CH_3 \qquad \text{Disubstituted}$$

with CH_3 CH_2 above.

$$CH_3\underset{|}{C}H\!-\!\underset{|}{C}\!=\!CHCH_3 \qquad \text{Trisubstituted}$$

with CH_3 CH_3 above.

PROBLEM

3.12. Complete the following reactions showing the major organic product:

$$\textbf{(a)} \quad CH_3\overset{\displaystyle CH_3}{\underset{\displaystyle OH}{CH}}CHCHCH_3 \xrightarrow{H_2SO_4}$$

$$\textbf{(b)} \quad CH_3\overset{\displaystyle CH_3}{\underset{\displaystyle Cl}{C}}\!\!-\!\!\overset{\displaystyle CH_3}{CH}CH_3 + KOH \longrightarrow$$

Panel 3

PETROLEUM AND ENERGY

For millions of years plants have been converting raw materials, especially carbon dioxide and water, and the energy they absorb from sunlight into the chemicals necessary to maintain their existence. The process by which this is accomplished, photosynthesis, results in the storage of energy as chemical bonds. With the normal life cycles of growth, death, and decay, or growth, consumption, and decay, the end products of prehistoric plant and animal life, especially that of plankton, settled to the bottoms of lakes, marshes, and oceans. Sediments of these organic wastes accumulated over the ages; under some conditions they were converted into a complex mixture of organic compounds called petroleum and under other conditions to massive deposits of coal. Coal is formed primarily from the decay of plant life, which occurs in stages. Initially, peat is formed, then lignite (texture of original wood is often still evident), then bituminous or soft coal, which contains volatile organic compounds (often aromatic), and finally anthracite or hard coal (a high-luster coal with a high elemental carbon content).

Coal had long been valued as an energy resource but it wasn't until 1859, when Edwin Drake drilled a 69.5-ft hole next to a surface oil seep, that the ancient legacy of uncountable generations of plants and animals was released—petroleum. Expert predictions set the first quarter of the twenty-first century as the probable depletion date for this resource.

Crude oil is usually a black, viscous, foul-smelling liquid, composed primarily of hundreds of different hydrocarbon molecules. Crude oil can be refined, that is, separated into different fractions that are then commercially developed into several thousand consumable products, from life-giving medicines to deadly pesticides. But most of a barrel (one barrel = 42 gallons) of crude oil is converted into fuels, over 40% into gasoline, 30%–35% into fuel oil for heating and other purposes, 7%–10% for jet fuel, and much of what remains is sold as aviation gasoline, liquefied petroleum gas, lubricating oils, greases, and asphalt. Thus the treasure of hydrocarbon energy originally obtained from ancient sunlight and stored in chemical bonds by prehistoric organisms is reconverted by combustion to the carbon dioxide and water whence it came.

$$\text{Combustion:} \quad \text{Hydrocarbon} + O_2 \longrightarrow CO_2 + H_2O + \text{energy}$$

Only a small portion, about 5%, of this rich supply of organic material is used in the petrochemical industry to manufacture the drugs, dyes, plastics, and other products common to our current society.

Fractionation of Crude Petroleum

[PANEL 3]   To convert crude oil into usable components, it must be separated into various fractions. Since the boiling points of hydrocarbons increase with increasing molecular weight (section 3.4), crude oil can be separated into its components by fractional distillation using gigantic distillation towers such as that schematically depicted in Figure 3.2. Low-boiling, low-molecular-weight compounds are collected high on the column, and those of increasing molecular weight and boiling point are separated at various stages lower on the column.

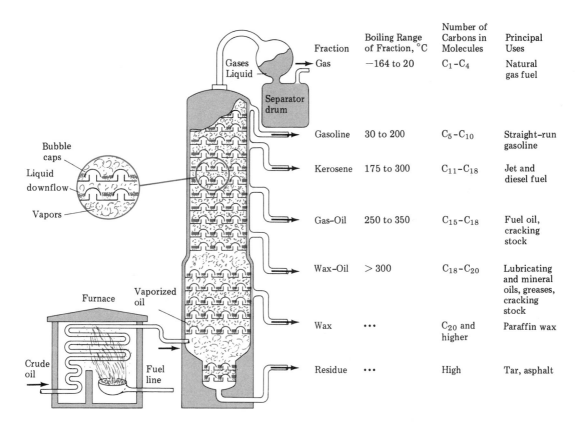

Fraction	Boiling Range of Fraction, °C	Number of Carbons in Molecules	Principal Uses
Gas	−164 to 20	C_1–C_4	Natural gas fuel
Gasoline	30 to 200	C_5–C_{10}	Straight-run gasoline
Kerosene	175 to 300	C_{11}–C_{18}	Jet and diesel fuel
Gas–Oil	250 to 350	C_{15}–C_{18}	Fuel oil, cracking stock
Wax–Oil	> 300	C_{18}–C_{20}	Lubricating and mineral oils, greases, cracking stock
Wax	⋯	C_{20} and higher	Paraffin wax
Residue	⋯	High	Tar, asphalt

FIGURE 3.2. Fractional distillation of crude petroleum. A bubble-cap distillation tower and the principal fractions of petroleum, their properties, and uses are illustrated. High-boiling, high-molecular-weight materials are collected at the bottom of the column and low-boiling, low-molecular-weight materials at upper regions of the column. Natural gas and liquified natural gas (LNG) are primarily methane, CH_4. Liquified petroleum gas (LPG)—used for heating in rural areas and in camping stoves, lighters, and torches—is mainly propane with lesser amounts of ethane and butane. Straight-run gasoline is neither of the quality or quantity for today's needs. Consequently, gasoline is also produced from the refining of other petroleum fractions (see Panel 6).

Alternative Energy Sources

Petroleum is a finite resource and the prospect of dwindling resources is causing political problems throughout the world. Coal could be the largest source

[PANEL 3] of energy in the future, but it is certainly not ideal. Unlike petroleum, it is not fluid and consequently it is difficult to handle and transport. Also, it is a dirty fuel and difficult to mine. Yet coal reserves in the United States are sufficient to meet the nation's energy needs for the next three to four centuries (in the early 1900s it supplied over 70% of the country's energy requirements; currently it provides about 20%). Current research is directed toward finding methods for recovering petroleum from noncontemporary sources and toward converting coal into liquid fuels for use in transportation.

1. Synthetic Fuels. Two methods for producing synthetic fuels from coal are coal gasification and coal liquefaction.

In coal gasification, solid coal is converted to methane, an extremely versatile, easily transported, nonpolluting fuel. In one method, called *hydrogasification,* coal is reacted with water (water supplies the hydrogen) at high temperatures ideally resulting in the formation of methane and carbon dioxide.

$$\underset{\text{Coal}}{2C} + 2H_2O \longrightarrow \underset{\text{Methane}}{CH_4} + CO_2$$

At the temperatures required to effect such a conversion, however, the methane is thermally unstable and some decomposes. A mixture of gases, called synthesis gas, including CH_4, CO_2, H_2, and CO results. In subsequent steps, the synthesis gas is catalytically oxygenated, with the carbon monoxide being converted to carbon dioxide. This continues until the CO and hydrogen are in the correct ratio for methanation.

$$CO + 3H_2 \xrightarrow{\text{Catalyst}} CH_4 + H_2O$$

Purification can yield synthetic natural gas that is 95%–98% methane. Research is underway to produce catalysts that allow the initial step to occur at lower temperatures so that methane is produced in higher yield without subsequent decomposition.

Coal liquefaction is not new. During World War II, Germany produced 100,000 barrels of gasoline daily by the uneconomical, but at the time necessary, Fischer-Tropsch process.

$$\textit{Fischer-Tropsch process} \quad \begin{aligned} &\text{Coke} + H_2O \xrightarrow{\text{Heat}} CO + H_2 \\ &6CO + 13H_2 \longrightarrow \underset{\text{Gasoline}}{C_6H_{14}} + 6H_2O \end{aligned}$$

Most coals are not pure carbon. They have a low hydrogen-to-carbon ratio. Thus, in coal liquefaction it is necessary to hydrogenate the coal and convert it into a liquid mixture resembling gasoline.

Recently, the alcohol methanol has received much attention in gasoline production. Methanol is commercially produced by hydrogenation of carbon monoxide.

$$CO + 2H_2 \longrightarrow CH_3OH$$

Recall that in coal gasification CO is the major product of the hydrogasification step and that measures can be taken to increase its yield. Methanol is therefore a potential derivative of coal. Methanol has been tested both as a replacement for gasoline and as a blend component of gasoline. More recently, research has de-

[PANEL 3] vised methods for converting methanol into conventional gasoline, providing the final line in the economically feasible conversion of coal to gasoline. The catalytic process converts methanol into water and hydrocarbon molecules with five to ten carbons.

$$n\mathrm{CH_3OH} \xrightarrow{\text{Catalyst}} \text{hydrocarbon gasoline molecules} + \mathrm{H_2O}$$

2. Shale Oil. Oil shale, such as the rich deposits on the western slopes of the Rocky Mountains, constitutes an important potential source of energy. Quality shale can yield 15–30 gallons of oil per ton of shale. Estimates range from 1800 billion to 3–7 trillion total barrels of oil buried in this region. (In comparison, Middle East crude oil reserves are about 400 billion barrels. Current United States petroleum production is about 4 billion barrels annually.) The actual amount of oil recoverable and the environmental problems associated with strip mining oil shale and disposing of processed shale (which has a high salinity) are immense problems that require solutions before a mature shale-oil industry can develop.

3. Tar Sands. Tar sands are sandy deposits containing highly viscous solid or semisolid hydrocarbons. They must be strip mined and the hydrocarbon mixture released and separated by flotation from the sand using hot water. Liquid hydrocarbons and gases are recovered from the organic layer, and the remainder is charred to coke which can be used to produce synthetic fuels. There is no significant commercial extraction of tar sands in the United States, although in Alberta, Canada, the world's largest tar-sand deposit is being mined successfully.

4. Enhanced Oil Recovery. By the beginning of 1976, 441 billion barrels of oil had been discovered in the United States, of which approximately 106 billion barrels had been already consumed. It was estimated that another 39 billion barrels could be recovered economically, leaving 300 billion barrels of unrecovered oil (for comparison, the Middle East was estimated at this time to have almost 400 billion barrels of oil that could be recovered coventionally). Research into secondary and tertiary recovery techniques using hot water, detergents, alcohols, and polymers is being pursued.

In addition, heavy crude oil, a thick, almost solid material, is not considered a conventional crude. Since it is more difficult to recover and does not give as much fuel after fractionation, it is not used in abundance. However, as oil prices rise heavy crude becomes a more economical energy source.

5. Biomass. Just over a century ago, firewood was the primary fuel in the United States, and it is still the main energy source in much of the developing world. Wood is the main energy source classified as biomass. *Biomass energy* is energy derived from biological matter that is renewable in a relatively short period of time. It is estimated that on a daily basis, plants through photosynthesis store 17 times as much energy as is used world-wide. Theoretically, if biomass resources were managed properly, enough land, water, sunshine, and plant nutrients exist to fulfill the world's needs in food and fuel.

The major biomass sources include urban and industrial wastes, agricultural and forestry wastes, land and fresh-water energy farming, and open-ocean farming. By the year 2000, experts estimate a biomass contribution to the total United States energy consumption at 10%, with an even higher future potential.

[PANEL 3]　　**6. Other Energy Sources.** The current national energy consumption in the United States is 75 quads (a quad is one quadrillion or 10^{15} Btu's; a Btu, British thermal unit, is the quantity of heat necessary to raise the temperature of one pound of water 1°F). In addition to petroleum and fossil fuels in general, other energy sources are in use or under consideration. These include nuclear fission; nuclear fusion; solar energy; geothermal energy; hydroelectric, tidal and wind power; and energy from thermal gradients in the ocean. Increased energy efficiency and conservation are also important. The human race has been fortunate in finding vast concentrations of important materials on this earth. If not careful, it will consume them and scatter the remains so widely that recovery will be impossible.

END-OF-CHAPTER PROBLEMS

3.13 Nomenclature of Alkanes: Name the following by the IUPAC system of nomenclature: **(a)** $CH_3CH_2CH_3$; **(b)** $CH_3(CH_2)_8CH_3$; **(c)** $CH_3(CH_2)_4CH_3$

3.14 Nomenclature of Alkanes: Name the following by the IUPAC system of nomenclature:

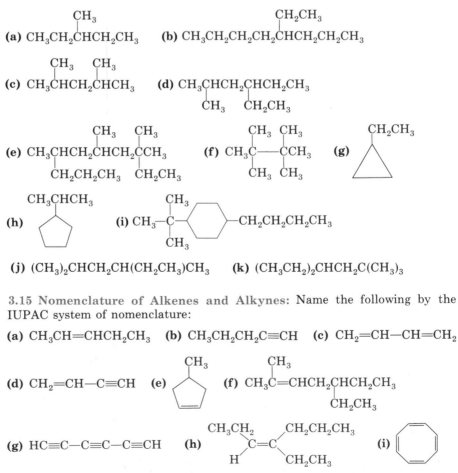

(j) $(CH_3)_2CHCH_2CH(CH_2CH_3)CH_3$　　**(k)** $(CH_3CH_2)_2CHCH_2C(CH_3)_3$

3.15 Nomenclature of Alkenes and Alkynes: Name the following by the IUPAC system of nomenclature:

(a) $CH_3CH{=}CHCH_2CH_3$　**(b)** $CH_3CH_2CH_2C{\equiv}CH$　**(c)** $CH_2{=}CH{-}CH{=}CH_2$

(d) $CH_2{=}CH{-}C{\equiv}CH$　**(e)**　**(f)** $CH_3C{=}CHCH_2CHCH_2CH_3$ with CH_3 above first carbon and CH_2CH_3 below

(g) $HC{\equiv}C{-}C{\equiv}C{-}C{\equiv}CH$　**(h)**　**(i)**

3.16 Nomenclature of Hydrocarbon Derivatives: Name the following by the IUPAC system of nomenclature:

(a) $CH_3\overset{\overset{\displaystyle CH_3}{|}}{C}H\underset{\underset{\displaystyle Br}{|}}{C}HCH_3$ (b) $ClCH_2CH{=}CHCH_2Br$ (c) CHI_3 (d) CH_2Br_2

3.17 IUPAC Nomenclature: Draw the following compounds:

(a) 3,7-dimethyl-1,3,6-octatriene (ocimene, component of basil)
(b) 2,2,4-trimethylpentane (a component of high-octane gasoline)
(c) tetrafluoroethene (precursor of Teflon)
(d) 2-chloro-1,3-butadiene (precursor of neoprene rubber)
(e) 1,2,3,4,5,6-hexachlorocyclohexane (an insecticide)
(f) 2,2,6-trimethyl-4-ethyl-5-propyloctane
(g) 10-bromo-9,9-dimethyl-1,3,5,7-decatetrayne
(h) *trans* 3-heptene

3.18 IUPAC Nomenclature: Alkynes occur only rarely in nature. The following compound with three double bonds is found in some plants and fungi. Name it by the IUPAC system (a thirteen-carbon chain is called tridecane).

$$CH_3CH{=}CHC{\equiv}C{-}C{\equiv}C{-}C{\equiv}CCH{=}CHCH{=}CH_2$$

3.19 Common Nomenclature: Draw the following compounds: (a) acetylene; (b) vinyl bromide; (c) ethylene; (d) allyl chloride; (e) isopropyl iodide; (f) ethylacetylene.

3.20 Geometric Isomerism: Draw the geometric isomers requested: (a) *cis-* and *trans*-3-methyl-2-pentene; (b) three geometric isomers of 2,4-hexadiene; (c) *cis, cis, cis*-2,4,6-octatriene.

3.21 Physical Properties: Explain the difference in melting points or boiling points, as indicated, for each of the following sets of compounds:

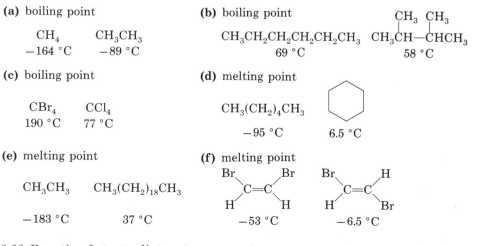

(a) boiling point

CH_4 CH_3CH_3
$-164\,°C$ $-89\,°C$

(b) boiling point

$CH_3CH_2CH_2CH_2CH_2CH_3$ $CH_3\overset{\overset{\displaystyle CH_3}{|}}{C}H{-}\overset{\overset{\displaystyle CH_3}{|}}{C}HCH_3$
$69\,°C$ $58\,°C$

(c) boiling point

CBr_4 CCl_4
$190\,°C$ $77\,°C$

(d) melting point

$CH_3(CH_2)_4CH_3$
$-95\,°C$ $6.5\,°C$

(e) melting point

CH_3CH_3 $CH_3(CH_2)_{18}CH_3$
$-183\,°C$ $37\,°C$

(f) melting point

$-53\,°C$ $-6.5\,°C$

3.22 Reactive Intermediates: Draw a tertiary butyl carbocation, free radical, and carbanion. Show how the electrons in the C—A bond in the following general compound must be distributed in the bond cleavage to form each:

$$(CH_3)_3C{-}A \longrightarrow$$

3.23 Halogenation of Alkanes: How many different monochlorination isomers could be formed by the light-induced monochlorination of the following compounds? Write the compounds formed.

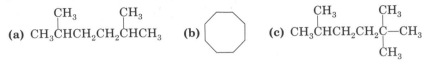

(a) $CH_3\overset{\displaystyle CH_3}{\underset{}{C}}HCH_2CH_2\overset{\displaystyle CH_3}{\underset{}{C}}HCH_3$ **(b)**   **(c)** $CH_3\overset{\displaystyle CH_3}{\underset{}{C}}HCH_2CH_2\overset{\displaystyle CH_3}{\underset{\displaystyle CH_3}{C}}{-}CH_3$

3.24 Halogenation of Alkanes: Write the structural formula for the alkane with each of the following molecular formulas that gives only one monobromination isomer: **(a)** C_5H_{12}; **(b)** C_8H_{18}.

3.25 Halogenation of Alkanes: Describe the reaction conditions that would favor the formation of bromoethane and hexabromoethane when ethane is treated with bromine in the presence of light.

3.26 Halogenation of Alkanes—Reaction Mechanism:
(a) Write a step-by-step reaction mechanism for the light-induced monobromination of methane.
(b) Tetraethyllead (an antiknock agent in gasoline) decomposes to elemental lead and ethyl free radicals at 140 °C.

$$Pb(CH_2CH_3)_4 \xrightarrow{\ 140\ °C\ } Pb + 4CH_3CH_2\cdot$$

Although methane and chlorine, in the absence of light, react at 250 °C, in the presence of minute quantities of tetraethyllead they can be made to react at 140 °C. Write a mechanism showing how tetraethyllead catalyzes the chlorination of methane.

3.27 Dehydration of Alcohols—Reaction Mechanism: Write a step-by-step reaction mechanism for the acid-catalyzed dehydration of $CH_3CHOHCH_3$ (rubbing alcohol) to form propene (precursor of polypropylene).

3.28 Elimination Reactions: Complete the following reactions, showing the major organic products:

(a) $CH_3CH_2CH_2Br + KOH \longrightarrow$ **(b)** $CH_3CH_2CH_2\overset{\displaystyle }{\underset{\displaystyle OH}{C}}HCH_3 \xrightarrow{\ H_2SO_4\ }$

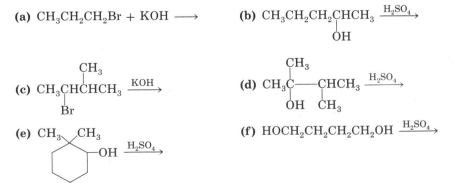

(c) $CH_3\overset{\displaystyle CH_3}{\underset{\displaystyle Br}{C}}HCHCH_3 \xrightarrow{\ KOH\ }$ **(d)** $CH_3\overset{\displaystyle CH_3}{\underset{\displaystyle OH}{C}}{-}\overset{\displaystyle }{\underset{\displaystyle CH_3}{C}}HCH_3 \xrightarrow{\ H_2SO_4\ }$

(e)

(f) $HOCH_2CH_2CH_2CH_2OH \xrightarrow{\ H_2SO_4\ }$

(g) $CH_3CH_2CHBr_2 + 2NaNH_2 \longrightarrow$

(h)
$$CH_3\overset{\overset{\displaystyle CH_3}{|}}{\underset{\underset{\displaystyle CH_3}{|}}{C}}\!\!-\!\!\overset{\overset{\displaystyle }{}}{\underset{\underset{\displaystyle Cl}{|}}{CH}}\!\!-\!\!\overset{}{\underset{\underset{\displaystyle Cl}{|}}{CH_2}} + 2NaNH_2 \longrightarrow$$

3.29 Preparation of Alkenes and Alkynes: Write the structure of a starting compound and necessary reagents for preparing the following in one step:

(a) $CH_3\overset{\overset{\displaystyle CH_3}{|}}{CH}CH\!=\!CH_2$ 　　**(b)** $CH_3\overset{\overset{\displaystyle CH_3}{|}}{C}\!=\!CHCH_2CH_3$ 　　**(c)** $CH_3CH_2C\!\equiv\!CH$

3.30 Homologous Series: Write a homologous series of at least five members, starting with each of the following compounds:

(a) $CH_3CH\!=\!CH_2$ 　　**(b)** CH_3OH 　　**(c)** HCO_2H 　　**(d)** $CH_3\overset{\overset{\displaystyle O}{\|}}{C}H$ 　　**(e)** CH_3NH_2

3.31 Combustion: Write balanced chemical reactions for the total combustion of each of the following hydrocarbons: **(a)** methane; **(b)** propane; **(c)** isooctane; **(d)** 1-pentene; **(e)** cyclohexane; **(f)** acetylene.

3.32 Petroleum Fractions: Write the structures of one or two molecules that could be representative of the following petroleum fractions: **(a)** gas; **(b)** gasoline; **(c)** kerosene; **(d)** gas-oil; **(e)** wax-oil; **(f)** wax.

Chapter 4

Nonaromatic Hydrocarbons II

In the previous chapter (section 3.5A), we saw that there are three major types of organic reactions—substitution, elimination, and addition. Halogenation of alkanes (section 3.6) gave us an example of substitution, and in the preparation of alkenes and alkynes (section 3.7) we saw examples of elimination reactions. In this chapter, we will continue examining the chemistry of hydrocarbons by looking at the characteristic reactions of alkenes and alkynes—addition reactions—and the commercial application of these reactions in producing addition polymers (plastics, fibers, resins).

ADDITION REACTIONS OF ALKENES AND ALKYNES

4.1   A. Introduction to Addition Reactions

Alkenes and alkynes are unsaturated because the carbons involved in a carbon-carbon double bond or triple bond do not have the maximum number of attached atoms, four (they do, of course, have four bonds). Because of this deficiency in attached atoms, addition reactions are characteristic of these unsaturated compounds. A double bond can undergo addition once, and a triple bond can undergo addition twice, given enough reagent.

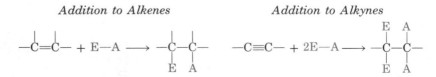

Note that these general reactions are exactly the reverse of the elimination reactions used to prepare alkenes and alkynes (section 3.7).

B. Addition Reactions of Alkenes

Alkenes add reagents once to form a saturated compound.

$$-\overset{|}{\underset{}{C}}=\overset{|}{\underset{}{C}}- \; + \; E-A \;\longrightarrow\; -\overset{|}{\underset{\underset{E}{|}}{C}}-\overset{|}{\underset{\underset{A}{|}}{C}}-$$

What exactly has happened? Considering the general reactant EA, E has bonded to one carbon of the double bond and A to the other. The double bond necessarily has been reduced to a single bond. The logical question now is, "What can add to carbon-carbon double bonds; specifically, what are E and A?" The most common reagents are simple inorganic compounds—hydrogen, halogens, hydrogen halides, water, and hypohalous acids. After examining the following general examples of these reactions, check your comprehension by working problem 4.1.

1. Hydrogenation. (E = H, A = H; nickel, platinum, or palladium metal necessary as catalysts; reaction carried out under pressure.)

$$-\overset{|}{\underset{}{C}}=\overset{|}{\underset{}{C}}- \; + \; H_2 \;\xrightarrow{\;Ni\;}\; -\overset{|}{\underset{\underset{H}{|}}{C}}-\overset{|}{\underset{\underset{H}{|}}{C}}-$$

The catalyst in hydrogenation acts by adsorbing both the hydrogen and alkene onto its surface, which is followed by transfer of the hydrogen to the alkene.

2. Halogenation. (E = X, A = X; X_2 = Cl_2 or Br_2; F_2 is too reactive, I_2 too unreactive.)

$$-\overset{|}{\underset{}{C}}=\overset{|}{\underset{}{C}}- \; + \; X_2 \;\longrightarrow\; -\overset{|}{\underset{\underset{X}{|}}{C}}-\overset{|}{\underset{\underset{X}{|}}{C}}-$$

3. Addition of Hydrogen Halides. (E = H, A = X; HX = HCl, HBr, HI.)

$$-\overset{|}{\underset{}{C}}=\overset{|}{\underset{}{C}}- \; + \; HX \;\longrightarrow\; -\overset{|}{\underset{\underset{X}{|}}{C}}-\overset{|}{\underset{\underset{H}{|}}{C}}-$$

4. Hydration. (E = H, A = OH; H_2SO_4 catalyst.)

$$-\overset{|}{\underset{}{C}}=\overset{|}{\underset{}{C}}- \; + \; H_2O \;\xrightarrow{\;H_2SO_4\;}\; -\overset{|}{\underset{\underset{HO}{|}}{C}}-\overset{|}{\underset{\underset{H}{|}}{C}}-$$

5. Addition of Hypohalous Acids. (E = X, A = OH; HOX = HOCl, HOBr, HOI.)

$$-\overset{|}{\underset{}{C}}=\overset{|}{\underset{}{C}}- \; + \; HOX \;\longrightarrow\; -\overset{|}{\underset{\underset{X}{|}}{C}}-\overset{|}{\underset{\underset{OH}{|}}{C}}-$$

PROBLEM

4.1. Write equations showing the reaction of 2-butene with each of the following reagents: **(a)** H_2/Ni; **(b)** Cl_2; **(c)** HBr; **(d)** HOI; **(e)** H_2O/H_2SO_4.

C. Addition Reactions of Alkynes

1. Addition of Hydrogen, Halogens, and Hydrogen Halides. Alkynes can add one mole of reagent, reducing the triple bond to a double bond, or two moles of reagent, converting the triple bond to a single bond.

$$E\text{—}A = H_2 \text{ (Ni catalyst)}; \ X_2 \text{ (Cl}_2, \text{Br}_2\text{)}; \ HX \text{ (HCl, HBr, HI)}$$

Hydrogen, halogen, and hydrogen halide readily add to alkynes. For example, one mole of HCl adds to one mole of acetylene to produce vinyl chloride (from which PVC is made), or two moles to form 1,1-dichloroethane.

$$HC\equiv CH \xrightarrow{\ HCl\ } \underset{\underset{Cl}{|}\quad\underset{H}{|}}{HC=CH} \xrightarrow{\ HCl\ } \underset{\underset{Cl}{|}\quad\underset{H}{|}}{\overset{\overset{Cl}{|}\quad\overset{H}{|}}{HC-CH}}$$

2. Addition of Water. Water and hypohalous acids add to alkynes but not in the way described above. The difference lies in the formation of an intermediate enol (a compound with both a double bond and an alcohol function) from the addition of one mole of reagent. Enols are unstable and rearrange to aldehydes and ketones. This can be illustrated by adding water to acetylene in the presence of sulfuric acid and mercury II sulfate as catalysts.

$$HC\equiv CH + H_2O \xrightarrow[\text{HgSO}_4]{\text{H}_2\text{SO}_4,} \underset{\underset{H}{|}\quad\underset{OH}{|}}{HC=CH} \longrightarrow \underset{\underset{H}{|}\quad\underset{O}{\|}}{\overset{\overset{H}{|}}{HC-CH}}$$
$$\text{Enol}\qquad\qquad\qquad\text{Aldehyde}$$

PROBLEM

4.2. Write equations showing the reaction of 2-butyne with each of the following reagents: **(a)** $1H_2$/Ni; **(b)** $2H_2$/Ni; **(c)** $1Br_2$; **(d)** $2Cl_2$; **(e)** 1HCl; **(f)** 2HBr; **(g)** H_2O/H_2SO_4, $HgSO_4$.

D. Mechanism of Electrophilic Addition

To introduce the mechanism of electrophilic addition, let us again consider the addition of a general reagent E—A to a carbon-carbon double bond.

$$\underset{}{-\overset{|}{C}=\overset{|}{C}-} + E\text{—}A \longrightarrow \underset{\underset{E}{|}\quad\underset{A}{|}}{-\overset{|}{C}-\overset{|}{C}-}$$

For E—A to react with an alkene, there must be some attractive force that could cause the two species to come together. Let us analyze each reactant.

E—A can be thought of as being divisible into a positive species, called an electrophile (E^+), and a negative species (A^-). *Electrophile* means "electron-loving." In the addition reactions we have covered, H^+ would be the electrophile when adding either hydrogen halide or water to an alkene, and X^+ would be the electrophile when adding either halogen or hypohalous acid (HOX). A of E—A is an anion (for example Cl^- in HCl) or a neutral Lewis base.

The carbon-carbon double bond of an alkene is composed of a sigma (σ) bond and a pi (π) bond. Recall that the electrons of a σ bond are concentrated between the bonded atoms and tightly held by them, and that the π bond is formed by the overlap of p orbitals on each carbon above and below the σ bond (Figure 4.1). Thus, the π bond can be thought of as a loosely held cloud of negative charge. Since opposite charges attract, the electron-deficient electrophile, E^+, is attracted to this negative cloud of π electrons and can be embedded in it. Now the reacting species have made contact and the reaction can commence. Since an electrophile initiated this contact, the reaction is termed *electrophilic addition*.

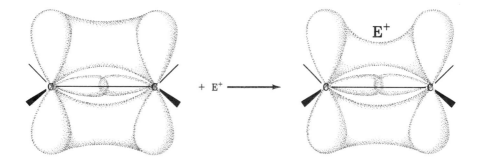

FIGURE 4.1. Bonding picture of an alkene showing σ and π bonds of the carbon-carbon double bond. The electrophile E^+ becomes embedded in the π cloud and thereby initiates the addition reaction.

Now that the electrophile is in contact with the alkene, it can actually form a bond to one of the carbons of the double bond. Since the electrophile is deficient in electrons, it must use the two π electrons to form a bond.

Carbocation

The two π electrons that were once shared between the two carbons are now shared between one carbon and the electrophile. The other carbon, having lost an electron, takes on a positive charge. It has six outer-shell electrons and is a carbocation (see section 3.5C).

The resultant carbocation is an unstable, short-lived intermediate and, being positive, it will be quickly neutralized by the anion $A:^-$.

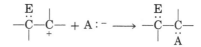

Basically, the mechanism of electrophilic addition is a two-step process: (1) attack by an electrophile to form a carbocation, and (2) neutralization of the carbocation.

Electrophilic Addition—General Mechanism

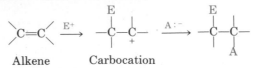

Alkene   Carbocation

Compare this general mechanism to the following specific examples of electrophilic addition.*

1. Mechanism for Addition of Hydrogen Halides (HX). (HX = HCl, HBr, HI.)

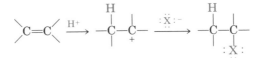

2. Mechanism for Addition of Halogen (X₂). (X₂ = Cl₂, Br₂.)

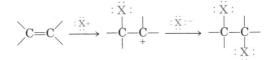

Although Cl_2 and Br_2 are nonpolar, as they approach the π cloud of the double bond the repulsion between the π bond and the nonbonding electrons in the outer shell of the halogen molecule momentarily polarizes the halogen molecule ($X^+—X^-$). The positive end of the polarized halogen (the electrophile, $:\overset{..}{X}{}^+$) bonds to the alkene forming a carbocation which is neutralized by $:\overset{..}{X}:^-$.

3. Mechanism for Addition of Water.

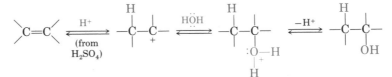

H^+ from the sulfuric acid catalyst bonds to the alkene forming a carbocation. Since there is no OH^- in an acid medium, the carbocation is neutralized by bonding to one of the lone pairs of electrons on a neutral water molecule. Loss of a hydrogen ion from the oxygen produces a neutral addition product and regenerates the catalyst. The reaction is completely reversible; the reverse has already been discussed as dehydration of alcohols (section 3.7B).

4. Mechanism for Addition of Hypohalous Acids (HOX).

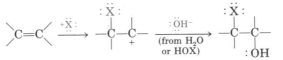

---

*Hydrogenation does not involve an electrophilic addition mechanism.

PROBLEMS

4.3. Write step-by-step electrophilic addition mechanisms for the reaction of 2-butene with: **(a)** HBr; **(b)** Cl_2; **(c)** H_2O (H_2SO_4 catalyst); **(d)** HOBr.

4.4. When HCl adds to propene, two products are possible, 1-chloropropane and 2-chloropropane. Write reaction equations and two-step electrophilic addition reaction mechanisms leading to each product. (After reading section 4.1E, come back to this problem and see if you can explain why 2-chloropropane is the predominant product.)

E. Orientation of Addition

When a symmetrical reagent adds to a symmetrical alkene only one product is possible.

$$CH_3CH=CHCH_3 + Br_2 \longrightarrow CH_3CH-CHCH_3$$

  Symmetrical   Symmetrical | |
                                                    Br   Br

A symmetrical reagent E—A is one in which E and A are identical, such as Br_2. Hydrogen bromide has atoms that are different and is therefore unsymmetrical. When the double bond of an alkene is bisected, the alkene is symmetrical if the two parts are the same (as in 2-butene) and unsymmetrical if they are different (as in 1-butene). If either the alkene or the added reagent is symmetrical, only one addition product is possible.

$$CH_3CH=CHCH_3 + HBr \longrightarrow CH_3CH_2CHCH_3$$

  Symmetrical   Unsymmetrical |
                                                   Br

$$CH_3CH_2CH=CH_2 + Br_2 \longrightarrow CH_3CH_2CH-CH_2$$

  Unsymmetrical   Symmetrical | |
                                                    Br   Br

If both the alkene and added reagent are unsymmetrical, however, two addition products are possible. They are usually formed in unequal amounts.

$$CH_3CH_2CH=CH_2 + HBr \longrightarrow CH_3CH_2CHCH_3 + CH_3CH_2CH_2CH_2Br$$

  Unsymmetrical   Unsymmetrical | Minor product
                                                  Br

                                          Major product

The proportion of each product depends on the relative stabilities of the intermediate carbocations formed in the reaction mechanism. To be able to predict which product will dominate, one must understand carbocation stability.

1. Carbocation Stability. Structurally, a carbocation is a carbon with three bonds, six outer-shell electrons (in the three bonds), and a positive charge. The simplest is the methyl carbocation, CH_3^+. Groups directly attached to the positive carbon that can partially neutralize or disperse the positive charge stabilize the carbocation. Alkyl groups (R groups: methyl, ethyl, propyl, etc.) are electron-releasing groups and stabilize carbocations. The electron density of the adjacent σ bonds "spreads over" to the positive carbon, partially neutralizing the charge. The more alkyl groups directly attached to the positive carbon, the more stable the carbocation. A tertiary (3°) carbocation has three attached alkyl groups and is more stable than a secondary (2°) carbocation with two, or a primary (1°) that has only one. To determine the number of directly attached alkyl groups, count the number of carbons directly bonded to the positive carbon. The order of carbocation stability follows:

Carbocation Stability

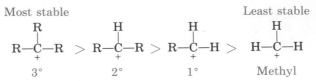

2. **Predicting Addition Products.** The reaction of 1-butene (an unsymmetrical alkene) with HBr (an unsymmetrical reagent) gives two addition products, 1-bromobutane and 2-bromobutane. To predict which product predominates, we write a reaction mechanism (electrophilic addition, section 4.1D) leading to each product and determine which carbocation is more stable.

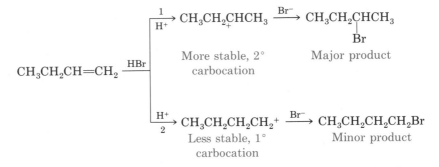

If the hydrogen ion from HBr bonds to carbon-1, the second carbon becomes positive (the two π electrons were pulled away from it). Since there are two attached alkyl groups, this is a secondary carbocation. Neutralization by bromide ion forms 2-bromobutane. Because this secondary carbocation is more stable than the primary carbocation formed if H+ bonds to the second carbon (leading to 1-bromobutane upon neutralization by Br⁻), it is formed more readily and 2-bromobutane is the predominant product.

Pioneering work in this field of organic chemistry was performed by the Russian chemist Vladimir Markovnikov. Appropriately, the rule for predicting orientation of addition is known as Markovnikov's rule. Stated in modern terms the rule is: *When an unsymmetrical reagent adds to an unsymmetrical alkene, the positive portion of the reagent adds to the carbon that results in the formation of the more stable carbocation.*

Similar reasoning applies to the addition of unsymmetrical reagents to unsymmetrical alkynes. For example:

$$CH_3C{\equiv}CH + 2HCl \longrightarrow CH_3\overset{\displaystyle Cl}{\underset{\displaystyle Cl}{C}}CH_3$$

PROBLEM

4.5. Predict the major product of addition in the following reactions by writing the two possible intermediate carbocations formed in the mechanism of electrophilic addition and determining which is more stable.

$$\textbf{(a)}\ CH_3\underset{\underset{CH_3}{|}}{C}{=}CH_2 + HCl \longrightarrow \qquad \textbf{(b)}\ CH_3\underset{\underset{CH_3}{|}}{C}{=}CHCH_3 + H_2O \xrightarrow{H_2SO_4}$$

ADDITION POLYMERS

4.2

An important application of addition reactions is in the formation of addition polymers, which make up a major portion of the gigantic plastics industry. The word *polymer* comes from two segments—*poly* meaning "many" and *mer* meaning "unit." Polymers are giant molecules composed of recurring structural units called *monomers* (single units).

$$nM \longrightarrow M{-}M{-}M{-}M{-}M \quad (\text{or } M_n)$$
$$\text{Monomer} \qquad \text{Polymer}$$

Polymers produced by the addition of alkene molecules to one another are called *addition polymers*.

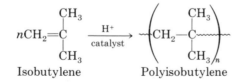

$$\text{Monomer} \qquad\qquad \text{Polymer}$$

Monomeric units are repeated hundreds or thousands of times. The general polymer formula usually does not show the ends of the molecule since they are variable and compose only a minute fraction of the total molecule.

A. Chemistry of Addition Polymerization—Mechanisms

1. Cationic Polymerization through Electrophilic Addition. We have seen that the most characteristic reaction of alkenes is electrophilic addition. Consider the acid-catalyzed polymerization of isobutylene to form the adhesive polyisobutylene.

$$n CH_2{=}\underset{\underset{CH_3}{|}}{\overset{\overset{CH_3}{|}}{C}} \xrightarrow[\text{catalyst}]{H^+} \left(CH_2{-}\underset{\underset{CH_3}{|}}{\overset{\overset{CH_3}{|}}{C}}\right)_n$$
$$\text{Isobutylene} \qquad\qquad \text{Polyisobutylene}$$

Suppose in this reaction we have a large amount of isobutylene relative to the amount of acid catalyst. In the first step of electrophilic addition (section 4.1D), the electrophile (H^+ in this case) is attracted to the π cloud of isobutylene's double bond and eventually bonds, forming the more stable tertiary carbocation (section 4.1E).

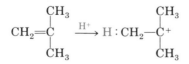

$$CH_2{=}\underset{\underset{CH_3}{|}}{\overset{\overset{CH_3}{|}}{C}} \xrightarrow{H^+} H:CH_2{-}\underset{\underset{CH_3}{|}}{\overset{\overset{CH_3}{|}}{C}}{}^+$$

In the absence of large concentrations of neutralizing agent, this carbocation will attack the double bond of another isobutylene molecule, forming another tertiary carbocation.

$$H : CH_2-\overset{\overset{\displaystyle CH_3}{|}}{\underset{\underset{\displaystyle CH_3}{|}}{C}}+ \quad \overset{\overset{\displaystyle CH_3}{|}}{\underset{\underset{\displaystyle CH_3}{|}}{CH_2{=}C}} \longrightarrow H : CH_2-\overset{\overset{\displaystyle CH_3}{|}}{\underset{\underset{\displaystyle CH_3}{|}}{C}} : CH_2-\overset{\overset{\displaystyle CH_3}{|}}{\underset{\underset{\displaystyle CH_3}{|}}{C}}+$$

The process can be repeated many times (as indicated by n in the following equation) until eventually one of the carbocations formed is neutralized—for example, by capturing an anion.

$$H : CH_2-\overset{\overset{\displaystyle CH_3}{|}}{\underset{\underset{\displaystyle CH_3}{|}}{C}}\left(CH_2-\overset{\overset{\displaystyle CH_3}{|}}{\underset{\underset{\displaystyle CH_3}{|}}{C}}\right)_n CH_2-\overset{\overset{\displaystyle CH_3}{|}}{\underset{\underset{\displaystyle CH_3}{|}}{C}}+ \xrightarrow{\ A:^-\ } H : CH_2-\overset{\overset{\displaystyle CH_3}{|}}{\underset{\underset{\displaystyle CH_3}{|}}{C}}\left(CH_2-\overset{\overset{\displaystyle CH_3}{|}}{\underset{\underset{\displaystyle CH_3}{|}}{C}}\right)_n CH_2-\overset{\overset{\displaystyle CH_3}{|}}{\underset{\underset{\displaystyle CH_3}{|}}{C}} : A$$

The net result is a polymer, an addition polymer, in which the double bonds of the alkene are essentially added to one another over and over again. Polymer chains produced by a polymerization reaction vary in length; that is, not all molecules of polyisobutylene have the same molecular weight.

2. Free-Radical Polymerization. Consider the polymerization of the monomer ethylene to form the polymer polyethylene, which, among other applications, is used in plastic packaging and squeeze bottles. The reaction can be induced by combining a small amount of peroxide (ROOR) with a large volume of ethylene. When peroxides are heated, they decompose readily to form free radicals.

Peroxide decomposition: $R\overset{..}{\underset{..}{O}} : \overset{..}{\underset{..}{O}}R \xrightarrow{\ Heat\ } 2R\overset{..}{\underset{..}{O}} \cdot$

$\qquad\qquad\qquad\qquad\quad$ A peroxide $\qquad\quad$ A free radical

These free radicals will immediately seek a source of electrons to complete their octets. The π bond in an ethylene molecule supplies an electron to pair with the radical, but in the process a new free radical is formed.

Initiation: $\qquad\qquad R\overset{..}{\underset{..}{O}} \cdot \quad CH_2{\overset{..}{=}}CH_2 \longrightarrow R\overset{..}{\underset{..}{O}} : CH_2-CH_2 \cdot$

As before, this new radical will seek a pairing electron by attacking another ethylene molecule, in turn producing yet another free radical. The process repeats itself many times as the polymer is built.

Propagation:

$$RO : CH_2-CH_2 \curvearrowright \ CH_2{\overset{..}{=}}CH_2 \ CH_2{\overset{..}{=}}CH_2 \ CH_2{\overset{..}{=}}CH_2 \cdots \longrightarrow$$

$$RO : CH_2-CH_2 : CH_2-CH_2 : CH_2-CH_2 : CH_2-CH_2 \cdot$$

The polymerization proceeds by a free-radical chain reaction analogous to the chlorination of alkanes (section 3.6C). Attack by the peroxide is the initiation step and the repeated additions are propagation steps. Occasionally, two propagating chains meet end to end to conclude the reaction and form a complete polymer molecule—a termination step.

Termination:

$$RO(CH_2CH_2)_x\cdot \ + \ \cdot(CH_2CH_2)_y OR \longrightarrow RO(CH_2CH_2)_x(CH_2CH_2)_y OR$$

3. Anionic Polymerization. In anionic polymerization the reaction is initiated by an anion, forming carbanion intermediates (section 3.5C) as in the production of Lucite (Plexiglas). The process will continue until the anion is neutralized by a cation such as H^+.

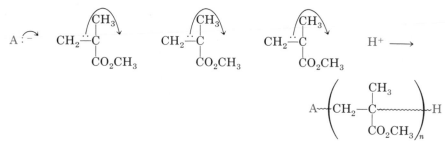

4.6. To aid your study of the mechanisms of polymerization, write a general mechanism (such as that shown for anionic polymerization) for the **(a)** cationic, **(b)** free-radical, and **(c)** anionic polymerizations of vinyl chloride

$$CH_2=\underset{\underset{Cl}{|}}{CH}$$

to polyvinyl chloride. Write them one directly under the other and compare the electron flow.

B. Copolymers

Copolymers are produced by polymerizing two or more different monomers, either simultaneously or consecutively. Vinyl acetate copolymerized with other monomers such as vinyl chloride produces copolymers used as adhesives, paint additives, and paper coatings and in textile treatment.

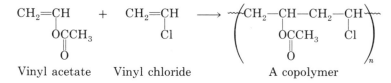

Vinyl acetate   Vinyl chloride   A copolymer

C. Types of Plastics and Plasticizers

Plastics can be classified as two basic types—thermosetting and thermoplastic. Thermosetting plastics must be molded into a shape as they are formed because they cannot be melted (they decompose before melting). Bakelite (light switches) and Melmac (plastic dishes) are examples of this type (see Panel 13 in Chapter 9). Thermoplastics, represented by most of the addition polymers in this chapter, can be repeatedly melted and reshaped.

Many polymers, particularly the addition polymers, are hard and brittle, and it is difficult to process them. Organic liquids of low volatility, when added to the polymer, act as internal lubricants between the millions of polymer chains. This action, by what are called plasticizers, allows greater molecular mobility and more flexibility in the plastic product. Dioctylphthalate (DOP), also called 2-ethylhexyl phthalate (DEHP), is an example of a plasticizer (Figure 4.2). As a plastic article ages, it may become brittle and crack since the plasticizer gradually diffuses and evaporates.

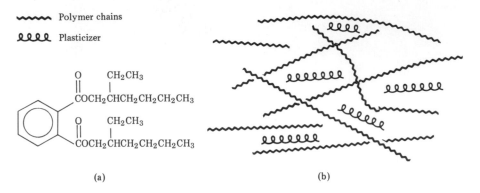

FIGURE 4.2. (a) The plasticizer dioctylphthalate (DOP); and (b) a diagram showing how its lubricating action allows polymer chains to slide across one another. Plasticizers allow physical flexibility in plastic products such as raincoats, shower curtains, and garden hoses.

Panel 4

COMMERCIAL APPLICATIONS
OF ADDITION POLYMERS

Baby bottles, children's toys, Styrofoam coolers, contact lenses, garden hose, credit cards, protective plastic wraps, Teflon-coated cookware, and many other articles common to our modern society are composed of addition polymers such as those discussed in this chapter. A general reaction for the production of addition polymers is

$$n \; \mathrm{C{=}C} \xrightarrow{\text{Catalyst}} \left(\mathrm{C-C} \right)_n$$

Alkene molecules add continuously to one another, producing giant molecules that are used to make a wide variety of commercial products. Following are some of the important addition polymers and their uses.

Polyethylene: $CH_2{=}CH_2 \longrightarrow (CH_2CH_2)_n$

Two types of polyethylene are commercially produced—a high-density polymer and a low-density polymer. High-density polyethylene is composed of continuous unbranched carbon chains that can pack closely together, giving extra strength and rigidity to the polymer (Figure 4.3). In low-density polyethylene, there is frequent chain branching along the molecules, which prevents them from coalescing. Because of this, low-density polyethylene is less rigid and softens at a lower temperature.

[PANEL 4]   High-density polyethylene is used for products that need a strong plastic, such as large drums, pipes and conduits, tanks, crates, and baby bottles that have to withstand sterilization temperatures. The low-density variety is made into films for packaging a variety of items—baked goods, frozen foods, meat, garments, dry cleaning—and for other uses, such as garbage bags and disposable diaper liners.

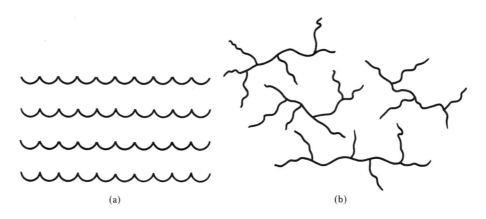

(a)   (b)

FIGURE 4.3. (a) High-density polyethylene. (b) Low-density polyethylene.

Polypropylene:

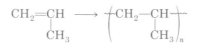

Polypropylene is a stronger and more rigid polymer than polyethylene. As such it is used in appliances, furniture, luggage, packing crates, and car and truck parts. The largest single use of polypropylene occurs in filaments and fibers used to manufacture products like indoor-outdoor carpeting.

Polystyrene:

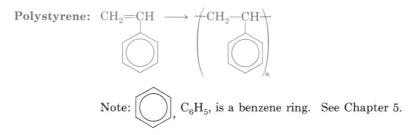

Note: ⬡ , C_6H_5, is a benzene ring. See Chapter 5.

Polystyrene is produced in two main forms: as a hard, often transparent, solid and as a foam (Styrofoam). Styrofoam is familiar because it is found in ice coolers, disposable hot drink cups, and protective packaging. Polystyrene can vary in texture from hard solid to foam, finding its way into electrical insulation, lighting fixtures, foundation and roofing insulation in buildings and homes, wall coverings and siding, plastic furniture, and disposable plastic knives, forks, and spoons. Polystyrene is also the most common plastic for toys (followed by polyethylene).

[PANEL 4]

Polymethyl Methacrylate:

Also called Lucite or Plexiglas, polymethyl methacrylate is often used as a substitute for glass because it is water-white and over 90% transparent. It is used to make safety glass (glass plus plastic) in automobile windshields and transparent pipes and tubing as well as Plexiglas. When polymethyl methacrylate is added to stucco and paints, it provides a hard, tough plastic coating on setting. One of the largest single uses is found in advertising signs and displays.

Because of its transparency, polymethyl methacrylate and modifications thereof are used in contact lenses. The basic optical principles of contact lenses were outlined by Leonardo da Vinci in 1508. In 1827 it was postulated that a gelatin disk superimposed on the cornea might correct astigmatism. In 1887 a glass shell was fitted over the eye of a patient whose upper eyelid had been destroyed. Glass contact lenses were available in the 1930s, but they often fit poorly. In the late 1930s (patent 1948), plastic contact lenses (polymethyl methacrylate) were introduced. An ideal contact lens should be optically transparent, mechanically stable, physiologically inert, and permeable to oxygen. Hard contact lenses are essentially polymethyl methacrylate and do not absorb water or transport oxygen. Soft contact lenses are often made from 2-hydroxy-ethyl methacrylate with or without other components.

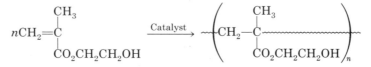

Because of the presence of the hydroxy groups, the polymer is able to absorb water thereby becoming more pliable, better lubricated, and more able to transport oxygen.

Orlon, Acrilan: $CH_2{=}CH \longrightarrow \big({-}CH_2CH{-}\big)$
                              $|$ $|$
                              CN CN $_n$

Two-thirds of Orlon is commonly spun into fibers for wearing apparel, and the remainder is used for home furnishings (carpeting, blankets, etc.).

Polyvinyl Chloride: $CH_2{=}CH \longrightarrow \big({-}CH_2CH{-}\big)$
                                    $|$ $|$
                                    Cl Cl $_n$

Polyvinyl chloride (PVC), along with polystyrene and polyethylene, is one of the top three plastics in terms of annual production figures. It has as wide a variety of applications as any other plastic. Intrinsically water-resistant, it can be manufactured into raincoats, shower curtains, garden hose, baby pants, swimming-pool liners, weather stripping, outdoor siding, and gutters. Two of the biggest uses of this polymer are in electrical conduit and in the PVC pipe found in plumbing and lawn-sprinkling systems. Polyvinyl chloride is also used to make blow-molded bottles, furniture, phonograph records, credit cards, toys, auto mats and upholstery, and also styled auto tops. Vinyl chloride, the monomer from

[PANEL 4] which PVC is formed, has been linked to the development of liver cancer and consequently worker exposure levels are regulated by federal standards.

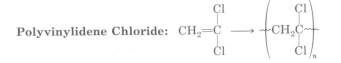

Polyvinylidene Chloride:

Films used for packaging foods, films such as saran, are made from polyvinylidene chloride.

Teflon: $CF_2{=}CF_2 \longrightarrow {+}(CF_2CF_2{)}_n$

Teflon, famous for its nonstick surface, has been applied practically as a surface coating in cooking utensils and in valves and gaskets. This polymer was accidentally discovered by Du Pont chemist Roy J. Plunkett in 1938. The morning after filling several cylinders with the gas tetrafluoroethylene ($CF_2{=}CF_2$), Plunkett found that the pressure gauge on one cylinder read zero, indicating that it was empty. But the seal was tight and the cylinder weighed the same as it had on the previous evening. When he opened the cylinder, a waxy white solid fell out. The tetrafluoroethylene had polymerized spontaneously. Teflon was born.

NATURAL AND SYNTHETIC RUBBER— RESONANCE STRUCTURES

4.3

A. Polyisoprene

1. Natural Polyisoprene Rubber. Natural rubber is produced from a milky-white colloidal latex found in the stems of some plants (even in the common dandelion and goldenrod). The commercial source is the rubber tree, which can yield as much as a ton of rubber per acre. The term *rubber* was coined by Joseph Priestley, who used it to "rub out" pencil marks. Structurally, natural rubber is a polyterpene (see Panel 6 on terpenes at the end of this chapter) composed of many recurring isoprene (2-methyl-1,3-butadiene) skeletons.

$$\underset{\text{Isoprene}}{CH_2{=}\overset{\overset{\displaystyle CH_3}{|}}{C}{-}CH{=}CH_2} \qquad \underset{\text{Natural rubber}}{\left(CH_2\overset{\overset{\displaystyle CH_3}{|}}{C}{=}CHCH_2\right)_n}$$

2. Synthetic Polyisoprene Rubber—1,4 Addition. Polyisoprene rubber can also be produced synthetically by the addition polymerization of isoprene (2-methyl-1,3-butadiene), a conjugated diene. In the peroxide-initiated polymerization, free radicals formed from decomposing peroxide attack isoprene molecules, forming another free radical.

Peroxide decomposition: $RO{:}OR \longrightarrow 2RO\cdot$

Initiation: $RO \,\,\, CH_2{\overset{..}{:}}\overset{\overset{\displaystyle CH_3}{|}}{C}{-}CH{=}CH_2 \longrightarrow RO{:}CH_2{-}\overset{\overset{\displaystyle CH_3}{|}}{\underset{\cdot}{C}}{-}CH{=}CH_2$

The free radical formed is called an *allylic free radical* and has a rather interesting structure. The unpaired electron occupies a p orbital that is coplanar with the p orbitals of the neighboring π bond (Figure 4.4). We can imagine this

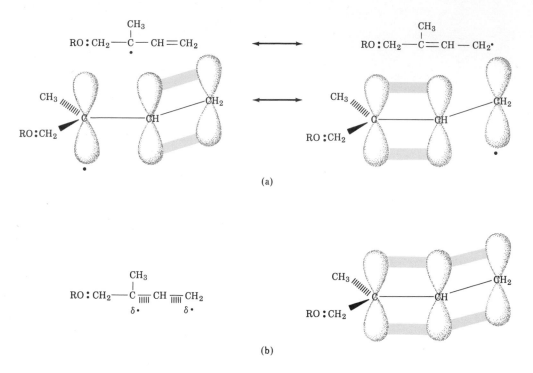

(a)

(b)

FIGURE 4.4. Free radical intermediate in the production of synthetic polyisoprene. (a) Resonance forms of the allylic free radical; (b) Resonance hybrid.

free-radical p orbital overlapping with the p orbital directly adjacent to it, forming a new π bond and a new locale for the lone electron. Figure 4.4a shows these two structures in terms of electron dot formulas and a molecular orbital picture. The two structures separated by a double-headed arrow are known as *resonance structures*. They describe the allylic free radical in terms of classical electron structures. In reality, neither of the two electron dot structures actually exists; rather, an "average" of the two, known as the *resonance hybrid,* is the true structure (Figure 4.4b). The resonance hybrid can be better comprehended by close examination of the π-bonding picture. Notice that the two resonance forms differ in the location of electrons, not atoms. Also notice that the middle p orbital is involved in the π bond no matter where the double bond is located. In reality, this middle p orbital overlaps unavoidably and simultaneously with the two p orbitals on either side. Consequently, there are no true double bonds or single bonds connecting these three carbons, but rather two bonds intermediate between double and single. The unpaired electron is spread across this system and can be "neutralized" at either carbon described by the resonance forms.

What does all of this mean in the production of synthetic rubber? After the free radical is formed, it will attack another isoprene molecule, thus propagating the chain reaction and continuing the growth of the addition polymer. If the radical is "neutralized" at carbon-2, then 1,2 addition has occurred. If neutralization occurs at carbon-4, 1,4 addition has occurred. Synthetic polyisoprene rubber is produced by a free-radical chain reaction involving mostly 1,4 addition to the conjugated diene isoprene.

Propagation:

$$
\begin{array}{ccc}
\text{CH}_3 & \text{CH}_3 & \text{CH}_3 \\
| & | & | \\
\text{RO}:\text{CH}_2-\text{C}=\text{CH}-\text{CH}_2 \;\curvearrowright\; \text{CH}_2-\text{C}-\text{CH}-\text{CH}_2, & \text{CH}_2-\text{C}-\text{CH}-\text{CH}_2, & \text{etc.}
\end{array}
$$

$$\downarrow$$

$$
\begin{array}{ccc}
\text{CH}_3 & \text{CH}_3 & \text{CH}_3 \\
| & | & | \\
\text{RO}:\text{CH}_2-\text{C}=\text{CH}-\text{CH}_2:\text{CH}_2-\text{C}=\text{CH}-\text{CH}_2:\text{CH}_2-\text{C}=\text{CHCH}_2\cdot
\end{array}
$$

Eventually, two propagating chains will come together in a termination step and we will have synthetic rubber.

$$
\begin{array}{c}
\text{CH}_3 \\
| \\
n\text{CH}_2=\text{C}-\text{CH}=\text{CH}_2
\end{array}
\xrightarrow[\text{peroxide}]{\text{ROOR}}
\text{RO}\!\left(\!\!\begin{array}{c}\text{CH}_3\\|\\\text{CH}_2\text{C}=\text{CHCH}_2\end{array}\!\!\right)_{\!\!n}\!\!\text{OR}
$$

Compare this free-radical chain-reaction mechanism to that of the production of polyethylene (section 4.2A.2) and the chlorination of methane (section 3.6C).

PROBLEM

4.7. Write a free-radical chain-reaction mechanism for the production of a rubber from 1,4-butadiene. Primarily 1,4 addition occurs.

B. Drawing Resonance Structures

Resonance theory is an important part of modern organic chemistry. We have used it in the previous section to explain 1,4 addition in the production of rubber, and we will employ resonance in subsequent chapters to explain orientation and rate of aromatic substitution and relative strengths of acids and bases. To effectively use the resonance theory, it is often helpful to be able to draw resonance structures and resonance hybrids. Resonance structures differ only in the positions of electrons, not atoms, as shown for the allylic carbocation, free radical, and carbanion in Figure 4.5. Drawing resonance structures for ions in essence involves manipulation of a three-atom allylic system. As you examine Figure 4.5, note the flow of electrons and how the resonance hybrid is drawn as an average of the resonance structures. Also note that there is a p orbital on each carbon of these allylic ions and radical.

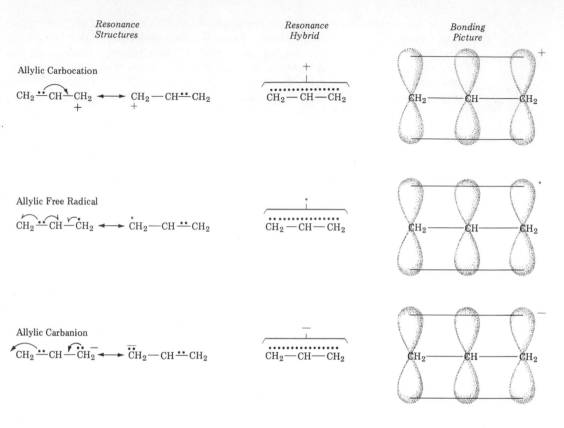

FIGURE 4.5. Drawing resonance structures for the allylic system ($CH_2{=}CH{-}CH_2{-}$).

Panel 5

TWO SIGNIFICANT EVENTS IN THE DEVELOPMENT OF THE RUBBER INDUSTRY

A. Vulcanization

Although rubber was introduced in Europe shortly after Columbus discovered the New World, it had limited use until 1839, when Charles Goodyear accidentally discovered vulcanization. Natural rubber tends to be sticky when warm, brittle when cold, and though elastic, does not regain its shape quickly or completely when stretched. For many years, Goodyear had unsuccessfully attempted to overcome these disadvantages. Then one day in 1839 he accidentally spilled one of his experiments, a mixture of latex rubber and sulfur, on a hot stove. After scraping the mixture off he found it was no longer sticky. Rather, it exhibited a greatly increased elasticity. The process is called *vulcanization*. Sulfur adds to the double bonds in rubber, constructing crosslinks between polymer chains (Figure 4.6). The sulfur bridges tend to maintain the conformations of the long poly-

[PANEL 5] mer chains so that they will be re-formed after the rubber is stretched. Stretching puts a strain on the polymer chains and sulfur bridges, and this is relieved only when the rubber is allowed to assume the original conformation.

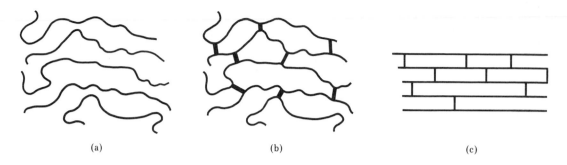

(a)   (b)   (c)

FIGURE 4.6. (a) Untreated rubber molecules are bent and convoluted. (b) Vulcanized rubber has many sulfur bridges linking the polymer chains thus providing a "chemical memory" of the original shape before stretching. (c) When stretched, the rubber molecules align. This puts a strain on the system since the sulfur bridges tend to hold the molecules in the original conformation. The strain is relieved when the rubber is allowed to snap back into the original conformation.

Soft rubber has 1%–3% sulfur by weight, whereas hard rubber (such as in rubber mallets) has as much as 20%–30% sulfur. Although Goodyear's discovery made possible the enormous growth of the rubber industry, he did not personally profit from it. He died in 1860 after years of court battles over patents, leaving his wife and six children with debts over $200,000.

B. Synthetic Rubber

On December 7, 1941, the United States became involved in World War II, following the Japanese attack on Pearl Harbor. Crucial to the subsequent war effort was the Japanese occupation of Malaya and the Dutch East Indies, source of virtually all crude rubber used in the United States. With a stockpile of less than a year's supply of natural rubber, it was essential that large-scale synthetic rubber production capacity be quickly established so that the armed forces, agriculture, and transportation could remain mobile. At the time, research on synthetic rubber production was in its infancy. Through cooperation and intellectual interchange among academic and industrial organizations, annual synthetic rubber production reached 20,000 tons in 1942, 215,000 tons in 1943, and almost 800,000 tons by 1945. Today, the United States produces over 3 million tons of rubber annually. Approximately 80% of it is synthetic.

Several synthetic rubbers play a large part in the present-day rubber industry. By far the most important of these is SBR, a styrene-butadiene copolymer with elastomer properties. This rubber, once called GRS (Government Rubber Styrene), was the first and most widely produced synthetic rubber of World War II. Composed of approximately 75% 1,3-butadiene and 25% styrene, it is similar in chemistry of formation to other addition polymers. Styrene undergoes normal 1,2 addition and the butadiene mainly 1,4 addition:

[PANEL 5] $nCH_2{=}CH{-}CH{=}CH_2 + nCH_2{=}CH \longrightarrow$

$$\left(CH_2CH{=}CHCH_2{-}CH_2CH\right)_n$$

The synthetic rubbers polybutadiene, polyisoprene, and neoprene are all made by the 1,4 addition polymerization of dienes:

$$\overset{\displaystyle G}{nCH_2{=}\underset{}{C}{-}CH{=}CH_2} \longrightarrow \left(CH_2\overset{\displaystyle G}{C}{=}CHCH_2\right)_{\!n}$$

where G = CH_3 for polyisoprene,
     = H for polybutadiene, and
     = Cl for neoprene.

Like most natural and synthetic rubbers, polybutadiene and polyisoprene have their predominant use in tire manufacture. Neoprene, however, is more resistant to oils, chemicals, heat, and air than most rubbers and thereby finds specialty uses, as in gasoline pump hoses and rubber tubing in automobile engines.

HYDROBORATION

4.4

Hydration of propene with H_2O/H_2SO_4 produces 2-propanol. This product predominates over 1-propanol because the carbocation leading to its formation is a more stable 2° carbocation, whereas 1-propanol is formed via a 1° carbocation.

$$CH_3CH{=}CH_2 + H_2O \xrightarrow[\text{(via } CH_3\overset{+}{C}HCH_3)]{H_2SO_4} CH_3\underset{\underset{\displaystyle OH}{|}}{C}HCH_3$$

How then can 1-propanol be prepared? Treatment of propene with diborane (B_2H_6) followed by hydrogen peroxide oxidation in aqueous sodium hydroxide will yield predominantly 1-propanol.

$$CH_3CH{=}CH_2 \xrightarrow{B_2H_6} \xrightarrow[H_2O/OH^-]{H_2O_2,} CH_3CH_2CH_2OH$$

This apparent reversal in the orientation of addition can be explained if one thinks of B_2H_6 as two BH_3 units. The boron-hydrogen bonds are polar, with hydrogen being partially negative and boron partially positive. Since the boron has only six outer-shell electrons, not an octet, it is a Lewis acid, that is, an electron-pair acceptor. These two factors cause the boron to be attracted to propene's π electron cloud and it bonds to the first carbon. As the carbon-boron bond forms, a secondary carbocation begins to form on carbon-2 (if the boron had bonded to carbon-2, a less stable primary carbocation would have formed on carbon-1). Actually, a carbocation never forms because the developing positive charge on carbon-2 is simultaneously neutralized by a negative hydrogen.

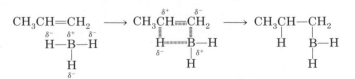

The two remaining boron-hydrogen bonds of BH_3 add to two more alkene molecules in an analogous manner.

$$CH_3CH_2CH_2BH_2 \xrightarrow{CH_3CH=CH_2} \xrightarrow{CH_3CH=CH_2} (CH_3CH_2CH_2)_3B$$

The trialkylborane is then oxidized to an alcohol.

$$(CH_3CH_2CH_2)_3B + 3H_2O_2 \longrightarrow 3CH_3CH_2CH_2OH + B(OH)_3$$

PROBLEM

4.8. Write equations showing the preparation of alcohols by hydroboration from the following alkenes: **(a)** 1-butene; **(b)** methylpropene.

OXIDATION OF ALKENES

4.5

A. Hydroxylation with Potassium Permanganate

Alkenes react with potassium permanganate to form 1,2-diols.

$$3-\overset{|}{\underset{}{C}}=\overset{|}{\underset{}{C}}- + 2KMnO_4 + 4H_2O \longrightarrow 3-\overset{|}{\underset{OH}{C}}-\overset{|}{\underset{OH}{C}}- + 2MnO_2 + 2KOH$$
$$\text{(Purple)} \qquad\qquad\qquad \text{(Brown)}$$

This reaction, known as the Baeyer test, is useful in distinguishing alkenes from alkanes. Alkanes do not undergo reaction with $KMnO_4$. A positive test is easy to detect visually because potassium permanganate solutions are deep purple. As they are added to alkenes, the purple quickly disappears, leaving a murky brown precipitate of manganese oxide.

PROBLEM

4.9. Write equations showing the reaction of the following with potassium permanganate: **(a)** ethene; **(b)** 2-butene; **(c)** cyclohexene.

B. Ozonolysis

Double bonds are easily cleaved oxidatively on reaction with ozone followed by hydrolysis in the presence of a reducing agent, zinc.

$$-\overset{|}{\underset{}{C}}=\overset{|}{\underset{}{C}}- + O_3 \longrightarrow \xrightarrow[\text{Zn}]{H_2O,} -\overset{|}{\underset{}{C}}=O + O=\overset{|}{\underset{}{C}}-$$

The reaction products are aldehydes and ketones. Ozone, prepared by passing oxygen gas through an electric discharge, is bubbled into a solution of the alkene in an inert solvent like carbon tetrachloride. Mechanistically, the ozone adds to the double bond, and a molozonide is formed. This rearranges to an ozonide in which the carbon-carbon bond is completely cleaved. Hydrolysis of the ozonide produces aldehydes and ketones.

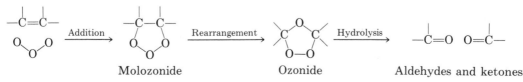

Molozonide   Ozonide   Aldehydes and ketones

Ozonolysis is particularly useful for elucidating the location of double bonds in alkenes. An unknown alkene is cleaved to smaller, more easily identifiable aldehydes and ketones. The aldehydes and ketones are then pieced back together like a puzzle. Wherever a carbon-oxygen double bond occurs, originally that carbon was involved in a carbon-carbon double bond. Suppose we have an unknown alkene with the molecular formula C_4H_8. After ozonolysis it is converted to the following compounds:

$$C_4H_8 \xrightarrow{O_3} \xrightarrow[\text{Zn}]{H_2O,} CH_3\overset{\underset{\|}{O}}{C}CH_3 + H\overset{\underset{\|}{O}}{C}H$$

Unknown alkene

The carbon-oxygen double bonds identify the carbons involved in the alkene linkage. Connecting these two carbons gives us the structure of the unknown alkene, 2-methylpropene.

$$CH_3\overset{\underset{\|}{CH_2}}{C}CH_3$$

The other two isomeric alkenes with the formula C_4H_8 give quite different ozonolysis products.

$$CH_3CH_2CH{=}CH_2 \xrightarrow{O_3} \xrightarrow[\text{Zn}]{H_2O,} CH_3CH_2\overset{\underset{\|}{O}}{C}H + H\overset{\underset{\|}{O}}{C}H$$

$$CH_3CH{=}CHCH_3 \xrightarrow{O_3} \xrightarrow[\text{Zn}]{H_2O,} 2CH_3\overset{\underset{\|}{O}}{C}H$$

PROBLEMS

4.10. Write products formed after ozonolysis of the following alkenes: **(a)** 2-methyl-2-pentene; **(b)** 3,7-dimethyl-1,3,6-octatriene (ocimene, a component of the spice basil); **(c)** cyclohexene.

4.11. Determine the structure of the alkene with the formula C_5H_{10} from the ozonolysis products.

$$C_5H_{10} \xrightarrow{O_3} \xrightarrow[H_2O]{Zn,} CH_3\overset{\underset{\|}{O}}{C}H + CH_3CH_2\overset{\underset{\|}{O}}{C}H$$

ACIDITY OF TERMINAL ALKYNES

4.6   Terminal alkynes, in which the triple bond is at the end of a carbon chain, are very weakly acidic. The hydrogen can be abstracted by strong bases such as sodium amide ($:NH_2^-$ is a stronger base than $:OH^-$).

$$RC{\equiv}CH + NaNH_2 \longrightarrow RC{\equiv}C:^-Na^+ + NH_3$$

Alkanes and alkenes are not acidic and do not undergo this reaction. These sodium salts of alkynes are useful in producing higher alkynes by nucleophilic substitution reactions (section 7.4A).

PROBLEM

4.12. Write a reaction equation showing the reaction between 1-butyne and sodium amide.

Panel 6

TERPENES

The odor of mint, the scent of cedar and pine, the fragrance of geraniums and roses, and the color of carrots and tomatoes are largely due to a class of compounds known as the terpenes. Besides imparting odor and flavor to plants, terpenes have commercial uses in perfume as well as medicines and food flavorings.

Terpenes are characterized by carbon skeletons constructed of isoprene units. Isoprene is a conjugated diene.

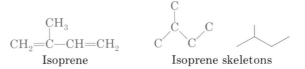

$$CH_2{=}\overset{\overset{\displaystyle CH_3}{|}}{C}{-}CH{=}CH_2$$
Isoprene

Isoprene skeletons

In fact, terpenes are classified according to the number of isoprene units in the molecule. The simplest terpenes have two isoprene units (10 carbons) and are called monoterpenes. Other classes are listed below.

Monoterpenes	Two isoprene units	C_{10}
Sesquiterpenes	Three isoprene units	C_{15}
Diterpenes	Four isoprene units	C_{20}
Triterpenes	Six isoprene units	C_{30}
Tetraterpenes	Eight isoprene units	C_{40}

Terpenes are further classified according to the number of rings in the molecule:

Acyclic	No rings	Bicyclic	Two rings
Monocyclic	One ring	Tricyclic	Three rings

And they are classified also on the basis of whether they are oxygenated (aldehydes, ketones, alcohols, ethers, and so on) or are pure hydrocarbons. The following is a brief survey of some common terpenes of each class. Note the isoprene skeletons in each compound.

A. Monoterpenes

Monoterpenes consist of two isoprene units usually connected in a head-to-tail fashion. Dashed lines are used in the following structures to show the isoprene skeletons. Myrcene (bayberry wax in candles) and ocimene (component of the spice basil) are examples of acyclic monoterpenes. Citral, oil of lemon, has a lemon odor, whereas the corresponding alcohol, geraniol, is a constituent of the essential oils of roses and geraniums. Both citral and geraniol are oxygenated monoterpenes. They also happen to be recruiting pheromones for the honeybee. Pheromones (see Panel 12 at the end of Chapter 8) are compounds produced and

[PANEL 6] secreted by insects (and have been found with increasing incidence in other animals) for "chemical communication."

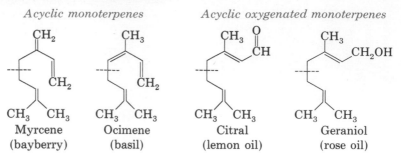

Acyclic monoterpenes   *Acyclic oxygenated monoterpenes*

Myrcene   Ocimene   Citral   Geraniol
(bayberry)   (basil)   (lemon oil)   (rose oil)

Limonene, oil of lemon and orange, and menthol and menthone from peppermint oil are examples of monocyclic monoterpenes. Menthol is the flavoring added to cigarette tobacco, shaving creams, and mouthwashes.

Monocyclic monoterpenes

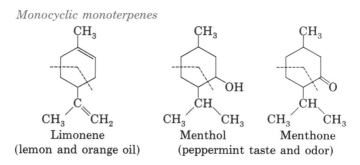

Limonene   Menthol   Menthone
(lemon and orange oil) (peppermint taste and odor)

Familiar examples of bicyclic monoterpenes include α-pinene, the principal constituent of pine oil; camphor, from the camphor tree; and camphene, oil of ginger. Pinene is used as a solvent and paint thinner (turpentine) and as the fragrance in room fresheners and pine cleaners. Camphor, a component of some chest rubs, has long been a valued medicinal ingredient although its therapeutic value has come into question in modern times. The number of rings in a polycyclic compound can be determined by counting the minimum number of scissions in the carbon skeleton that would be necessary to make it an open-chain structure.

Bicyclic monoterpenes

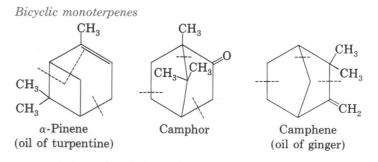

α-Pinene   Camphor   Camphene
(oil of turpentine)   (oil of ginger)

PROBLEM

4.13. To ensure that you understand the structures presented in this section, draw (a) citral, (b) menthol, (c) α-pinene and (d) farnesol (next section) showing all carbons and hydrogens.

[PANEL 6]   **B. Sesquiterpenes**

Farnesol (scent of lily of the valley), zingiberene (oil of ginger), selinene (oil of celery), and cedrol (cedar oil) are examples of acyclic, monocyclic, bicyclic, and tricyclic sesquiterpenes, respectively.   Each has three isoprene units in its carbon skeleton.

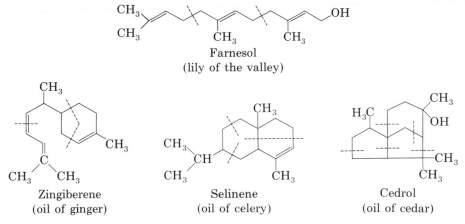

Farnesol
(lily of the valley)

Zingiberene   Selinene   Cedrol
(oil of ginger)   (oil of celery)   (oil of cedar)

C. Diterpenes

Diterpenes have 20 carbons arranged in skeletons of four isoprene units. Vitamin A, whose deficiency can cause night blindness in adults and total blindness in children (100,000 annually world-wide), is an example of a diterpene.   Its

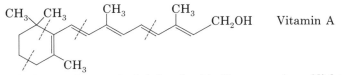

Vitamin A

aldehyde oxidation product, retinal, is involved in the conversion of light impulses to nerve transmissions in the eye (see Panel 1, section 2.7).

Vitamin A is used medicinally in the treatment of acne.   Recently, a vitamin A derivative, 13-*cis*-retinoic acid, has shown promising test results on individuals suffering disfiguring acne that has resisted other treatments such as oral and topical (externally applied) antibiotics and unmodified vitamin A therapy.   The drug is also being tested as an anticancer agent.

D. Triterpenes

Squalene, a triterpene (six isoprene units), is found in large amounts in shark-liver oil and in smaller amounts in livers of higher animals.   As illustrated in the following structures, it is a precursor in the biosynthesis of cholesterol.

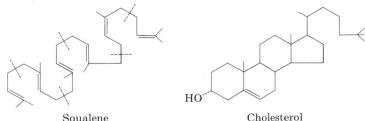

Squalene   Cholesterol

[PANEL 6]   E. Tetraterpenes

Carotene, $C_{40}H_{56}$, is a bicyclic tetraterpene responsible for the orange color of carrots. Carrots are a source of vitamin A since cleavage of carotene at the central double bond produces two identical diterpenes with the vitamin A carbon skeleton (thus the claim that carrots are beneficial to vision). Carotene itself has no vitamin A activity. Its cleavage is effected by enzymes in the intestines and liver.

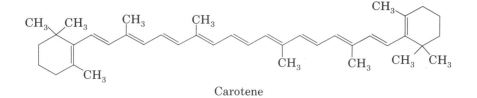

Carotene

F. Polyterpenes

Natural rubber is a good example of a polyterpene. It is extracted as a milky liquid, called latex, from the rubber tree and is composed of thousands of isoprene units (section 4.3).

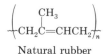

Natural rubber

PROBLEM

4.14. The following are structures of common terpenes. Classify each compound as a mono-, di-, tri-, or tetraterpene and as acyclic, mono-, bi-, tri-, or tetracyclic. Using dashed lines or circles, identify each isoprene unit.

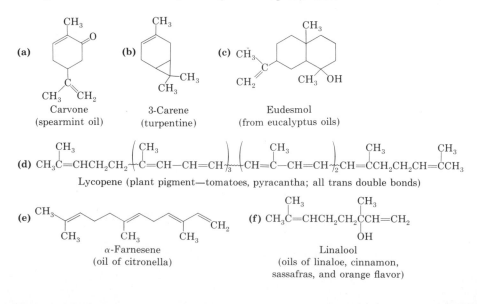

(a) Carvone (spearmint oil)

(b) 3-Carene (turpentine)

(c) Eudesmol (from eucalyptus oils)

(d) Lycopene (plant pigment—tomatoes, pyracantha; all trans double bonds)

(e) α-Farnesene (oil of citronella)

(f) Linalool (oils of linaloe, cinnamon, sassafras, and orange flavor)

[PANEL 6]

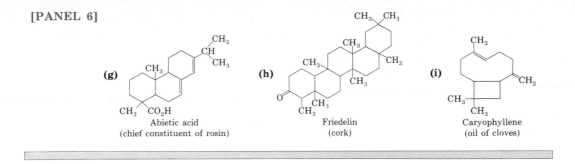

(g) Abietic acid
(chief constituent of rosin)

(h) Friedelin
(cork)

(i) Caryophyllene
(oil of cloves)

END-OF-CHAPTER PROBLEMS

4.15 IUPAC Nomenclature: Name the following by the IUPAC system of nomenclature:

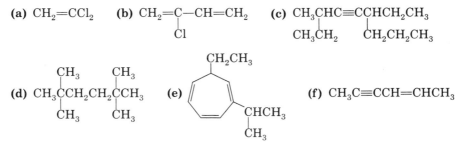

(a) $CH_2{=}CCl_2$ **(b)** $CH_2{=}\underset{\underset{Cl}{|}}{C}{-}CH{=}CH_2$ **(c)** $CH_3\underset{\underset{CH_3CH_2}{|}}{CHC}{\equiv}C\underset{\underset{CH_2CH_2CH_3}{|}}{CHCH_2CH_3}$

(d) $CH_3\underset{\underset{CH_3}{|}}{\overset{\overset{CH_3}{|}}{C}}CH_2CH_2\underset{\underset{CH_3}{|}}{\overset{\overset{CH_3}{|}}{C}}CH_3$ **(e)** [structure]   **(f)** $CH_3C{\equiv}CCH{=}CHCH_3$

4.16 IUPAC Nomenclature: Draw the following from the IUPAC name: **(a)** 1-chloro-4-t-butylcyclohexane; **(b)** 2-methylpentane; **(c)** 3,4-diethylhexane; **(d)** 2-hexyne; **(e)** 5-propyl-3-octyne; **(f)** 2-bromo-3-heptene; **(g)** 6-ethyl-4,8-dimethyl-2,4,6-nonadiene; **(h)** 1-buten-3-yne.

4.17 IUPAC Nomenclature of Terpenes: Name the following terpenes by the IUPAC system of nomenclature. See Panel 6 in this chapter for structures. **(a)** ocimene; **(b)** squalene; **(c)** isoprene; **(d)** α-farnesene.

4.18 Reactions of Alkenes and Alkynes: Complete the following reactions showing the major organic products:

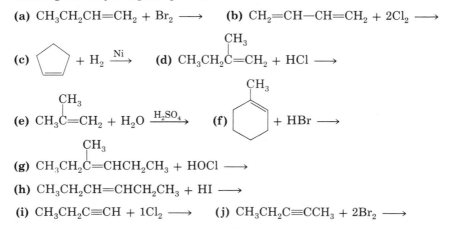

(a) $CH_3CH_2CH{=}CH_2 + Br_2 \longrightarrow$ **(b)** $CH_2{=}CH{-}CH{=}CH_2 + 2Cl_2 \longrightarrow$

(c) [structure] $+ H_2 \xrightarrow{Ni}$ **(d)** $CH_3CH_2\underset{\underset{CH_3}{|}}{C}{=}CH_2 + HCl \longrightarrow$

(e) $CH_3\underset{\underset{CH_3}{|}}{C}{=}CH_2 + H_2O \xrightarrow{H_2SO_4}$ **(f)** [structure] $+ HBr \longrightarrow$

(g) $CH_3CH_2\underset{\underset{CH_3}{|}}{C}{=}CHCH_2CH_3 + HOCl \longrightarrow$

(h) $CH_3CH_2CH{=}CHCH_2CH_3 + HI \longrightarrow$

(i) $CH_3CH_2C{\equiv}CH + 1Cl_2 \longrightarrow$ **(j)** $CH_3CH_2C{\equiv}CCH_3 + 2Br_2 \longrightarrow$

(k) $CH_3C \equiv CH + 1HCl \longrightarrow$ **(l)** $CH_3C \equiv CH + 2HBr \longrightarrow$

(m) $CH_3\overset{CH_3}{\underset{|}{C}}HC \equiv CCH_3 + 2H_2 \xrightarrow{Ni}$ **(n)** $CH_3C \equiv CH + H_2O \xrightarrow{H_2SO_4}$

(o) $CH_3CH_2\overset{CH_3}{\underset{|}{C}}=CH_2 \xrightarrow{B_2H_6} \xrightarrow[OH^-]{H_2O_2}$ **(p)** ⬡$=CH_2 \xrightarrow{B_2H_6} \xrightarrow[OH^-]{H_2O_2}$

(q) $CH_3CH=CH_2 + KMnO_4/H_2O \longrightarrow$ **(r)** ⬠ $+ KMnO_4/H_2O \longrightarrow$

(s) $CH_3\overset{CH_3}{\underset{|}{C}}=\overset{CH_3}{\underset{|}{C}}CH_3 \xrightarrow{O_3} \xrightarrow[H_2O]{Zn,}$ **(t)** (bicyclic structure) $\xrightarrow{O_3} \xrightarrow[H_2O]{Zn,}$

(u) $CH_3CH=CHCH_2CH=CHCH_3 \xrightarrow{O_3} \xrightarrow[H_2O]{Zn,}$ **(v)** $CH_3C \equiv CH + NaNH_2 \longrightarrow$

4.19 Ozonolysis: Write the products of ozonolysis of the following compounds with O_3 followed by Zn/H_2O:

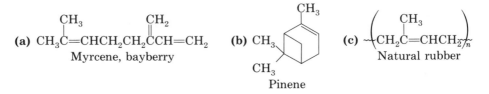

(a) $CH_3\overset{CH_3}{\underset{|}{C}}=CHCH_2CH_2\overset{CH_2}{\underset{\|}{C}}CH=CH_2$

Myrcene, bayberry

(b) Pinene structure

(c) $\left(CH_2\overset{CH_3}{\underset{|}{C}}=CHCH_2\right)_n$

Natural rubber

4.20 Ozonolysis: Write molecular structures for the following compounds based on the molecular formulas and ozonolysis products shown:

(a) $C_8H_{16} \xrightarrow{O_3} \xrightarrow[H_2O]{Zn,} 2CH_3\overset{O}{\underset{\|}{C}}CH_2CH_3$

(b) $C_7H_{14} \xrightarrow{O_3} \xrightarrow[H_2O]{Zn,} CH_3CH_2CH_2\overset{O}{\underset{\|}{C}}H + CH_3\overset{O}{\underset{\|}{C}}CH_3$

(c) $C_{10}H_{18} \xrightarrow{O_3} \xrightarrow[H_2O]{Zn,} H\overset{O}{\underset{\|}{C}}CH_2CH_2\overset{O}{\underset{\|}{C}}H + 2CH_3\overset{O}{\underset{\|}{C}}CH_3$

(d) $C_{10}H_{16} \xrightarrow{O_3} \xrightarrow[H_2O]{Zn,} CH_3\overset{O}{\underset{\|}{C}}CH_2\overset{O}{\underset{\|}{C}}CH_2\overset{O}{\underset{\|}{C}}CH_3 + 3HCH$

(e) $C_8H_{14} \xrightarrow{O_3} \xrightarrow[H_2O]{Zn,} CH_3\overset{O}{\underset{\|}{C}}(CH_2)_4\overset{O}{\underset{\|}{C}}CH_3$

(f) $C_9H_{14} \xrightarrow{O_3} \xrightarrow[H_2O]{Zn,} CH_3\overset{O}{\underset{\|}{C}}-$(cyclopentane)$-\overset{O}{\underset{\|}{C}}CH_3$

4.21 Hydration: Write a reaction equation showing how the following alcohols can be prepared from alkenes: **(a)** 1-butanol; **(b)** 2-butanol; **(c)** cyclohexanol; **(d)** 2-methyl-1-propanol; **(e)** 2-methyl-2-propanol.

4.22 Reaction Mechanisms: Write step-by-step reaction mechanisms for the following reactions:

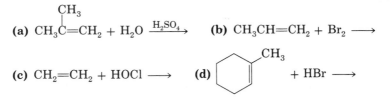

(a) $CH_3\overset{\underset{|}{CH_3}}{C}{=}CH_2 + H_2O \xrightarrow{H_2SO_4}$ **(b)** $CH_3CH{=}CH_2 + Br_2 \longrightarrow$

(c) $CH_2{=}CH_2 + HOCl \longrightarrow$ **(d)** [cyclohexene with CH$_3$ structure] $+ HBr \longrightarrow$

4.23 Reaction Mechanisms—Addition Polymers: Write step-by-step reaction mechanisms for the following polymerization reactions: **(a)** cationic polymerization of propene to polypropylene; **(b)** free radical polymerization of 1,1-dichloro-ethene to saran; **(c)** anionic polymerization of acrylonitrile to Orlon; **(d)** polymerization of chloroprene (2-chloro-1,3-butadiene) to neoprene rubber by a free-radical chain reaction involving 1,4 addition.

4.24 Addition Polymers: Write structures for the polymers produced from the following monomers:

(a) $CH_2{=}CHF$ Vinyl fluoride polymerizes to Tedlar, which is used as a coating.

(b) $CH_2{=}CF_2$ Its polymer is polyvinylidene fluoride, used to make rubber-type articles.

(c) $CH_2{=}\underset{\underset{\overset{\|}{O}}{OCCH_3}}{CH}$ Polyvinyl acetate is used in paints, paper, and adhesives.

(d) $CH_2{=}CH{-}CH{=}CH_2$ Polybutadiene rubber.

(e) $CH_2{=}CHBr$ Vinyl bromide is used to produce flame-retardant polymers.

(f) $CF_2{=}CF_2$, $CF_3CF{=}CF_2$ A copolymer called Teflon FEP is made from tetrafluoroethylene and hexafluoropropylene.

(g) $CH_2{=}CH{-}CH{=}CH_2$,   ABS (acrylonitrile, butadiene styrene) copolymers are made from these three monomers.
$CH_2{=}\underset{\underset{CN}{|}}{CH}$, $CH_2{=}CH$ [benzene ring]

(h) $CH_2{=}CH{-}CH{=}CH_2$,   Nitrile rubber is a copolymer resistant to oil that is used to make hoses, seals, gaskets, and O-rings.
$CH_2{=}\underset{\underset{CN}{|}}{CH}$

(i) CH_2=$\overset{\overset{\displaystyle CH_3}{|}}{C}$—CH=$CH_2$,   Butyl rubber is a copolymer of isoprene and iso-butylene.

CH_2=$\overset{\overset{\displaystyle CH_3}{|}}{C}$—$CH_3$

4.25 Synthesis: In formulating a synthesis for an organic compound, one often works backward from the product desired to a starting material that is available. Try some of the following by working backwards and proposing structures for the unknown compounds A, B, C, ... The reactions you will need are the elimination reactions to prepare alkenes and alkynes (section 3.7) and the electrophilic addition reactions of alkenes and alkynes (section 4.1).

(a) A + H_2 $\xrightarrow{\text{Ni}}$ $CH_3CH_2CH_2CH_2CH_3$ **(b)** B $\xrightarrow{\text{2KOH}}$ CH_3C≡CH

(c) C $\xrightarrow{\text{H}_2\text{SO}_4}$ $CH_3\overset{\overset{\displaystyle CH_3}{|}}{C}HCH$=$CHCH_3$ **(d)** D $\xrightarrow{\text{HCl}}$ $CH_3\overset{\overset{\displaystyle CH_3}{|}}{C}H\underset{\underset{\displaystyle Cl}{|}}{C}HCH_3$

(e) E $\xrightarrow{\text{H}_2\text{SO}_4}$ F $\xrightarrow[\text{Ni}]{\text{H}_2}$ ⬡   **(f)** G $\xrightarrow{\text{KOH}}$ H $\xrightarrow{\text{Br}_2}$ $CH_3CHBrCH_2Br$

(g) I $\xrightarrow{\text{HBr}}$ J $\xrightarrow{\text{KOH}}$ K $\xrightarrow{\text{Cl}_2}$ $CH_3CHClCHClCH_3$

(h) L $\xrightarrow{\text{H}_2\text{SO}_4}$ M $\xrightarrow{\text{Br}_2}$ N $\xrightarrow{\text{2KOH}}$ O $\xrightarrow{\text{2HCl}}$ CH_3CHCl_2

4.26 Industrial Reactions: Processes for making industrial chemicals are described in the following paragraphs. Write reaction equations illustrating the descriptions.

(a) beverage alcohol: hydration of ethene
(b) rubbing alcohol: hydration of propene
(c) vinyl chloride, monomer from which PVC is made: addition of chlorine to ethene followed by dehydrochlorination
(d) chloroprene, monomer from which neoprene rubber is made: addition of one equivalent HCl to 1-butene-3-yne (triple bond reacts)
(e) triclene, dry cleaning agent: addition of 2 equivalents chlorine to ethyne followed by dehydrochlorination with one equivalent of base
(f) vinylidene chloride, monomer from which saran is made: one equivalent HCl added to acetylene followed by addition of one equivalent of chlorine; this product is dehydrochlorinated with one equivalent of base

4.27 Gasoline Octane Boosters: TBA (tertiary butyl alcohol, 2-methyl-2-propanol) can be made industrially by the hydration of isobutylene (methylpropene) with acid catalyst (sections 4.1B4 and 4.1D3), and MTBE (methyl tertiary butyl ether, 2-methyl-2-methoxypropane) is prepared by treating isobutylene with methanol (CH_3OH) and acid catalyst. Both are octane boosters that are becoming increasingly important as lead is phased out of gasoline. Write chemical reactions for each process. Then write a reaction mechanism for each. Both involve electrophilic addition.

4.28 Reaction Mechanisms: If bromine dissolved in a concentrated sodium chloride solution is added to cyclohexene, the product is largely 1-bromo-2-chloro-cyclohexane. Explain this using a reaction mechanism.

4.29 Geometric Isomers: There are two forms of citral (lemon oil, Panel 6) that are geometric isomers. Draw each.

4.30 Qualitative Analysis: Suggest a reagent that would chemically distinguish between each member of the following pairs of compounds. Write chemical equations for all positive tests.

(a) 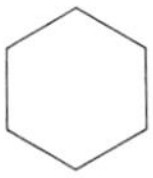 ⬡ and ⬡   **(b)** $CH_3CH_2C{\equiv}CH$ and $CH_3C{\equiv}CCH_3$

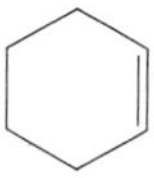

(c) $CH_3C{\equiv}CCH_3$ and $CH_3CH_2CH_2CH_3$ **(d)** $CH_3CH{=}CH_2$ and $CH_3C{\equiv}CH$

4.31 Electron Dot Formula, Production of Acetylene: Acetylene (ethyne) can be made from coke and limestone. In the final step, calcium carbide is treated with two equivalents of water to yield acetylene and calcium hydroxide. Write a chemical equation for this reaction and an electron dot formula for calcium carbide (CaC_2).

4.32 Resonance Structures: Draw the resonance structures and resonance hybrid for each of the following ions and radicals:

(a) $CH_3\overset{+}{C}HCH{=}CH_2$ **(b)** $CH_3\overset{-}{C}HCH{=}CH_2$ **(c)** $CH_3\overset{\cdot}{C}HCH{=}CH_2$

(d) 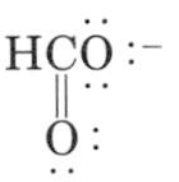 $H\overset{\displaystyle \underset{\|}{O:}}{\underset{:}{C}}\overset{..}{O}{:}^{-}$ **(e)** $CH_2{=}CH\overset{+}{C}HCH{=}CH_2$ **(f)** 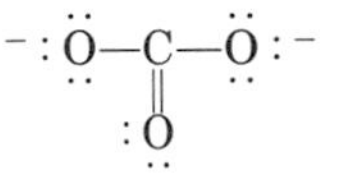 $^{-}{:}\overset{..}{O}{-}\overset{\displaystyle \underset{\|}{:O}}{C}{-}\overset{..}{O}{:}^{-}$

(g) 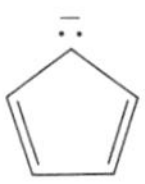 ⬠⁻   **(h)**  ⬡·   **(i)** 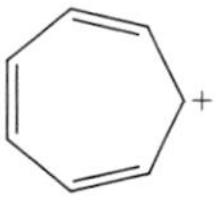 ⬡⁺

Chapter 5

Aromatic Hydrocarbons

INTRODUCTION TO AROMATIC COMPOUNDS

5.1   *Aromatic compounds* are defined as benzene and compounds that are similar to benzene in chemical behavior.

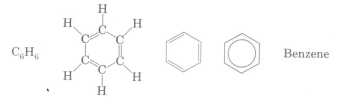

C_6H_6 .... Benzene

Although benzene and many other aromatic compounds are extracted from foul-smelling coal tar, they have a fragrant odor, hence the term *aromatic*. In fact, some that contain other additional functional groups are responsible for the characteristic fragrances of wintergreen, cinnamon, cloves, vanilla, almonds, and roses.

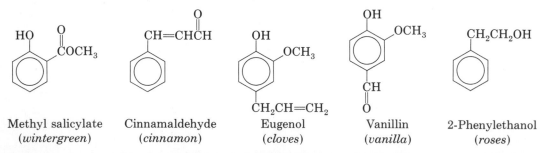

| Methyl salicylate | Cinnamaldehyde | Eugenol | Vanillin | 2-Phenylethanol |
| *(wintergreen)* | *(cinnamon)* | *(cloves)* | *(vanilla)* | *(roses)* |

Benzene was first isolated by Michael Faraday in 1825. He obtained it from the illuminating gas that was manufactured in England from whale oil. In 1834, Mitscherlich established the formula C_6H_6, and in 1865 August Kekule proposed a structure. Benzene is a cyclic compound commonly written as a hexagon with alternating double and single bonds or with a circle in the middle to describe the bonding. Each corner of the hexagon represents a carbon with one bonded hydrogen. The structure of benzene is not as simple as it appears and because of some unusual characteristics, its real structure eluded chemists for many years.

BENZENE: STRUCTURE AND BONDING

5.2   A. Unusual Characteristics of Benzene

1. Stability. Benzene is an unusually stable compound as evidenced by its characteristic reactions. Most unsaturated compounds characteristically undergo addition reactions in which a double or triple bond is altered to a single bond (see addition reactions of alkenes and alkynes, section 4.1).

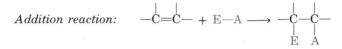

Addition reaction:

Benzene, on the other hand, engages predominantly in substitution reactions, in which the integrity of the benzene ring is preserved—testimony to its unusual stability.

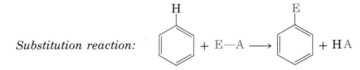

Substitution reaction:

2. Resonance Energy. Another indication of benzene's unusual stability is something called *resonance energy*. When a chemical reaction occurs, energy is either evolved, an exothermic reaction, or absorbed from the surroundings, an endothermic reaction. The hydrogenation of cyclohexene, for example, is an exothermic reaction, with 28.6 kcal of energy evolved for each mole of cyclohexene hydrogenated.

$$\text{cyclohexene} + H_2 \xrightarrow{\text{Ni}} \text{cyclohexane} + \text{heat (28.6 kcal/mole)}$$

For 1,3-cyclohexadiene, we should predict a heat of hydrogenation of 57.2 kcal/mol, twice that of cyclohexene, since there are twice as many double bonds. The actual value, 55.4 kcal/mol, is very close to our prediction. Following this

$$\text{1,3-cyclohexadiene} + 2H_2 \xrightarrow{\text{Ni}} \text{cyclohexane} + \text{heat (55.4 kcal/mole)}$$

same logic, we could treat benzene as a 1,3,5-cyclohexatriene, predicting a heat of hydrogenation three times that of cyclohexene, or 85.8 kcal/mol. The observed value, 49.8 kcal/mol, is vastly different from our prediction.

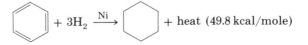

$$\text{benzene} + 3H_2 \xrightarrow{\text{Ni}} \text{cyclohexane} + \text{heat (49.8 kcal/mole)}$$

The difference, 36 kcal/mol, is called the resonance energy. Benzene contains 36 kcal/mol less energy than would be predicted and thus is 36 kcal/mol more stable than would be expected.

3. Carbon-Carbon Bond Lengths.   Only one form of 1,2-dibromobenzene is known.   The following structures represent the same compound even though as written, the bromines are separated by a single bond in one case and a double bond in the other.

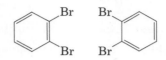

Physical measurements show that all the carbon-carbon bond lengths are identical and intermediate in length between normal carbon-carbon single and double bonds.

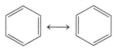

C—C	C⬚C	C=C
Single bonds	Bonds in benzene	Double bonds
1.54 Å	1.40 Å	1.34 Å

As a result, benzene is often depicted as a hexagon with a circle drawn inside rather than with alternating double and single bonds.   The circle within a hexagon is descriptively more accurate, and the alternating-bond model is better for electron bookkeeping.

B. Bonding in Benzene

1. Resonance Hybrid Picture.   Benzene can be described by two resonance forms.

   Resonance forms of benzene

Resonance forms (indicated by double-headed arrows) are used to describe unusual molecules such as benzene.   They themselves do not actually exist.   Instead, the resonance forms are classical electronic structures used to describe the structure of the actual molecule.   They differ only in the position of electrons, not in atoms.   In the two resonance forms of benzene, the positions of the carbon-carbon double bonds have been changed but the carbon atoms remain stationary.

Benzene does not alternate between the two resonance structures, nor are some benzene molecules of one form and the rest of the other.   The true structure of benzene is an average of the resonance forms called the *resonance hybrid*. Wherever there is a double bond in one resonance form, there is a single bond in the other.   Averaging these, we get a resonance hybrid with six identical carbon-carbon bonds all intermediate in length between a carbon-carbon double and single bond.

Resonance hybrid of benzene

Describing benzene by using two resonance forms to depict the resonance hybrid is analogous to describing a mule as a hybrid of a horse and a donkey.*

---

*Analogy by G. W. Wheland, University of Chicago.

The mule is not a horse part of the time and a donkey the rest but an individual creature with characteristics of both. The analogy fails in that horses and donkeys actually exist whereas contributing resonance structures do not exist. Another analogy describing a rhinoceros as a hybrid of the fictional dragon and unicorn is better.* The rhinoceros is real but the dragon and unicorn are not.

2. Molecular Orbital Picture of Benzene. A molecular orbital description of benzene most satisfactorily explains the structure of the resonance hybrid. Since each carbon in the resonance forms is involved in a double bond, and we know that a double bond is composed of a σ bond and a π bond, each carbon must possess a p orbital (Figure 5.1a). The only difference in the two resonance forms is in which p orbitals are shown overlapping (Figure 5.1b). However, if you could put yourself in the position of a p orbital, you would find the two adjacent p orbitals on either side of you to be identical and equidistant. Consequently, the p orbital would necessarily and unavoidably overlap with *both* of the adjacent p orbitals. This is the situation with each of the p orbitals in benzene. There is continuous overlap of the six p orbitals around the ring in the resonance hybrid (Figure 5.1c). This explains the fact that all carbon-carbon bond lengths in benzene are equivalent and intermediate between single and double bonds.

C. Structure of Benzene—A Summary

The following summary statements describe benzene, the parent of aromatic compounds (see Figure 5.1).

1. The molecular formula is C_6H_6.
2. The carbons exist in a flat six-membered ring with a cloud of six π electrons overlapping above and below the ring.
3. All six carbons are equivalent.
4. All carbon-carbon bond lengths are equivalent and intermediate between single and double bonds.
5. All six hydrogens are equivalent.
6. Each carbon is trigonal, sp^2 hybridized, and has 120° bond angles.

AROMATICITY: STRUCTURAL AND BONDING REQUIREMENTS

5.3

A. Properties of Aromatic Compounds

Aromatic compounds are compounds which resemble benzene in chemical behavior. They are characterized by: (1) a significant amount of resonance energy, and (2) a tendency toward substitution rather than addition reactions even though all have a high degree of unsaturation.

B. Structural Requirements for Aromaticity

Following are three simple structural characteristics that are required for a compound to be aromatic:

---

*Analogy by J. D. Roberts, California Institute of Technology.

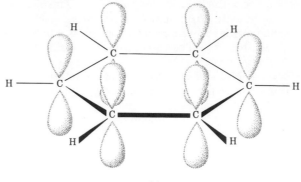

(a)

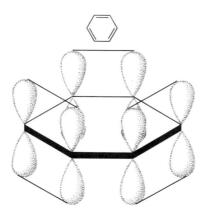

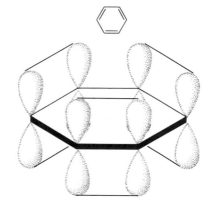

(b)

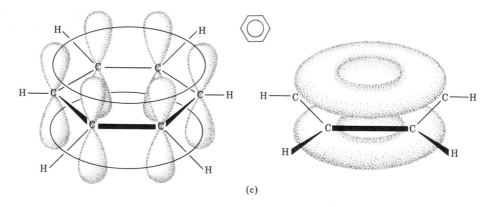

(c)

FIGURE 5.1. Molecular Orbital Representation of Benzene: (a) Each carbon in a benzene ring has a p orbital. Solid lines represent σ bonds. (b) Resonance forms of benzene. (c) Resonance hybrid of benzene showing continuous overlap of p orbitals. The total π cloud represents six electrons and overlaps above and below the ring.

1. Aromatic compounds, or at least the aromatic portion of a compound, must be cyclic and planar.
2. Each atom in an aromatic ring must have a p orbital so that there can be continuous overlap of parallel p orbitals around the ring. It will be necessary to determine the number of electrons occupying these p orbitals. Each atom in a double bond has a p orbital with one electron. A nitrogen, oxygen, or sulfur not involved in a double bond contributes one p orbital, containing two electrons, that is parallel with the rest.

$$A=A \qquad -\overset{|}{\underset{..}{N}}- \qquad -\overset{..}{\underset{..}{O}}- \qquad -\overset{..}{\underset{..}{S}}-$$

$$2\pi e \qquad\qquad 2\pi e \qquad\qquad 2\pi e \qquad\qquad 2\pi e$$

3. Hückel $4n + 2$ Rule: The cyclic clouds of parallel p orbitals must contain $4n + 2$ pi electrons (where $n =$ an integer 0, 1, 2, 3, ...). In other words, there must be a total of 2, 6, 10, 14, ... pi (π) electrons.

Applying these rules to benzene, we find that benzene is cyclic and planar, has a p orbital on every carbon because each carbon is involved in a double bond, and has a total of six π electrons because there are three double bonds.

PROBLEM

5.1. Apply the rules for aromaticity to the following molecules. Determine which are aromatic and which are not. Explain your answers.

(a) $CH_2=CH-CH=CH-CH=CH_2$

(b) ⬡   (c) ⬡   (d) ⬡⬡

C. Heterocyclic Aromatics

Heterocyclic compounds are cyclic compounds in which one or more of the ring atoms are not carbons. Many of the compounds are aromatic and are found prolifically as structural units in biological molecules. Two of the simpler of these are pyridine, part of the structure of nicotine, and pyrimidine, a structural unit in DNA and RNA (Chapter 15).

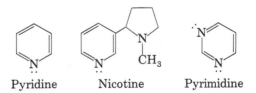

Pyridine   Nicotine   Pyrimidine

In both pyridine and pyrimidine, a nitrogen has merely been substituted for a C—H in benzene. Consequently, these molecules, like benzene, are aromatic. Each is cyclic and planar and has a p orbital on each ring atom and six π electrons. Because the nitrogen is already involved in a double bond, the remaining unshared pair of electrons on the nitrogen cannot be coplanar with the other p orbitals and is not part of the aromatic system.

Pyrrole, a structural unit of hemoglobin and chlorophyll, and imidazole, a structural feature in proteins and histamine (responsible for symptoms of allergies such as hay fever) are both five-membered ring nitrogen heterocycles.

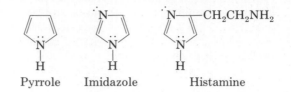

Pyrrole   Imidazole   Histamine

In pyrrole we have a cyclic, planar molecule with a p orbital on each atom. The p orbital on the nitrogen possesses the unshared electron pair and this added to the four π electrons from the two double bonds accounts for the six π electrons needed to satisfy the Hückel $4n + 2$ rule. The corresponding nitrogen in imidazole likewise provides its unshared pair to the "aromatic sextet," whereas the unshared pair on the other nitrogen, that involved in a double bond, is not included.

There are many polynuclear aromatic rings that are structural components of biological molecules. These include purine, a base in DNA and RNA and the structural feature in caffeine; indole, part of LSD and of the amino acid analogue serotonin, important in nerve transmission; quinoline, a structural unit in the antimalarial drug quinine; and isoquinoline, a chemical modification of quinoline which characterizes morphine, codeine, and heroin.

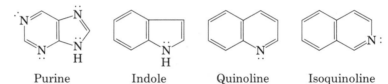

Purine   Indole   Quinoline   Isoquinoline

PROBLEM

5.2. Explain why (a) purine, (b) indole, (c) quinoline, and (d) isoquinoline are aromatic.

TABLE 5.1. MELTING POINTS AND BOILING POINTS OF AROMATIC HYDROCARBONS

Name	Formula	Molecular Weight	Melting Point, °C	Boiling Point, °C
Benzene	C_6H_6	78	5.5	80
Toluene	$C_6H_5CH_3$	92	− 95	111
o-Xylene	$1,2\text{-}C_6H_4(CH_3)_2$	106	− 25	144
m-Xylene	$1,3\text{-}C_6H_4(CH_3)_2$	106	− 48	139
p-Xylene	$1,4\text{-}C_6H_4(CH_3)_2$	106	13	138
Ethylbenzene	$C_6H_5CH_2CH_3$	106	− 95	136
Naphthalene	$C_{10}H_8$	128	81	218
Anthracene	$C_{14}H_{10}$	178	216	340
Phenanthrene	$C_{14}H_{10}$	178	101	340

NOMENCLATURE OF AROMATIC COMPOUNDS

5.4

A. Aromatic Hydrocarbon Ring Systems

Benzene, C_6H_6, is the most common aromatic ring. There are a variety of fused-ring aromatic hydrocarbons of which naphthalene, anthracene, and phenanthrene are the most common. The numbering system shown is used to name derivatives of these three compounds.

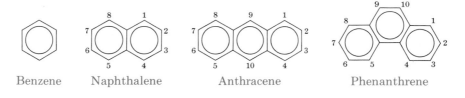

Benzene   Naphthalene   Anthracene   Phenanthrene

PROBLEM

5.3. Draw all the positional isomers in which a bromine can replace a hydrogen on **(a)** naphthalene, **(b)** anthracene, and **(c)** phenanthrene. Do not repeat any structures.

B. Monosubstituted Benzenes

Monosubstituted benzenes are named as derivatives of benzene.

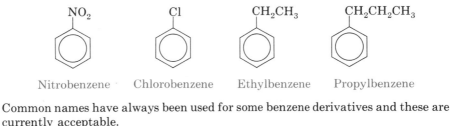

Nitrobenzene   Chlorobenzene   Ethylbenzene   Propylbenzene

Common names have always been used for some benzene derivatives and these are currently acceptable.

Toluene   Benzaldehyde   Benzoic   Benzene sulfonic   Phenol   Aniline
                           acid   acid

PROBLEM

5.4. Name the following compounds:

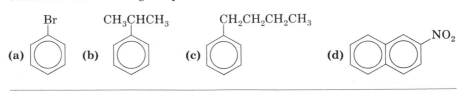

C. Disubstituted Benzenes

To name a disubstituted benzene, both groups and their relative positions must be identified. Every disubstituted benzene will have three positional isomers as illustrated by the xylenes (dimethylbenzenes used in high-octane gasoline). If the groups are adjacent, in a 1,2 relation they are termed ortho, if 1,3, meta and if 1,4, para.

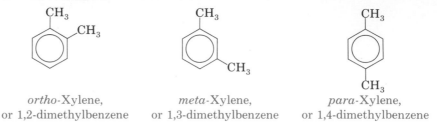

ortho-Xylene,   *meta*-Xylene,   *para*-Xylene,
or 1,2-dimethylbenzene   or 1,3-dimethylbenzene   or 1,4-dimethylbenzene

When two substituents are different, they are usually put in alphabetical order. If the compound is a derivative of a monosubstituted benzene designated by an accepted common name, it can be named as such.

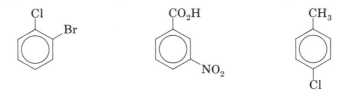

o-Bromochlorobenzene   *m*-Nitrobenzoic acid   *p*-Chlorotoluene

Substituents on the nitrogen of aniline are designated by an *N*.

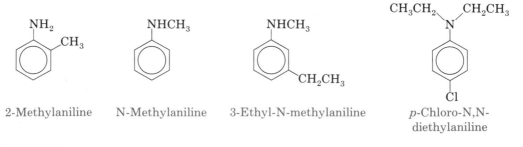

2-Methylaniline   N-Methylaniline   3-Ethyl-N-methylaniline   *p*-Chloro-N,N-diethylaniline

PROBLEM

5.5. Name the following disubstituted benzenes:

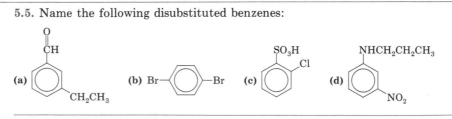

D. Polysubstituted Benzenes

When more than two groups are on a benzene ring, their positions must be numbered. Ortho, meta, and para designations are not acceptable. If one of the groups is associated with a common name, the molecule can be named as a derivative of the monosubstituted compound numbering from the group designated in the common name.

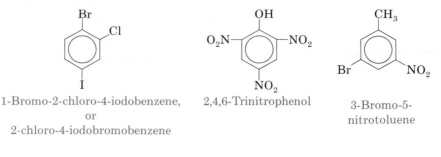

1-Bromo-2-chloro-4-iodobenzene,
or
2-chloro-4-iodobromobenzene

2,4,6-Trinitrophenol

3-Bromo-5-
nitrotoluene

PROBLEM

5.6. Name the following polysubstituted benzenes:

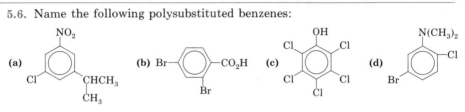

E. Aromatics Designated By Prefixes

Occasionally, the substituents on an aromatic ring are too complex to name conveniently with a prefix. In these cases, the aromatic ring is named with a prefix. The terms *phenyl* and *benzyl* are commonly used.

Phenyl   *Benzyl*

In the examples, the longest carbon chain is used as the base of the name and the aromatic portion is identified as a substituent on the chain.

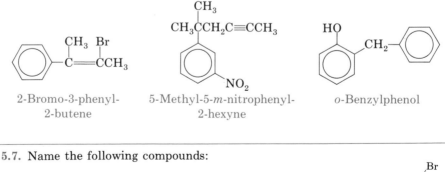

2-Bromo-3-phenyl-
2-butene

5-Methyl-5-*m*-nitrophenyl-
2-hexyne

o-Benzylphenol

PROBLEM

5.7. Name the following compounds:

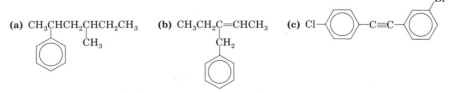

(a) $CH_3CHCH_2CHCH_2CH_3$
    CH_3

(b) $CH_3CH_2C{=}CHCH_3$
    CH_2

(c) Cl—⟨◯⟩—C≡C—⟨◯⟩
    Br

USES OF AROMATIC COMPOUNDS

5.5   Aromatic compounds can be obtained naturally from coal and synthetically from petroleum. They have a tremendous variety of uses and applications, some of which are summarized in Table 5.2.

TABLE 5.2.   TYPICAL USES OF SOME AROMATIC COMPOUNDS

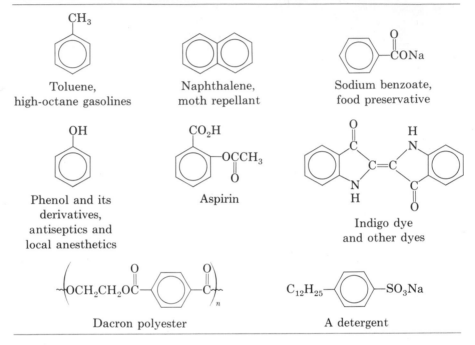

Toluene,
high-octane gasolines

Naphthalene,
moth repellant

Sodium benzoate,
food preservative

Phenol and its
derivatives,
antiseptics and
local anesthetics

Aspirin

Indigo dye
and other dyes

Dacron polyester

A detergent

ELECTROPHILIC AROMATIC SUBSTITUTION

5.6   The characteristic reaction of aromatic compounds is substitution.   One might have predicted addition because of the high degree of unsaturation in aromatic compounds. However, addition would destroy one or more of the "double bonds" and thus the aromaticity. No longer would there be a p orbital on every carbon of the ring allowing continuous overlap of the six π electrons. With substitution, however, the integrity of the benzene ring is preserved and the unusually stable bonding pattern is still intact.

We shall introduce electrophilic aromatic substitution the same way we did electrophilic addition to alkenes and alkynes. First we will discuss what happens in these reactions and then how it happens, that is, the reaction mechanism. Using these concepts we will develop a method for predicting the most predominant product when more than one is possible (compare with section 4.1).

A. Electrophilic Aromatic Substitution: The Reaction

Substitution reactions on a benzene ring are basically very simple—a hydrogen on the ring is replaced by another atom or group.

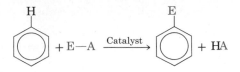

The catalyst acts by freeing the electrophile, E, from E—A. The electrophile replaces the hydrogen, which forms a by-product with the anion, A. The logical question now is: What specifically are EA and the catalyst? They are outlined in the following paragraphs. The reagent is EA, the catalyst is written over the arrow, and E is the atom or group that ends up replacing the hydrogen.

1. Halogenation. (X = Cl, Br; F_2 is too reactive, I_2 too unreactive.)

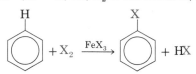

2. Alkylation, Acylation: Friedel-Crafts Reaction. (R = alkyl group.)

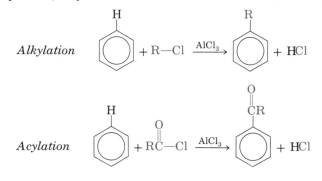

3. Nitration. (HNO_3, H_2SO_4 are both concentrated.)

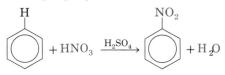

4. Sulfonation. (Fuming sulfuric acid with dissolved SO_3 is used.)

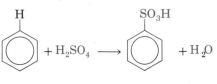

PROBLEM

5.8. Write reaction equations illustrating the reaction of *p*-xylene (1,4-dimethyl-benzene) with the following reagents. All reactions are electrophilic aromatic substitution: **(a)** Cl_2, $FeCl_3$; **(b)** Br_2, $FeBr_3$; **(c)** CH_3CH_2Cl, $AlCl_3$; **(d)** CH_3CCl, $AlCl_3$; **(e)** HNO_3, H_2SO_4; **(f)** H_2SO_4.

$$\overset{\parallel}{O}$$

B. Electrophilic Aromatic Substitution: The Mechanism

To introduce the mechanism of electrophilic aromatic substitution, let us look at the general reaction of benzene with an electrophilic reagent and a catalyst.

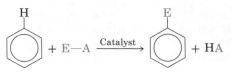

We have seen from bonding considerations that benzene has a large cloud of π electrons above and below the plane of the ring. This cloud attracts positive or electron-seeking species (electrophiles). The catalyst functions by converting the reactant EA into an electrophile.

$$E—A + \text{catalyst} \longrightarrow \quad E^+$$
$$\text{Electrophile}$$

Once formed, the electrophile is strongly attracted to the electron-rich π cloud of benzene (Figure 5.2).

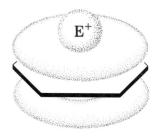

FIGURE 5.2. Attraction of electrophile to benzene's π cloud.

As the electrophile penetrates the π cloud and approaches bonding distance, it acquires two electrons from the π system and actually bonds to the benzene ring.

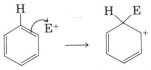

Carbocation

A carbocation results because the electrons now used in the bond between a carbon and the electrophile were previously shared by two carbons. We have the electrophile attached. Next the hydrogen must be removed to complete the substitution. If the hydrogen leaves as H^+, the two bonding electrons can enter the π system, neutralize the carbocation, and restore the aromaticity.

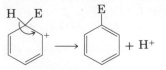

The H⁺ can react with the catalyst-anion complex to regenerate the catalyst and form the by-product HA. The essential elements of the mechanism of electrophilic aromatic substitution are summarized as follows:

Generation of the electrophile: $E-A + \text{catalyst} \longrightarrow E^+ + A:^-$

Two-step substitution:

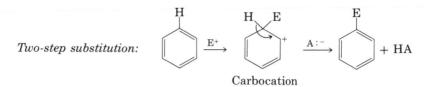

Carbocation

In the following sections this mechanism is applied to the specific reactions we have considered.

1. Halogenation.

Generation of the electrophile: $X_2 + FeX_3 \longrightarrow X^+ + FeX_4^-$
$$(X = Cl, Br)$$

Two-step substitution:

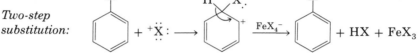

Iron III halide is a Lewis acid since the iron does not have a complete octet of electrons. To complete its octet, it abstracts a halide ion from X_2 leaving a positive halogen electrophile.

$$:\ddot{X}:\ddot{X}: \quad \begin{matrix} :\ddot{X}: \\ Fe:\ddot{X}: \\ :\ddot{X}: \end{matrix} \longrightarrow :\ddot{X}^+ \quad :\ddot{X}:Fe:\ddot{X}:^- \\ :\ddot{X}:$$

After X⁺ bonds to the ring forming a carbocation, an X⁻ from FeX_4^- combines with the hydrogen ion to form the by-product and regenerate the catalyst.

$$FeX_4^- + H^+ \longrightarrow FeX_3 + HX$$

2. Alkylation and Acylation: Friedel-Crafts Reaction. Let us use methylation of benzene as an example.

Generation of the electrophile: $CH_3Cl + AlCl_3 \longrightarrow CH_3^+ + AlCl_4^-$

Two-step substitution:

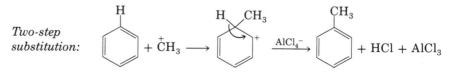

Like FeX_3, $AlCl_3$ is a Lewis acid. To complete its octet, aluminum abstracts a chloride ion from CH_3Cl generating the electrophile, a methyl carbocation.

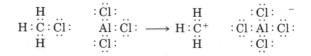

The catalyst is regenerated when a Cl^- from $AlCl_4^-$ combines with the hydrogen ion eliminated in the last step.

3. Nitration.

Generation of the electrophile: $HNO_3 + H_2SO_4 \longrightarrow \overset{+}{N}O_2\ H\bar{S}O_4 + H_2O$

Two-step substitution:

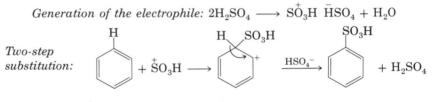

Since sulfuric acid is a stronger acid than nitric acid, one can imagine an acid-base reaction occurring to generate the electrophile in which nitric acid is acting as the base.

$$HO—NO_2 + H^+H\bar{S}O_4 \longrightarrow H_2O + NO_2^+HSO_4^-$$

The catalyst, H^+, is returned in the last step of the mechanism.

4. Sulfonation.

Generation of the electrophile: $2H_2SO_4 \longrightarrow \overset{+}{S}O_3H\ \bar{H}SO_4 + H_2O$

Two-step substitution:

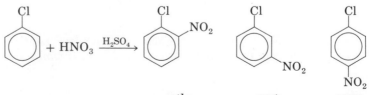

PROBLEMS

5.9. Write electrophilic aromatic substitution reaction mechanisms for the reaction of benzene with **(a)** $Br_2/FeBr_3$; **(b)** $CH_3CH_2Cl/AlCl_3$; **(c)** $CH_3\overset{O}{\overset{\|}{C}}Cl/AlCl_3$ and for the reaction of 1,4-dimethylbenzene with **(d)** $Cl_2/FeCl_3$; **(e)** HNO_3/H_2SO_4; **(f)** H_2SO_4.

5.10. The mechanism of electrophilic addition (section 4.1D) and electrophilic substitution (section 5.6B) are almost identical in the first of the two steps. As an aid to study, write general reaction mechanisms one below the other for comparison.

C. Orientation of Substitution

1. Directive Effects. Since benzene is a symmetrical molecule, electrophilic substitution gives only one substitution product no matter which of the six hydrogens is replaced. Most benzene derivatives, however, are not symmetrical, and more than one substitution isomer is usually possible. For example, the nitration of chlorobenzene can give three positional isomers of chloronitrobenzene.

The ortho and para isomers are formed almost to the exclusion of the meta product.

What determines the orientation of substitution, and how does one predict the predominant products? *The atom or group already present on the benzene ring directs the orientation of substitution of the incoming electrophile.* For example, in the nitration of chlorobenzene, the chlorine directs the nitro group primarily to the ortho and para positions. However, in the chlorination of nitrobenzene, the nitro group directs the incoming chlorine almost exclusively to the meta position.

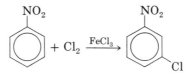

Groups already present on the benzene ring either direct incoming electrophiles to the ortho and para positions or to the meta position. Table 5.3 lists ortho-para directors and meta directors.

TABLE 5.3. ORIENTATION OF SUBSTITUTION

Ortho, Para Directors		*Meta Directors*	
—OH	Hydroxy-	$-\overset{\text{O}}{\overset{\|}{\text{C}}}\text{OH}$	Carboxylic acid
—OR	Alkoxy-		
—NH$_2$	Amino-	$-\overset{\text{O}}{\overset{\|}{\text{C}}}\text{H}$	Aldehyde
—NHR	Alkylamino-		
—NR$_2$		$-\overset{\text{O}}{\overset{\|}{\text{C}}}\text{R}$	Ketone
—X	Halogens	$-\text{C}\equiv\text{N}$	Cyano-
—R	Alkyl-	—NO$_2$	Nitro-
		—SO$_3$H	Sulfonic acid

2. Predicting Substitution Products. To predict the orientation of substitution, analyze the effect of the groups already bonded to the benzene ring. In the sulfonation of toluene, the methyl group (an alkyl group) is an ortho-para director (Table 5.3).

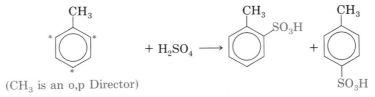

(CH$_3$ is an o,p Director)

A carboxylic acid group is a meta director, as illustrated by the bromination of benzoic acid.

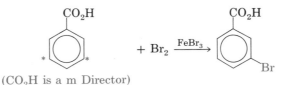

(CO$_2$H is a m Director)

If two or more groups are already present on the benzene ring, analyze the directive effects of each group individually and predict the product based on the complete analysis. In the following example, the bromine as an o,p director and the sulfonic acid group as a m director both direct to the same positions.

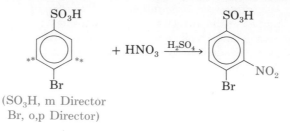

(SO₃H, m Director
Br, o,p Director)

PROBLEM

5.11. Complete the following reactions and predict the principal substitution products:

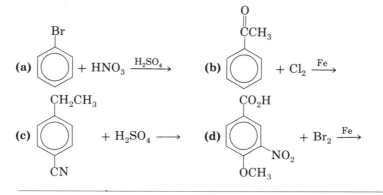

---

3. Synthesis. In synthesizing an aromatic compound, not only do we need to know the reagents necessary for putting on a particular group but we also need to determine the order in which to add the reagents. For example, consider the synthesis of *meta*-bromonitrobenzene from benzene.

from

If the bromine were put on the ring first, it would direct the nitration to the ortho and para positions because halogens are o,p directors. Very little of the desired product would result.

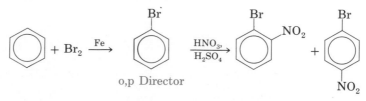

This would be an acceptable synthesis of either *ortho-* or *para*-bromonitrobenzene. If the nitro group is placed on the ring first, however, since it is a meta director, bromination would give almost exclusively *m*-bromonitrobenzene, the desired product.

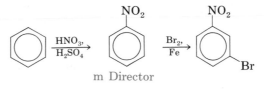

PROBLEM
_____

5.12. Using reaction equations, show how the following compounds could be synthesized from benzene: **(a)** *m*-chlorobenzenesulfonic acid; **(b)** *p*-ethylnitrobenzene.
_____

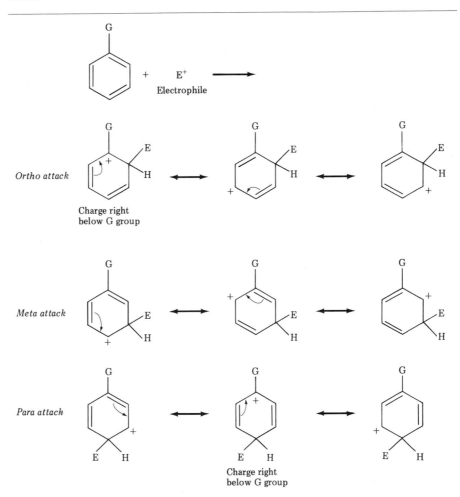

FIGURE 5.3. Resonance contributors resulting from ortho, meta, and para attack on a substituted benzene.

4. Theory of Directive Effects.   A group on a benzene ring directs incoming electrophiles to ortho and para positions simultaneously or to meta positions. The electrophile bonds in such a way as to form the most stable intermediate carbocation.   Figure 5.3, on page 121, illustrates the first step in the mechanism of electrophilic aromatic substitution—the formation of a carbocation.   The benzene derivative is monosubstituted, so there are three different positions for electrophile attachment and three possible carbocations.   By movement of electron pairs as shown in Figure 5.3, three resonance forms of each carbocation can be drawn.   Through resonance, the charge is dispersed and the carbocation is stabilized.

For ortho and para attack, notice that one of the resonance forms allows placement of the positive charge on the carbon bearing the group G (the one already bonded to the ring).   This is not the case in meta attack.   Whether G stabilizes or destabilizes carbocations, its effect will be greatest if the electrophile attacks ortho or para.

The groups in Table 5.3 labeled ortho, para directors are electron-releasing groups and are capable of stabilizing carbocations.   Therefore, if one of these carbocation-stabilizing groups is already on the ring it will direct the incoming electrophile ortho and para for its most effective stabilization.   The groups labeled meta directors in Table 5.3 are electron-withdrawing groups and destabilize carbocations.   Thus, if one of. these is present, the incoming electrophile bonds meta where the group G will have the least effect.

D. Activating and Deactivating Groups

A substituent already present on a benzene ring not only directs the orientation of substitution of an incoming group (electrophile) but also influences the rate of reaction.   A group that increases the rate of electrophilic aromatic substitution is called an *activating group,* whereas those that decrease the rate are termed *deactivating groups.*

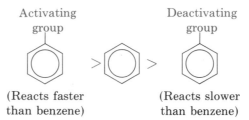

Activating group > > Deactivating group

(Reacts faster than benzene)   (Reacts slower than benzene)

The rate of electrophilic substitution depends on the availability of the π cloud of electrons above and below the benzene ring to the attacking electrophile. The more electron-rich (the more negative) the cloud, the faster the electrophilic attack.   Electron-releasing groups increase the electron density of the ring and are activating groups.   Electron-withdrawing groups, on the other hand, decrease the electron density of the π cloud, decreasing its availability to the attacking electrophile.   Therefore electron-releasing groups are activating and electron-withdrawing groups are deactivating toward electrophilic aromatic substitution.

All ortho, para directors (Table 5.3) with the exception of the halogens are activating groups.   Except for alkyl groups, all have a lone pair of electrons that is partially donated to the ring through resonance.   This makes the ring more negative and consequently more attractive to the positive electrophiles.   The hydroxy group on phenol is a good example of an activating group.

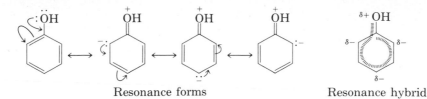

Resonance forms   Resonance hybrid

All meta directors (Table 5.3) and the halogens are deactivating groups. They withdraw electrons from the benzene ring, making it less attractive to an incoming electrophile. Because of bond polarity, many of these groups have either a full or partial positive charge on the atom bonded to the benzene ring.

By their greater electronegativity or by resonance (as with the nitro group) they withdraw electron density from the benzene ring.

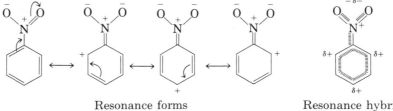

Resonance forms   Resonance hybrid

The halogens represent a dual effect. Owing to their strong electronegativity, they withdraw electrons from the benzene ring, thus deactivating it. However, once an electrophile has bonded and formed a carbocation, the halogen releases a lone pair of electrons by resonance, stabilizing the positive charge. Although the halogens are deactivating, they are ortho, para directors.

OXIDATION OF ALKYLBENZENES

5.7

When alkylbenzenes are treated with an oxidizing agent such as potassium permanganate, alkyl groups on a benzene ring are oxidized to carboxylic acids. All primary and secondary alkyl groups are oxidized despite their size or number (given enough reagent).

$$CH_3CH_2 - \bigcirc - CH_3 \xrightarrow{\text{KMnO}_4} HO_2C - \bigcirc - CO_2H$$

Note in this reaction that alkyl groups, which are o,p directors, are changed to acid groups, which are m directors. This must be considered in synthesis problems. For example, the following sequence is effective for producing *m*-nitrobenzoic acid from toluene. The opposite sequence would produce ortho and para nitrobenzoic acid.

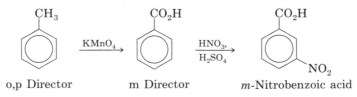

o,p Director   m Director   *m*-Nitrobenzoic acid

PROBLEM

5.13. Write a reaction equation illustrating the oxidation of propylbenzene with potassium permanganate.

Panel 7

GASOLINE

Gasoline is primarily a mixture of hydrocarbon molecules with 5-10 carbons obtained from petroleum. In an internal combustion engine, a gasoline-air mixture is injected into a cylinder. A piston compresses the vaporized fuel mixture, which is ignited by a spark plug at the point of maximum compression (Figure 5.4). Combustion of the fuel generates hot expanding gases, which force the piston to the bottom of the cylinder, eventually turning the wheels.

The efficient operation of an engine requires the smooth, even, controlled combustion of the fuel to produce a steady pressure against the piston. Sudden explosive detonation, or autoignition of the fuel without a spark, especially contrary to engine timing, causes pinging, knocking (rattling of piston against cylinder walls), and afterburn, all of which result in power loss and potential damage to the engine.

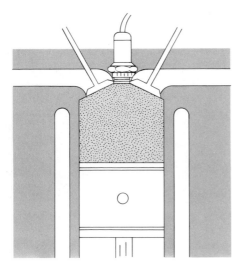

OCTANE NUMBER REQUIREMENTS
OF ENGINES ACCORDING TO
COMPRESSION RATIO

Compression Ratio	Research Octane Number for Knock-Free Performance
4:1	60
5:1	73
6:1	81
7:1	87
8:1	91
9:1	95
10:1	98
11:1	100
12:1	102

FIGURE 5.4. The internal combustion engine and octane ratings necessary for knock-free performance.

Structure and Quality of Gasoline Molecules

A good gasoline must be easily vaporized in the carburetor, but not so volatile that it boils out of the gas tank. It must burn in a controlled fashion to give knock-free performance. Hydrocarbon molecules containing approximately 5 to 10 carbons meet the volatility requirements for gasoline but show a wide variation

[PANEL 7] in combustion quality. The following hydrocarbon structural characteristics produce quality gasolines with good antiknock properties:

1. low molecular weight (conventional gasolines, however, must be in the C_5–C_{10} carbon range)
2. branching of hydrocarbon chains
3. unsaturated, cyclic, and in particular, aromatic molecules

The ability of a gasoline to burn smoothly without knocking is described by the octane number. The octane scale, as originally conceived, spanned from 0 to 100, with straight-chained heptane occupying the bottom of the scale and the highly branched isooctane (2,2,4-trimethylpentane) the top.

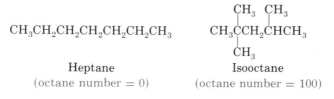

$$CH_3CH_2CH_2CH_2CH_2CH_2CH_3 \qquad\qquad CH_3CCH_2CHCH_3$$

Heptane   Isooctane

(octane number = 0)   (octane number = 100)

For a test gasoline, the burning characteristics are compared with those of standard mixtures of isooctane and heptane. For example, a gasoline that burns like a 90:10 mixture of isooctane and heptane has an octane rating 90.

Octane numbers are determined by two laboratory methods, which are called the research octane number (RON) and the motor octane number (MON). In general, the values found by the research method correlate with the low speeds and frequent accelerations of city driving, whereas the more severe motor octane test relates to highway performance. Research octane numbers are usually higher, and an average of the two determination methods are posted on gasoline pumps.

The octane requirements of an engine increase with increasing compression ratio. The compression ratio of an engine is the ratio of the maximum cylinder volume (cylinder volume with the piston at the bottom of the cycle) to cylinder volume at maximum compression. Figure 5.4 compares octane requirements to compression ratios.

In 1935, the research octane number for regular gasoline was 72 and for premium it was 78. Octane numbers peaked around 1968 with regular gasoline rated at 94 and premium at 100. Since then octane numbers have decreased as fuel conservation has gained importance in engine design.

The octane number of a particular gasoline component varies with molecular structure as previously described. Table 5.4 illustrates the effects of molecular size, chain branching, unsaturation, cyclization, and aromaticity on octane number.

Types of Gasoline

1. Leaded Gasoline. Addition of small amounts of tetraethyllead, $Pb(CH_2CH_3)_4$, to gasoline can produce a substantial increase in octane number. Addition of 3 g of tetraethyllead (TEL) to a gallon of gasoline can increase the octane rating 5 to 15 points depending on the source of the gasoline. TEL moderates the combustion of fuel so that it burns in a slow and controlled manner instead of detonating contrary to engine timing. As a leaded gasoline is burned, lead metal deposits form that could foul an engine. Ethylene bromide

TABLE 5.4. RESEARCH OCTANE NUMBERS FOR HYDROCARBON MOLECULES

Hydrocarbon	Unleaded RON	Hydrocarbon	Unleaded RON
$CH_3CH_2CH_3$	97	(toluene: benzene ring with CH_3)	120
$CH_3(CH_2)_4CH_3$	25		
$CH_3(CH_2)_6CH_3$	−19		
$CH_3(CH_2)_2CH{=}CHCH_3$	93		
$CH_3\overset{\displaystyle CH_3}{\underset{}{C}}HCH_2CH_2\overset{\displaystyle CH_3}{\underset{}{C}}HCH_3$	65	(o-xylene: benzene ring with two adjacent CH_3)	107
(cyclopentane ring)	101		
(benzene ring)	106	(m-xylene: benzene ring with two meta CH_3)	118

[PANEL 7] (BrCH$_2$CH$_2$Br) or ethylene chloride is added to react with this lead metal to produce lead II bromide (or chloride), which is volatile at engine temperatures and escapes with the exhaust.

$$Pb \quad + \quad \underset{\underset{Br}{|}\ \underset{Br}{|}}{CH_2CH_2} \longrightarrow \quad PbBr_2 \quad + \quad CH_2{=}CH_2$$

Nonvolatile ... Volatile

Because lead compounds are often toxic, the amount of tetraethyllead in gasoline has undergone a regulated decrease for environmental and health reasons. In 1970, leaded gasolines (and the total gasoline pool) averaged 2.5 g per gallon with some gasolines containing 4–6 g. The 1980 requirement was that the total gasoline pool (leaded and unleaded) average no more than 0.5 g per gallon (leaded gasolines will probably average 1–2 g in the 1980's).

Currently the future of leaded gasolines is dim. In 1973, regulations required refiners to provide an unleaded gasoline, and in the 1975 model year, automakers began putting catalytic converters on cars as standard equipment to meet Environmental Protection Agency (EPA) emission standards. Catalytic converters are made ineffective by lead-containing exhaust. Consequently, unleaded gasolines must be used in these automobiles. Whereas unleaded gasolines comprised less than 2% of the market before the 1975 model year, by the mid-1980s their share will approach 80% at the current pace.

2. Unleaded Gasoline. Octane boosters must be added to gasolines without tetraethyllead to achieve the necessary octane ratings. One of the common methods for increasing gasoline performance is to increase the proportion of aromatic hydrocarbons in the blend, notably benzene, xylene, and toluene (commonly called BXT), all of which have high octane numbers (Table 5.4). Two of the newer blending agents are tertiary butyl alcohol (TBA) and methyl tertiary butyl ether (MTBE), which have octane ratings of 108 and 115, respectively. A lead substitute, methylcyclopentadienyl manganese tricarbonyl (MMT), has been used in small amounts ($\frac{1}{16}$ g per gallon gives a 0.9 point octane boost), but it appears that the EPA will not allow its continued use.

[PANEL 7]

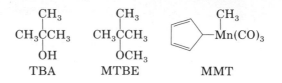

TBA   MTBE   MMT

3. Gasohol. Gasohol is a blend of 90% unleaded gasoline and 10% ethanol (beverage alcohol). Because ethanol can be distilled from almost any type of crop or crop waste—corn, grain, wood chips, grapefruit rinds, sugar cane stalks, even garbage—it can be a cash crop for farmers and a renewable energy source. Successfully tested in Nebraska and Illinois, it appears that higher concentrations of ethanol are possible with no major problems. Availability has now been extended to other states. Not only is ethanol a good fuel extender but it increases the octane rating of the gasoline with which it is blended. Brazil, with a large sugar cane industry, is well on the way to fuel self-sufficiency. Its vehicles use pure alcohol or gasohol as fuel. Sugar cane is fermented to produce alcohol and the refuse is burned for energy to distill the alcohol.

Gasoline Production

American drivers burn more than 110 billion gallons of gasoline annually driving over 1 trillion, 315 billion miles. Straight-run gasoline obtained directly from petroleum fractionation is neither of sufficient quality nor quantity to meet this tremendous demand. In oil refineries, lighter and heavier crude oil fractions are converted into gasoline, and the quality of low-octane fuels is enhanced.

1. Cracking. In the cracking process, high-boiling, high-molecular-weight petroleum fractions (such as the gas-oil and wax-oil) are broken down into molecules of a size appropriate for use as gasoline, that is, 5 to 10 carbons. For example, an 18-carbon hydrocarbon, when heated with silica-alumina catalysts, breaks down, or cracks, into smaller molecules.

Cracking

$$C_{18}H_{38} \xrightarrow[\text{catalyst}]{\text{Heat,}} C_9H_{20} + C_9H_{18}$$

$$C_{18}H_{38} \xrightarrow[\text{catalyst}]{\text{Heat,}} C_{10}H_{22} + 4CH_2{=}CH_2$$

Molecules too large C_5-C_{10} *gasoline*
to be used as gasoline   *range molecules*

Cracking is specifically used for other purposes such as the production of ethylene ($CH_2{=}CH_2$) from higher-molecular-weight feedstocks. The ethylene is used to make polyethylene and antifreeze among other products. A coproduct of cracking (and some other refinery operations) is hydrogen gas, which is used with nitrogen to make ammonia by the Haber process ($N_2 + 3H_2 \longrightarrow 2NH_3$). For this reason the price of fertilizer is related to the price of crude oil.

2. Polymerization and Alkylation. Catalytic cracking produces substantial amounts of gaseous hydrocarbons as well as gasoline. These compounds, too light and volatile to be used as gasoline, can be recombined into compounds of higher molecular weight by polymerization and alkylation.

In polymerization, two alkenes, each with less than 5 carbons, are combined

[PANEL 7] to form a larger molecule useful in gasoline. This process is similar in concept to addition polymerization (section 4.2).

Alkylation, also a buildup process, involves adding an alkane to an alkene. It has largely replaced polymerization. For example, isobutane can be added to isobutene in the presence of a catalyst (such as sulfuric acid, hydrofluoric acid, or aluminum chloride) to produce isooctane.

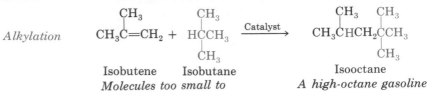

Alkylation

Isobutene   Isobutane   Isooctane
Molecules too small to   *A high-octane gasoline*
be useful as gasoline

3. Reforming: Isomerization and Aromatization. Reforming alters the molecular structures of the components of low-octane gasolines to form a higher-octane fuel. The two main reforming processes are isomerization and aromatization.

Branched-chain hydrocarbons have a higher octane rating than unbranched ones. In isomerization, a straight-chained hydrocarbon is catalytically isomerized with aluminum chloride or platinum to a branched paraffin. The isomerization of octane to isooctane, for example, causes a dramatic increase in octane rating.

Isomerization $CH_3(CH_2)_6CH_3$ $\xrightarrow{\text{Catalyst}}$

Octane   Isooctane
(octane number = −19)   (octane number = 100)

In aromatization, platinum-containing catalysts (platforming) convert paraffins into aromatic hydrocarbons, as in the aromatization of hexane.

Aromatization $CH_3CH_2CH_2CH_2CH_2CH_3$ $\xrightarrow{\text{Catalyst}}$

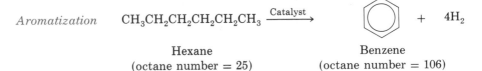

Hexane   Benzene
(octane number = 25)   (octane number = 106)

Panel 8

CANCER AND CARCINOGENS

"Cancer is a scourge that strikes two families in three, or one person out of every four Americans. It kills more children between the ages of three and 14 than any other disease. Cancer killed more Americans in 1971 (the year Congress passed the National Cancer Act) than in four years of World War II."

Those are the words of former Senator Hubert H. Humphrey shortly before he succumbed to cancer. Cancer certainly is unavoidably visible to the average

[PANEL 8] person either through personal tragedy or by media coverage of common substances that are now suspected carcinogens. *Carcinogens* are cancer-producing substances. Consider the following list of widely used chemicals that have been withdrawn from the market or are still embroiled in controversy as cancer-causing suspects: the artificial sweeteners, cyclamate and saccharin; the insecticides DDT, dieldrin, aldrin, chlordane, and heptachlor; chloroform, once found in over 2000 products (particularly cough medicines); red dyes Nos. 2 and 40, added to many food products; nitrites, added as preservatives and antibotulism agents to packaged meats; asbestos, used as insulation; diethylstilbestrol, used at one time by women with hormonal difficulties (higher incidences of vaginal cancer were found in their daughters) and to promote fattening of livestock; TRIS, a flame retardant in children's sleepwear; estrogens in birth control pills and drugs used by menopausal women; and methapyriline, a very common antihistamine used in nonprescription nasal decongestants, allergy relief medications, and sleeping pills. Recently, it has been suggested that black pepper and excessively browned foods may contain carcinogens. And even sexual activity may increase cervical cancer in women through a dormant virus transferred in seminal fluid.

Most cancer experts agree that cigarette smoking is the single biggest cause of cancer, possibly accounting for as many as 40% of all cancers. American cigarette makers produce 600–700 billion cigarettes annually, and these are smoked regularly by about 37.5% of the men and 29.6% of the women in the United States. Cigarette smoke contains some 4000 chemical compounds.* In addition to cancer, smoking can cause other health problems including cardiovascular disease, pulmonary disease, increased risk of miscarriage, and increased risk of fetal birth defects in the children of smoking women. There is some evidence that alcohol and cigarettes have a synergistic effect in inducing several forms of cancer.

Cancer is the second leading cause of death in the United States today (heart disease is number one). Approximately 375,000 people die of it annually— over 1000 each day. There is no question that some chemicals are carcinogens and can cause cancer. The fact that the mortality rate from cancer is not evenly distributed around the United States but is more highly concentrated in areas heavy in chemical industry and chemical pollution such as Houston, southern Louisiana, the northeastern Atlantic coast states (particularly New Jersey), and some locations around the Great Lakes is evidence of this. What types of chemicals cause cancer? Obviously, the question has not yet been completely answered. But some structural features in organic compounds are common in carcinogens such as some polynuclear aromatic compounds and some chlorinated hydrocarbons. Among the most common of the polynuclear aromatics is 3,4-benzpyrene found in coal dust, cigarette smoke, automobile exhaust gases, and on the outside of charcoal-broiled steaks. Benzene itself, one of the most heavily produced organic chemicals, has been implicated in the onset of leukemia, and it is now used in academic and industrial situations only under restricted conditions. Among other uses, benzene is a component of high-octane gasolines. Structures of some suspected carcinogens are presented in Table 5.5.

---

*One omnipresent component of cigarette smoke is carbon monoxide. Carbon monoxide binds much more readily to hemoglobin than does oxygen and the blood of a heavy smoker could contain 5%–15% carboxyhemoglobin. The background level of carboxyhemoglobin is 0.5%. Carbon monoxide is also in excessive concentrations around freeways, industrial areas, and airports. A New York City cab driver was found to have 13% carboxyhemoglobin in his blood.

[PANEL 8]   **TABLE 5.5.** SUSPECTED CARCINOGENS

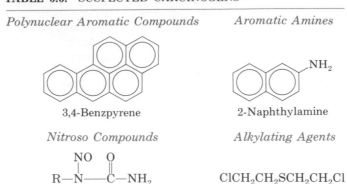

Polynuclear Aromatic Compounds   *Aromatic Amines*

3,4-Benzpyrene   2-Naphthylamine

Nitroso Compounds   *Alkylating Agents*

$$\underset{\text{Alkyl nitroso urea}}{\text{R—N——C—NH}_2}$$

Alkyl nitroso urea $ClCH_2CH_2SCH_2CH_2Cl$
(R = alkyl group)   Mustard gas

What is cancer and how do carcinogens act? There are a variety of cancers having in common the abnormal growth and multiplication of cells. Many cancerous growths show some loss of normal physiological activity. An abnormal growth that remains localized at its original site and that is not likely to recur once removed is termed *benign*. If the growth or tumor invades neighboring tissue creating new tumors in other parts of the body, it has metastasized and is termed *malignant* or cancerous. Carcinogens initiate cancers by reacting with important biological molecules such as DNA, RNA, or proteins. Some carcinogens are electrophiles and as such seek electrons. Genetic material (nucleic acids) and proteins have many nucleophilic or basic sites that can attract the carcinogen. The resulting Lewis acid-Lewis base type reaction chemically changes the molecule, and normal biological activity is altered. In another type of activity, carcinogen molecules of the correct shape and size become inserted in the DNA helix and distort or partly uncoil the molecule, again impairing its normal biological activity. A cancerous cell may result. As the cells divide and multiply, cells or clusters of cells may enter the blood and lymphatic systems and become trapped by small blood vessels of organs. At these new locations they develop into new tumors. As the cancer spreads and develops, normal body chemistry is strained or altered and eventually ceases.

A cancer cell has different metabolic properties and may have different nutritional requirements or different enzymatic processes than normal cells. In chemotherapy, these differences are taken advantage of by finding substances that will preferentially react with and destroy the cancerous cells. The process is analogous to chemical treatment of a lawn to destroy weeds. Further, because many cancerous growths are fast growing, they may preferentially absorb the anticancer drug. Unfortunately, chemotherapy can also destroy many normal, rapidly dividing cells such as hair follicles, cells lining the gastrointestinal tract, and bone marrow cells that are involved in immunity. This causes the common side effects of chemotherapy—nausea, vomiting, hair loss, and increased susceptibility to infection.

Testing a substance for carcinogenic properties is time-consuming and controversial. Because many years are sometimes necessary for the development of a malignancy, much testing involves subjecting mice and rats to extremely heavy

[PANEL 8] doses of the suspected carcinogen to develop a quick response. Among the most controversial actions are those involving the artificial nonnutritive sweeteners, cyclamates and saccharin. Cyclamates, which had been on the market since the 1950s, were the most widely used nonnutritive sweeteners in 1969 when they were banned. Within weeks of the ban, soft-drink manufacturers had diet drinks sweetened with saccharin on the shelves. In March, 1977, the Food and Drug Administration announced a ban on saccharin based on a Canadian study that showed large doses of saccharin fed to rats (5% of their diet) caused bladder tumors. In terms of human weight, the dose was equivalent to the consumption of 800 cans of soft drinks daily for a lifetime. Public outcry and additional reports ordered by Congress have delayed the ban. However, a National Academy of Sciences report has tagged saccharin as a moderate-to-high-risk (considering how many people use it) carcinogen, and it is likely that its use will be banned or greatly curtailed.

Research towards devising quicker and more reliable tests for chemical carcinogens is underway. Recent work in this area, specifically the Ames test, has shown promise. The test, developed by Dr. Bruce N. Ames of the University of California, Berkeley, involves testing the ability of a substance to cause mutations in selected strains of salmonella bacteria. Since bacteria have very short life cycles and propagate frequently, biological mutations caused by foreign substances may be evident in a few days. The assumptions that mutagens are always carcinogens or that experimental results on bacteria can be extrapolated to humans are yet to be completely accepted by the scientific community.

END-OF-CHAPTER PROBLEMS

5.14 Bonding Pictures: Draw a bonding picture showing all π bonds for **(a)** naphthalene, and **(b)** anthracene (structures in section 5.4A).

5.15 Resonance Energy: Predict for biphenyl the calculated heat of hydrogenation (if the substance is not aromatic), the observed energy (if it is aromatic), and the resonance energy. See section 5.2.A.2 for hydrogenation values.

Biphenyl

5.16 Resonance Forms: Draw the resonance forms and resonance hybrid for **(a)** naphthalene, and **(b)** anthracene.

5.17 Aromaticity: Determine whether the following compounds are aromatic or not. Explain your answers.

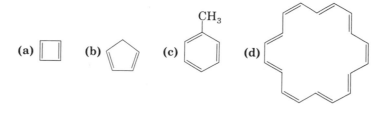

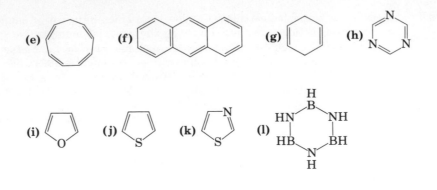

(e)   (f)   (g)   (h)

(i)   (j)   (k)   (l)

5.18 Nomenclature: Name the following compounds:

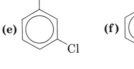

(a)   (b)   (c)   (d)

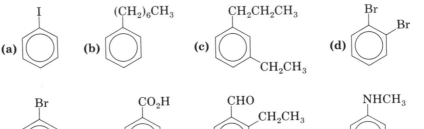

(e)   (f)   (g)   (h)

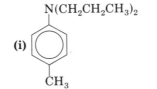

(i)   (j)   (k)

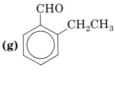

(l)   (m)   (n)

    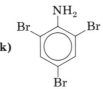

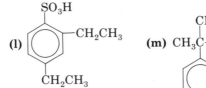

(o)   (p)

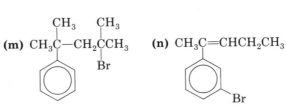

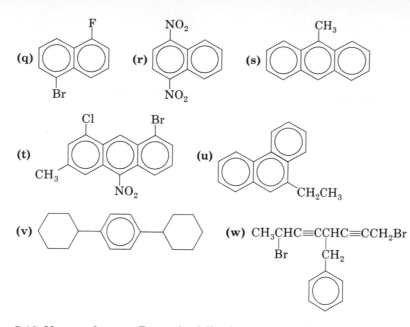

5.19 Nomenclature: Draw the following compounds:
(a) *p*-dichlorobenzene (mothballs)
(b) *m*-xylene (component of high-octane gasoline)
(c) 1,3,5-trinitrobenzene (an explosive, TNB)
(d) *o*-phenylphenol (a disinfectant in household deodorizers)
(e) 2,6-di-*t*-butyl-4-methylphenol (antioxidant used in gasoline)
(f) benzaldehyde (oil of bitter almonds)
(g) 2-methylnaphthalene (found in coal tar)
(h) pentachlorophenol (ant and termite killer)
(i) 2,4,6-trinitrophenol (picric acid, an explosive chemical being removed from old high school and college chemistry stockrooms)

5.20 Positional Isomers: Draw the positional isomers of the following compounds:
(a) tribromobenzenes
(b) chlorodibromobenzenes
(c) bromochlorofluorobenzenes
(d) dibromonaphthalenes
(e) dinitroanthracenes
(f) dinitrophenanthrenes

5.21 Positional Isomers: There are three dibromobenzenes. Their melting points are 87 °C, 6 °C, and −7 °C. Nitration of the isomer with the 87 °C mp results in only one mononitrated dibromobenzene. The isomer with mp = +6 °C gives two mononitro isomers, and the one with mp = −7 °C gives three. Write the structure of each isomer.

5.22 Reaction Mechanisms: Write step-by-step reaction mechanisms for the reaction of toluene with each of the following reagents:

(a) $Cl_2/FeCl_3$ **(b)** $CH_3CH_2\overset{\displaystyle O}{\overset{\displaystyle \|}{C}}Cl/AlCl_3$ **(c)** $CH_3\overset{\displaystyle Cl}{\underset{\displaystyle |}{C}}HCH_3/AlCl_3$

(d) HNO_3/H_2SO_4 **(e)** H_2SO_4

5.23 Reaction Mechanisms: Toluene can react with bromine in two different ways depending on the reaction conditions. When treated with Br_2 and $FeBr_3$, electrophilic aromatic substitution occurs on the benzene ring. If treated with bromine alone in the light, a free-radical chain reaction (section 3.6) occurs involving bromination of the methyl group. These happen because toluene is both an aromatic hydrocarbon (benzene ring) and an alkane (methyl group). Write step-by-step reaction mechanisms for both reactions.

5.24 Reaction Mechanisms: Ethylbenzene can be made by the Friedel-Crafts reaction in three ways. In section 5.6.A.2, we see that treating benzene with chloroethane and aluminum chloride will give ethylbenzene. Treating benzene with either ethanol and sulfuric acid or ethene and sulfuric acid will likewise produce ethylbenzene. Write step-by-step reaction mechanisms for each process. See sections 5.6.B.2, 3.7.B.2, and 4.1.D.1 for assistance. Note that all three processes involve the same reaction intermediates.

5.25 Reactions of Aromatic Compounds: Predict the major product(s) of the following reactions:

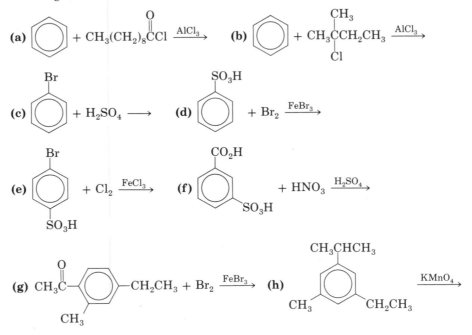

5.26 Reactions of Aromatic Compounds: Write structures for each product indicated by a letter.

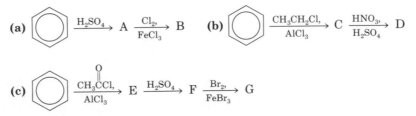

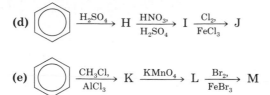

5.27 Synthesis: Outline the steps in the synthesis of the following compounds from benzene (assume that ortho and para isomers can be separated):

(a) *p*-bromochlorobenzene   **(b)** *p*-isopropylbenzenesulfonic acid
(c) *m*-bromobenzenesulfonic acid   **(d)** *m*-chloronitrobenzene
(e) *p*-chloronitrobenzene   **(f)** 2-bromo-4-nitroethylbenzene
(g) *m*-nitrobenzoic acid   **(h)** *p*-nitrobenzoic acid

5.28 Synthesis: From the word descriptions, write reactions illustrating the preparation of the following familiar substances:

(a) Mothballs: treatment of benzene with 2 moles chlorine and $FeCl_3$ as the catalyst.

(b) TNT: trinitration of toluene with 3 moles nitric acid and sulfuric acid (conc.) as a catalyst.

(c) Pentachlorophenol: a wood preservative (prevents attack by fungi and termites) used on fence posts, telephone poles. Produced by pentachlorination of phenol.

(d) Synthetic detergents: 2-chlorododecane ($C_{12}H_{25}Cl$) and benzene in the presence of $AlCl_3$ catalyst react by a Friedel-Crafts reaction. The product is sulfonated with fuming sulfuric acid. The sulfonic acid group is neutralized with sodium hydroxide (simple acid-base reaction) to give the detergent.

(e) Food preservative, sodium benzoate: oxidation of toluene to benzoic acid followed by neutralization of the acid with sodium hydroxide.

5.29 Activating and Deactivating Groups: Arrange the compounds within the following sets in order of reactivity (least to most reactive) toward electrophilic aromatic substitution:

(a) benzene, phenol, nitrobenzene
(b) benzene, chlorobenzene, aniline
(c) *p*-xylene, *p*-methylbenzoic acid, benzoic acid
(d) benzene, toluene, *p*-chloronitrobenzene, *p*-nitrotoluene, *p*-xylene

5.30 Activating and Deactivating Groups: Which benzene ring would you expect to be nitrated (HNO_3/H_2SO_4) in the following compounds? Explain your answer.

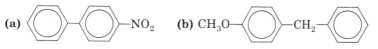

5.31 Qualitative Organic Analysis: Describe what you would do and see in distinguishing between cyclohexane, cyclohexene, and benzene using bromine.

5.32 Physical Properties: Explain the difference in melting points or boiling points, as indicated, for the compounds of the following sets:

(a) boiling point of methylbenzene (111 °C), and ethylbenzene (136 °C)
(b) melting point of xylenes, components of high-octane gasoline

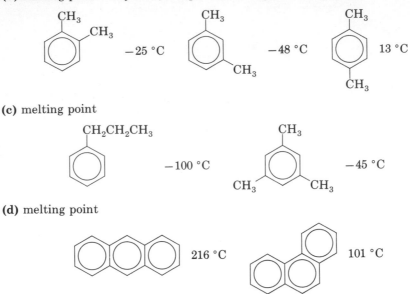

(c) melting point

(d) melting point

5.33 Gasoline: Write three structural characteristics of high-octane gasoline molecules. Draw representative molecules that possess each of the characteristics.

5.34 Production of Gasoline: What petroleum refinery method could improve the suitability of the following compounds as gasoline components?
(a) $CH_3(CH_2)_3CH_3$ **(b)** $CH_3(CH_2)_{18}CH_3$ **(c)** $CH_3(CH_2)_5CH_3$

(d) $CH_3CH{=}CH_2$ **(e)** $CH_3\underset{\underset{\displaystyle CH_3}{|}}{CH}CH_3$ **(f)**

5.35 Gasoline: Although alkenes have high octane ratings, they tend to form gums when blended into gasoline. What might be happening to the alkenes?

5.36 Methanol: Write a balanced equation showing the complete combustion of methanol. Why does methanol as a gasoline deliver only about half the miles per gallon of conventional hydrocarbon gasolines?

5.37 Pollution: Nitrogen oxides are serious air pollutants produced by automobiles. How do these arise?

Chapter 6

Optical Isomerism

ISOMERISM

6.1   You will recall that isomers are chemical compounds with identical molecular formulas but different structural formulas; that is, the arrangement of atoms in the molecules is in some way different. There are two principal classes of isomerism—structural isomerism (skeletal, positional, and functional) and stereoisomerism (geometrical, conformational, and optical). We have already examined all of the types of isomerism (Chapter 2) except optical isomerism, the most subtle.

Structural isomers differ in the bonding arrangement of atoms in the molecules; stereoisomers have identical bonding arrangements. Stereoisomers differ in how the atoms are oriented in space. *Optical isomers* are stereoisomers that interact with a type of light known as plane-polarized light. A simple definition of an *optically active compound* is a compound that rotates plane-polarized light.

PROBLEM

6.1. For the formula $C_6H_{12}O$, draw a representative pair of isomers, illustrating each of the following types of isomerism: **(a)** skeletal; **(b)** positional; **(c)** functional; **(d)** geometric; **(e)** conformational.

PLANE-POLARIZED LIGHT AND THE POLARIMETER

6.2   ### A. Plane-Polarized Light

Light can be described as a wave vibrating perpendicular to its direction of propagation. There are an infinite number of planes in which vibration can occur at right angles to the direction of propagation. Light vibrating in all possible planes is said to be unpolarized. Light vibrating in only one of the possible planes is *plane-polarized*. Plane-polarized light can be produced by passing unpolarized light through a Nicol prism (Iceland spar, a form of calcite, $CaCO_3$, developed by the Scottish physicist William Nicol) or through a Polaroid sheet (specially oriented crystals embedded in plastic; invented by E. H. Land). In either case, light vibrating in only one plane is allowed to pass. All other planes are rejected (Figure 6.1). A polarizer can be compared to a picket fence and the vibrating light waves could be depicted as two people, on opposite sides of the fence, oscillating a rope between two pickets. The only oscillation allowed is that parallel to the pickets. All other oscillations would be destroyed as they tried to pass through the fence.

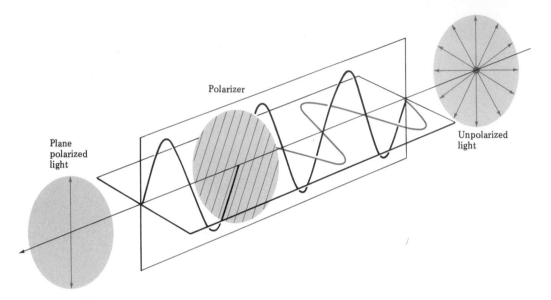

FIGURE 6.1. As unpolarized light encounters a polarizer, all but one plane is rejected. The resulting light that is transmitted is plane polarized.

B. The Polarimeter

The rotation of plane-polarized light by an optically active compound is detected and measured with an instrument called a *polarimeter,* shown diagramatically in Figure 6.2. A polarimeter has a monochromatic (single-wavelength) light source on one end, which produces unpolarized light vibrating in all possible planes perpendicular to the direction of propagation. As this light encounters the stationary polarizer, all planes but one are rejected. The light passing through is plane-polarized. The polarized light continues on through the sample tube (which for now we shall assume is empty) and reaches the variable analyzing polarizer. If this polarizer is lined up with the stationary polarizer, the polarized light will be allowed to pass and will be visible to the observer. If the variable polarizer is rotated, however, so that its allowable planes of light transmission are 90° to the stationary polarizer, the polarized light will be rejected and the observer will perceive darkness. This can be demonstrated with two pairs of Polaroid sunglasses. If one pair is placed in front of the other so that the lenses are lined up, light will pass through both and the pair will be transparent. If now, one pair is rotated 90°, the two lenses will have their planes of allowable light transmission out of phase and the pair will appear opaque. (Try this at a local store.)*

---

* Three-D movies and slides make use of polarized light. To see in three dimensions, each of our eyes visualizes a scene from slightly different perspectives. These are combined by the brain to give a 3-D visual

Assume now that the two polarizers of the polarimeter are aligned so that there is maximum light transmission, and that an optically active compound is placed in the sample tube. The plane of light transmitted by the stationary polarizer will be rotated in the sample tube and will not be maximally transmitted by the variable polarizer. The operator, however, can rotate the variable polarizer until light transmission is again maximized (when the analyzer's allowable transmission planes are the same as the light). In this way not only can optical activity be detected but the angle of rotation can also be measured.

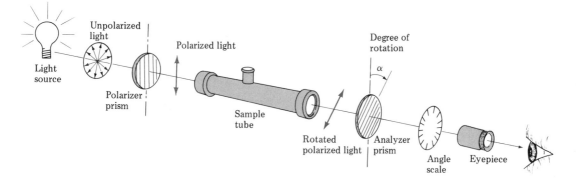

FIGURE 6.2. Schematic representation of a polarimeter containing an optically active sample.

C. Specific Rotation

A compound that does not rotate plane-polarized light is optically inactive, whereas a compound that does is optically active. A compound that rotates polarized light to the right, or clockwise, is termed *dextrorotatory,* represented by *d* or + (the variable polarizer is turned to the right to obtain maximum light transmission). If the rotation is to the left (counterclockwise), the substance is *levorotatory, l* or −. The degree of rotation is used to calculate the specific rotation according to the following equation:

$$\alpha = \frac{\text{specific}}{\text{rotation}} = \frac{\text{observed rotation, degrees}}{\underset{\text{tube, dm}}{\text{length of sample}} \times \underset{\text{sample, g/cm}^3}{\text{concentration of}}}$$

Like density, melting point, and boiling point, specific rotation is a physical property of a compound. The following are some specific rotations measured at about 20 °C, using light of wavelength 5893 Å:

Menthol	$\alpha = -50°$	Sucrose	$\alpha = +66.5°$
Cholesterol	$\alpha = -31.5°$	Vitamin C	$\alpha = +21.5°$
α-D-glucose	$\alpha = +112.2°$	Nicotine	$\alpha = -169°$

image. Two cameras are used in making 3-D movies to get two different views. The two images are projected on the screen with two different beams of polarized light (polarized in different planes). Without glasses, both of our eyes see both images and the picture appears blurred. Using Polaroid glasses with the lenses unsynchronized, however, we get a 3-D view. Since the lenses are oriented differently, each allows a different plane of polarized light to be transmitted; the other of the two planes is rejected by each lens. Thus, each eye sees only one of the originally filmed images and the two are then combined by the brain to give the 3-D effect.

STRUCTURE OF OPTICALLY ACTIVE COMPOUNDS

6.3   A. Historical Development

In 1815, the French physicist Jean Baptiste Biot found that plates of two different kinds of quartz rotated plane-polarized light in opposite directions but to equal degrees. Earlier, René Haüy, a French mineralogist, had determined that these two types of quartz crystals differed only in the position of two facets but that this difference caused the crystals to be nonidentical mirror images. They were called *enantiomorphs,* from the Greek *enantios,* "opposite" and *morph,* "form."

Louis Pasteur (1822–1895) made a discovery in 1848 that was very important to the development of stereochemistry. Pasteur is known primarily for his research on fermentation, the basis for microbiology; and pasteurization, the process carried out on milk, is named for him. At the time, Pasteur, a young man of 26 years, was studying the crystal structure of the sodium ammonium salt of tartaric acid at the Ecole Normale in Paris. Two isomeric forms of this acid were being deposited in wine barrels during fermentation. One, called tartaric acid, was dextrorotatory, whereas the other, then called *racemic acid,* was optically inactive. By slow crystallization of a solution of the sodium ammonium salt of the optically inactive racemic acid, Pasteur obtained two different types of crystals that were subtly different—they were mirror images. Using a magnifying glass and a pair of tweezers, Pasteur carefully separated the two types of crystals. This separation was like separating a barrel of gloves into right-handed and left-handed gloves. Of course, Pasteur's task required much greater concentration and dexterity. Although a solution of the mixture of crystals did not rotate plane-polarized light, solutions of equal concentration of each separate crystal form rotated plane-polarized light to an equal degree but in opposite directions. One form was dextrorotatory and the other levorotatory. If the two solutions were combined, the mixture became optically inactive. Pasteur concluded that racemic acid was then a mixture of *d* and *l* tartaric acids. Although each form separately rotated plane-polarized light, their combined effects were mutually cancelling, resulting in an optically inactive mixture.

By 1874, the structures of several optically active compounds were known. In that year, two chemists, Jacobus Hendricus van't Hoff (Dutch, 1852–1911) and Jules Achille Le Bel (French, 1847–1930), working independently, published papers pointing out that every optically active compound whose structure was known at that time had at least one carbon that was bonded to four different groups. Both van't Hoff and Le Bel also showed that if carbon were tetrahedral, two arrange-

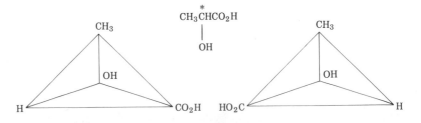

FIGURE 6.3. Three-dimensional tetrahedral representations of lactic acid, showing two nonsuperimposable mirror-image forms. Assume that the carbon with the asterisk lies in the center of the tetrahedron.

ments of these four groups could be possible and these arrangements would be related as mirror images.

Consider lactic acid, for example, in Figure 6.3. The indicated carbon has four different bonded groups, which can be arranged in two mirror-image forms. Two nonidentical forms of lactic acid are actually known: one, found in sour milk, is levorotatory; and the other, found in muscle tissue, is dextrorotatory. They have identical physical properties (densities, melting points, boiling points, refractive indices, and so on) except for the way they rotate plane-polarized light. One rotates plane-polarized light to the right and the other, an equal amount to the left.

B. Chiral Carbons and Enantiomers

1. Chiral Carbons. A carbon with four different bonded groups is called a *chiral carbon*. The term *chiral* is derived from the Greek *cheir* for "hand" and is used because a chiral carbon is similar to a hand. Our two hands are mirror images, yet in no way superimposable. Similarly, a carbon with four different bonded groups can exist in two nonsuperimposable mirror-image arrangements (Figure 6.4). A compound that is not superimposable on its mirror image is termed *chiral*.

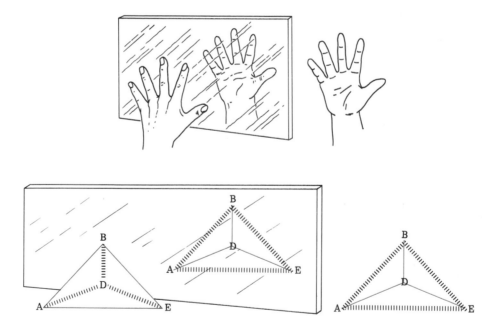

FIGURE 6.4. Mirror-image relations of a right and a left hand and the two arrangements of a chiral carbon. The chiral carbon (not shown) would be at the center of the tetrahedron.

Let us examine the chiral carbon shown in Figure 6.4 more closely. Using ball-and-stick models, we can construct a mirror-image model of the chiral carbon (Figure 6.5a). The mirror images can be turned and rotated so that any two groups on the chiral carbon can be superimposed at one time but the other two will always be in conflict. Try this yourself with ball-and-stick models or with gumdrops and toothpicks. If any two groups are identical, then the mirror images are easily superimposable, as shown in Figure 6.5b.

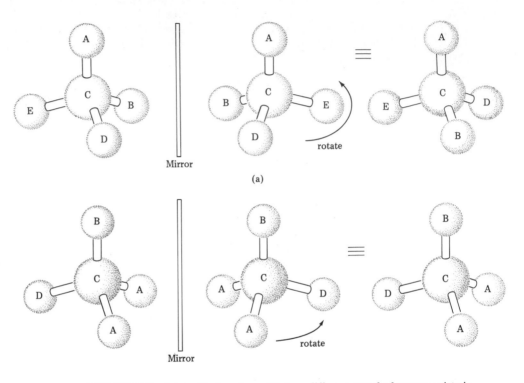

FIGURE 6.5. (a) A chiral carbon with four different attached groups exists in two nonsuperimposable mirror-image forms. Any two groups (in this picture A and E) can be superimposed by rotation but the other two (B and D) will always be in conflict. (b) If a carbon has two or more identical groups, it is superimposable on its mirror image.

2. Enantiomers. Optical isomers that are nonsuperimposable mirror images are called *enantiomers*. Enantiomers obviously come in pairs and each member of the pair has physical properties identical to the other's except in the way they rotate plane-polarized light. One rotates light to the right (dextrorotatory) and the other an equal magnitude to the left (levorotatory). Lactic acid (Figure 6.3) is an example of a pair of enantiomers. A 50:50 mixture of two enantiomers is called a *racemic mixture*. Such a mixture is optically inactive (does not rotate plane-polarized light) because the two components rotate plane-polarized light equally in opposite directions and cancel one another.

3. Representation of Enantiomers. How can the three-dimensional structure of an enantiomeric pair be expressed effectively on a two-dimensional surface (such as a chalkboard or paper)? One can draw a ball-and-stick molecular model from a perspective in which horizontal bonds are protruding out of the plane of the paper and vertical bonds are behind the plane (see Figure 6.6). This

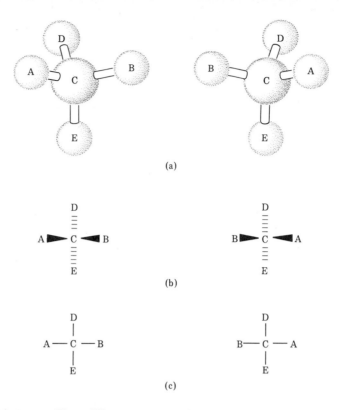

FIGURE 6.6. Three different representations of a chiral carbon. (a) With a tetrahedral carbon placed in this way, the horizontal bonds are in front of the plane and the vertical bonds behind. (b) The three-dimensional nature is shown by flared lines (in front of plane) and dashed lines (behind plane). (c) No attempt is made in these formulas (called Fischer projections) to show three dimensions. Vertical bonds are understood to be behind the plane, however, and horizontal bonds in front.

can also be represented on paper using flared lines for the horizontal bonds in front of the plane and dashed lines for the vertical bonds behind the plane (Figure 6.6b). The molecule can even be drawn with no indication of its three-dimensional nature (Figure 6.6c), so long as it is remembered that horizontal bonds are in front and vertical bonds behind the plane of the paper. Therefore, though the structures in Figure 6.6c appear to be superimposable by flipping one structure over onto the other, they really are not, because groups A and B would be transposed from the front to the back of the plane of the page. In other words, using this completely two-dimensional representation, we must not rotate the structure out of the plane of the page to test for superimposability. If we wish to rotate the structure out of the plane, representations such as that in Figure 6.6b should be used. To illustrate the manipulation of these figures, work through the accompanying example, which shows the superimposability of two structures. The same result can be obtained by interchanging groups in pairs (for avoiding configuration change, see page 152).

Representations such as those described in Figure 6.6c are called *Fischer projections* and are frequently used in biological chemistry. In representing a

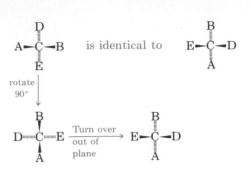

biological molecule, the structure is drawn with the most highly oxidized carbon (usually a carboxylic acid, aldehyde, or ketone) at the top followed vertically by the chiral carbons. Consider, for example, the amino acid phenylalanine (the disease phenylketonuria is caused by the inability to metabolize phenylalanine) and the carbohydrate glucose (blood sugar, the main source of body energy).

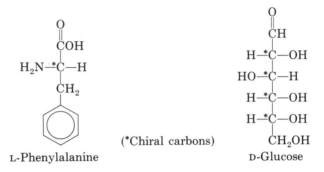

Naturally occurring amino acids (the structural units of proteins) are of the L-series in which the amino group (NH_2) is on the left when the molecule is written in a Fischer representation. Carbohydrates, however, are of the D-series in which the bottom chiral carbon has the OH oriented on the right. Note that there are four chiral carbons in glucose. Of the 16 possible optical isomers with the same chemical formula, the human body selectively produces and consumes mainly the one shown.

PROBLEMS

6.2. Which of the following have a chiral carbon and can exist as a pair of enantiomers? Using figures such as those in Figure 6.6b, draw the pair of enantiomers.

(a) 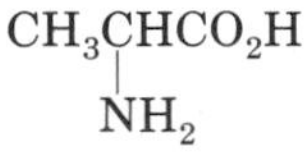 CH_3CHCO_2H
         |
         NH_2

(b) 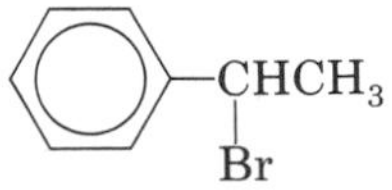 ⬡—$CHCH_3$
              |
              Br

(c) 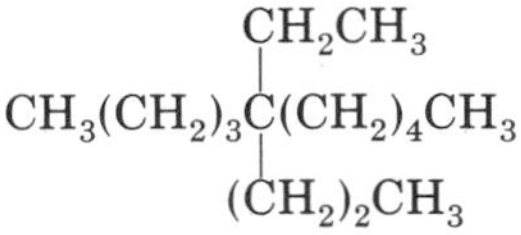 $CH_3(CH_2)_3\overset{\displaystyle CH_2CH_3}{\underset{\displaystyle (CH_2)_2CH_3}{C}}(CH_2)_4CH_3$

6.3. Determine which of the structures below (a–e) are identical and which are mirror images of the following compound:

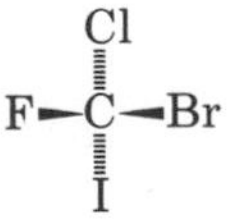

$$F-\overset{Cl}{\underset{I}{C}}-Br$$

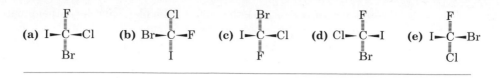

OPTICAL ISOMERS WITH TWO CHIRAL CARBONS

6.4   A compound possessing one chiral carbon can exist in two mirror-image forms called enantiomers. A compound with two chiral carbons can have a maximum of four optical isomers, because each chiral carbon can exist in two configurations that are mirror images. The maximum number of optical isomers possible for a compound is 2^n, where n is the number of chiral carbons. This is sometimes called the *van't Hoff rule*.

A. Molecules with Two Dissimilar Chiral Carbons

Consider the carbohydrate molecule 2-deoxyribose, which is a structural component of the genetic material deoxyribonucleic acid (DNA).

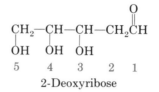

2-Deoxyribose

This molecule has two chiral carbons, carbons 3 and 4. The four different groups bonded to carbon-3 are —CH$_2$CH (with =O), —H, —OH, and —CHOHCH$_2$OH, and the four bonded to carbon-4 are —CHOHCH$_2$CH (with =O), —H, —OH, and —CH$_2$OH. Carbon-1 has three different bonded groups but needs four to be chiral. Each of carbons 2 and 5 has two identical bonded groups (hydrogens) and thus cannot be chiral.

There is a maximum of four optical isomers possible for 2-deoxyribose ($2^n = 2^2 = 4$). These should be drawn in a systematic fashion, with the chiral carbons emphasized. It is also convenient to draw the isomers as pairs of mirror images for comparison purposes. Following are the four isomers. Remember that horizontal bonds are in front of the plane of the paper and vertical bonds behind. Only the two chiral carbons and the bond between them are in the plane.

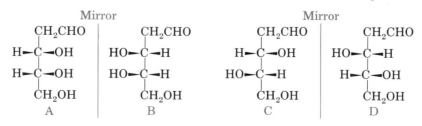

Structures A and B are mirror images of one another yet in no way are they superimposable. We cannot superimpose A and B by turning either out of the plane nor can we rotate either 180° in the plane and have them match since the top and bottom of the molecule are different. Thus, A and B are an enantiomeric pair. All physical properties are identical except the rotation of plane-polarized light; A and B rotate light in equal magnitudes but in opposite directions. Likewise, structures C and D are nonsuperimposable mirror images and are a pair of enantiomers.

How then are structures A and C, A and D, B and C, and B and D related? Each represents a pair of optical isomers of 2-deoxyribose, but none of the pairs are mirror images. Optical isomers that are not related as enantiomers are called *diastereomers*. In other words, diastereomers are optical isomers that are not mirror images. Whereas enantiomers differ only in the rotation of plane-polarized light, diastereomers can differ in all physical properties. Their melting points, boiling points, densities, refractive indices, and, if they are chiral, specific rotations can differ and usually do.

B. Molecules with Two Similar Chiral Carbons

Let us consider the compound tartaric acid, which Pasteur studied extensively in his research on optical isomerism.

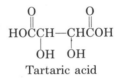

Tartaric acid

Tartaric acid is found in grape juice and cream of tartar and is also the acid component of some baking powders. It has two similar chiral carbons, carbon-2 and carbon-3. Each has four different bonded groups ($-CO_2H$, $-H$, $-OH$, $-CHOHCO_2H$).

Remember that one should draw the structures systematically, in pairs of mirror images, with emphasis on the chiral carbons. In the following, structures E and F are mirror images. However, by rotating either molecule 180° in the plane of the page, we can superimpose one on the other. Thus, structures E and F are identical and F should be eliminated from the list of optical isomers. Although E has chiral carbons, the overall molecule is not chiral since it is superimposable on its mirror image. Such compounds are called *meso compounds* and are optically

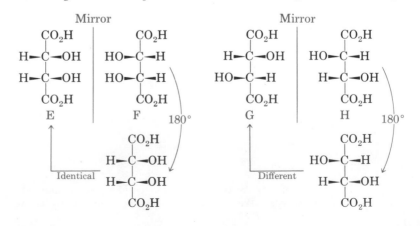

inactive; they do not rotate plane-polarized light. Actually, meso compounds probably interact with plane-polarized light but the rotation is undetectable due to internal compensation. Close inspection of *meso*-tartaric acid (E) reveals that the top half of the molecule is a mirror image of the bottom. Their individual effects on plane-polarized light cancel each other and the compound is optically inactive. Note then that possession of chiral carbons is not necessarily sufficient for optical activity, but the compound must also be chiral overall in its structure.

Structures G and H are related as mirror images and are not superimposable even if rotated 180°. Thus G and H constitute an enantiomeric pair. There are two pairs of diastereomers—EG and EH. Table 6.1 lists some properties of the various forms of tartaric acid, and Table 6.2 summarizes the definitions of terms used in describing optical isomers.

TABLE 6.1. PROPERTIES OF THE OPTICAL ISOMERS OF TARTARIC ACID

	Dextrorotatory Form	Levorotatory Form	Racemic Mixture	Meso Form
Rotation (α)	$+12$ °	-12 °	0 °	0 °
Melting point	168–170 °C	168–170 °C	206 °C	140 °C
Water-solubility 20 °C, 100 ml H_2O	139 g	139 g	20.60 g	125 g
pK_{A_1} (acidity)	2.93	2.93	2.96	3.11
Density	1.7598	1.7598	1.697	1.666

TABLE 6.2. TERMS USED TO DESCRIBE OPTICAL ISOMERS

Optically active, or chiral, compound A compound that is not superimposable on its mirror image. Such compounds rotate plane-polarized light.

Chiral carbon A carbon bonded to four different groups.

Van't Hoff rule The maximum number of optical isomers a compound may have is 2^n; n represents the number of chiral carbons.

Enantiomers Optical isomers that are mirror images. Enantiomers have identical physical properties except for the rotation of plane-polarized light. One of the pair is levorotatory and the other dextrorotatory, to equal extents.

Racemic mixture A mixture of equal parts of enantiomers. Racemic mixtures are optically inactive.

Diastereomers Optical isomers that are not mirror images (not related as enantiomers). All physical properties of diastereomers are usually different.

Meso compound A compound that has more than one chiral center and that is superimposable on its mirror image. *Meso* compounds are optically inactive.

PROBLEM

6.4. Draw the optical isomers of **(a)** 2-bromo-3-chlorobutane, and **(b)** 2,3-dibromo-butane. Identify pairs of enantiomers, pairs of diastereomers, and meso compounds.

OPTICAL ISOMERISM IN CYCLIC COMPOUNDS

6.5   Cyclic compounds can exhibit optical isomerism as well as geometric isomerism (section 2.8C).   Using 1,2-dibromocyclopropane as an illustration, we find that the cis isomer has its two bromines on the same side of the planar ring while the trans isomer has one above and one below the ring.   Since 1,2-dibromocyclopropane has two chiral carbons (the carbons bonded to bromines), it should have a maximum of four optical isomers.   Let us draw the mirror images of the cis and trans isomers and test for superimposability.

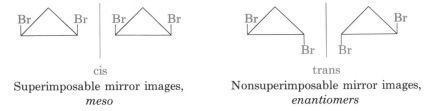

cis

Superimposable mirror images,
meso

trans

Nonsuperimposable mirror images,
enantiomers

cis-1,2-Dibromocyclopropane is superimposable on its mirror image.   Therefore, the molecule is not chiral (it is achiral) and it is an optically inactive meso structure.   The trans isomer is not superimposable on its mirror image; it exists as an enantiomeric pair.

The compound 1-bromo-2-chlorocyclopropane has two dissimilar chiral carbons and thereby two pairs of enantiomers.

cis **Enantiomers**

trans **Enantiomers**

PROBLEM

6.5. Draw the optical isomers of **(a)** 1,2-dibromocyclopentane, and **(b)** 1-bromo-3-chlorocyclopentane.   Identify pairs of enantiomers, pairs of diastereomers, and meso compounds.

SPECIFICATION OF CONFIGURATION

6.6   **A. R, S Designation of Chiral Carbons**

We have seen that a chiral carbon is bonded to four different groups and that the four groups can be arranged in two different ways that are related as mirror images.   The specific arrangement of the groups characterizes a particular stereoisomer and is known as the *configuration*.   This configuration can be described by actually drawing the compound.   But how can we describe the configuration more conveniently?

A most effective method was developed by R. S. Cahn, C. Ingold, and V. Prelog.   It involves two steps:

Step 1. By a set of sequence rules, described in Table 6.3, the groups connected to the chiral carbon are assigned a priority sequence.

Step 2. The molecule is then visualized so that the group of lowest priority is directed away from the observer.   The remaining three groups are in a plane facing the observer. If the eye travels clockwise as we look from the group of highest priority to the groups of second priority and third priority, the configuration is designated R (Latin, *rectus,* "right"). If we look in a counterclockwise direction, the configuration is designated S (Latin, *sinister,* "left").

TABLE 6.3. SEQUENCE RULES FOR R,S CONFIGURATIONS

Rule 1: If all four atoms directly attached to the chiral carbon are different, priority depends on atomic number, with the atom of highest atomic number getting the highest priority. The priority order of some atoms commonly found in organic compounds is:

High Priority									Low Priority
	$I > Br > Cl > S > F > O > N > C > H$								
	53	35	17	16	9	8	7	6	1
				Atomic numbers					

Rule 2: If two or more of the atoms directly bonded to the chiral carbon are identical, the priority of these groups is determined by comparing the next atoms of the groups and so on, working outward until a difference is found (as in the following examples).

Example

$$\underset{\underset{CH_2CH_2CH_3}{|}}{\overset{\overset{Cl}{|}}{ICH_2CH_2CCH_2Br}}$$

Priority sequence

$$Cl > CH_2Br > CH_2CH_2I > CH_2CH_2CH_3$$

Connected directly to the chiral carbon are a chlorine and three carbons. Chlorine has the highest atomic number and the highest priority. Connected, in each case, to the three carbons are 2 H's and a Br, 2 H's and a C, and 2 H's and a C. Bromine has the highest atomic number of C, H, and Br and thus CH_2Br is the highest priority of these three. The other two carbons are still identical. Connected to the second carbon of these groups are 2 H's and an I and 2 H's and a C. Iodine has the highest priority of these atoms (C, H, and I), so that $-CH_2CH_2I$ is next in the priority list and $-CH_2CH_2CH_3$ is last.

Example

$$\underset{\underset{CH_2Cl}{|}}{\overset{\overset{CF_3 \quad CH_3}{| \quad |}}{HSCH_2CH_2C—CHCH_3}}$$

Priority sequence

$$CH_2Cl > CF_3 > \overset{\overset{CH_3}{|}}{CH_3CH} > CH_2CH_2SH$$

The atoms directly bonded to the chiral carbon are all carbons, and it will be necessary to analyze the atoms bonded to these. Considering each individual carbon, we find that the bonded atoms are 3 F's; 2 C's; and an H, 2 H's and a Cl; and 2 H's and a C. Of these bonded atoms (F, Cl, C, and H), chlorine has the highest atomic number and fluorine the next. Thus $-CH_2Cl$ has the highest priority, followed by CF_3; it makes no difference that 3 F's add up to more than 1 Cl and 2 H's. The remaining carbons both contain only carbon and hydrogen.

(continues on page 150)

TABLE 6.3   (continued)

However, 2 C's and 1 H (CH_3CHCH_3) take precedence over 1 C and 2 H's ($-CH_2CH_2SH$), making CH_3CHCH_3 next highest in priority, and leaving $-CH_2CH_2SH$ as the lowest-priority group.

Rule 3: If a double or triple bond must be considered, the involved atoms are treated as being duplicated or triplicated, respectively.

$$-C{=}A \ \text{equals} \ -\overset{\displaystyle A}{\underset{\displaystyle \ }{C}}-A \qquad -C{\equiv}A \ \text{equals} \ -\overset{\displaystyle A}{\underset{\displaystyle A}{C}}-A$$

Example

$$H_2NCH_2\overset{\displaystyle C{\equiv}N}{\underset{\displaystyle H}{C}}-CH_2OH$$

Priority sequence

$$CH_2OH > C{\equiv}N > CH_2NH_2 > H$$

Three carbons and a hydrogen are directly bonded to the chiral carbon. Hydrogen has the lowest priority. If we consider the three carbons, the bonded atoms are 3 N's (in $C{\equiv}N$ the nitrogens are triplicated), 2 H's and an O, and 2 H's and an N. Of these atoms (N, O, H), oxygen has the highest atomic number and $-CH_2OH$ the highest priority. Of the remaining two groups, 3 N's take precedence over 1 N and 2 H's, so that $-C{\equiv}N$ is next in priority, followed by $-CH_2NH_2$ and H.

Let us apply this procedure in a general way. First, we put the four groups bonded to the chiral carbon in a priority order (we shall use the numbers 1–4 as groups here).

Highest Priority ① > ② > ③ > ④ Lowest Priority

Now put the group of lowest priority back away from the observer.

If groups 1, 2, 3 are arranged in a clockwise fashion, the configuration is R. If they occur in a counterclockwise fashion, it is S.

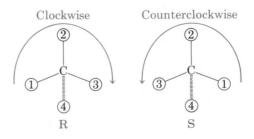

By drawing configurations in this manner, we are picturing the tetrahedral carbon as an inverted pyramid. The peak of the pyramid (tetrahedron) is behind the plane of the paper and the three-cornered base is in the plane of the paper. To describe this molecule by the Fischer projections (in which vertical bonds are behind the plane and horizontal bonds in front of the plane), we need only tilt the molecule a bit.

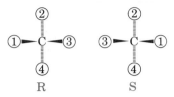

Now let us draw the R and S configurations of bromochlorofluoromethane. The priority sequence (Table 6.3) by atomic number is Br > Cl > F > H.

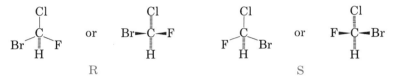

Suppose we wish to assign an R,S configuration to a structure that is not conveniently drawn as to its configuration. For example, what is the configuration of the following molecule?

For the configuration to be determined, the molecule must be positioned so that the lowest priority group (H) is down and away from the observer.

Now tilting the molecule forward and proceeding from the highest to lowest priority group (Br → Cl → F), we see that this is the R configuration.

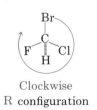

Clockwise
R configuration

The same result can be accomplished by interchanging groups bonded to the chiral carbon. If two groups are switched, the configuration of the chiral atom is changed. If two more groups are interchanged, however, the molecule assumes its original configuration. Thus, the interchange operation must be done in pairs to avoid a configuration change. Using the example at hand, we should exchange the

hydrogen and chlorine so as to get the lowest priority group down and back. Then
to retain the original configuration, interchange any two other groups, say the
bromine and fluorine.

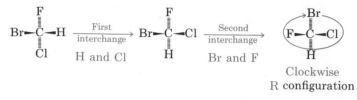

Clockwise
R configuration

For further practice, rationalize the configurations given for the following
compounds. (Priority sequences for these specific examples are explained in
Table 6.3.)

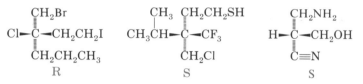

PROBLEM

6.6. **(a)** Draw and label the R and S configurations of 4-bromo-1-chloro-2-methyl-
butane; **(b)** specify the configuration of the following molecule as R or S:

B. Configuration of Geometric Isomers

The configuration of geometric isomers can be designated as cis or trans.
If the two identical groups are on the same side of a double bond, the isomer is cis;
and if they occur on opposite sides, trans. This method is not convenient when
there are no identical groups, as in 1-bromo-1-chloro-2-fluoro-2-iodoethene
(BrClC=CFI). The configuration of these and all other geometric isomers can be
specified using the letters Z and E. To do this, first determine the group of highest
priority on each carbon. If the two high-priority groups are together on the same
side, the configuration is Z (*zusammen,* German, "together"). If they are on opposite sides, the configuration is E (*entgegen,* German, "opposite").

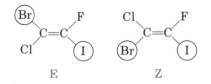

Here, Br > Cl and I > F in priority, and the higher-priority groups are circled. Of
course, the same procedure can be used on molecules that can be designated as cis
and trans.

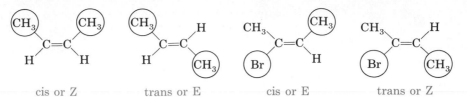

cis or Z   trans or E   cis or E   trans or Z

This method can also be used in specifying the configuration of cyclic compounds.

Z-1,3-Dibromocyclobutane   E-1,3-Dibromocyclobutane

PROBLEM

6.7. Draw the Z and E isomers of **(a)** 1,2-dichloroethene, and **(b)** 1,2-dichloro-cyclopropane.

RESOLUTION OF ENANTIOMERS

6.7   Recall that optical isomers that are nonsuperimposable mirror images are called enantiomers. As such they have completely identical physical properties except for the rotation of plane-polarized light. How can a racemic mixture be separated (resolved) into pure R and S enantiomers? One cannot take advantage of differences in boiling point, melting point, or solubility, since enantiomers are identical in each of these aspects. Following are three methods for resolving an enantiomeric pair.

A. Mechanical Resolution

Pasteur's original separation of the sodium ammonium salt of tartaric acid, known as *mechanical resolution,* is mainly of historical interest. By slow crystallization of this salt, he obtained crystals that existed as two mirror-image forms (section 6.3A). Mechanical separation (with tweezers) of the two crystalline forms resulted in the resolution of the racemic mixture. This method is not generally practical because most racemic mixtures do not readily form enantiomorphic crystals. Furthermore, the separation would be extremely tedious.

B. Biological Resolution

A second method for resolution of enantiomers takes advantage of the varying rates of metabolism of the components of a racemic mixture by microorganisms. Microorganisms produce enzymes that are themselves chiral and consequently react differently with the two enantiomeric forms. Pasteur found that *Penicillium glaucum,* the green mold found on aging fruit and cheese, selectively consumes the dextrorotatory form of tartaric acid. Although this method has a better chance of being successful than mechanical resolution and can be applied to a large variety of compounds, it suffers in that one enantiomeric form is lost (metabolized) and the other is often obtained in poor yield.

C. Resolution through Diastereomers

The most generally useful (and theoretically successful) method for separating enantiomers involves converting them into diastereomers. Whereas enantiomers are identical in all physical properties except the rotation of plane-polarized light, diastereomers differ in all properties, including melting point, boiling point, and solubility. If an enantiomeric pair can be converted into a pair of diastereomers, which are separable, and the separated diastereomers converted back into the enantiomers, resolution would be accomplished.

Take the physically inseparable pair of enantiomers in Figure 6.7. If group X on each enantiomer reacts with group O of a single enantiomer (optically active) from another source, a pair of diastereomers (optical isomers that are not mirror images) is formed. The diastereomers are separated by physical means such as distillation or fractional crystallization. The separated diastereomers are then individually converted back to the original enantiomers, which are now isolated from one another.

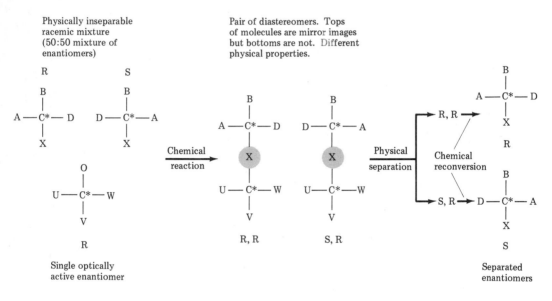

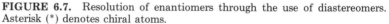

FIGURE 6.7. Resolution of enantiomers through the use of diastereomers. Asterisk (*) denotes chiral atoms.

Panel 9

OPTICAL ISOMERISM IN THE BIOLOGICAL WORLD

Most organic compounds produced or consumed by living organisms are chiral and thus rotate plane-polarized light. Although the differences between optical isomers appear very subtle on the surface, they are very significant in biological systems. Epinephrine (adrenalin) has one chiral carbon and has two enantiomeric forms, for instance. Although these forms are merely mirror images,

[PANEL 9] the levorotatory isomer is twenty times as active in raising blood pressure as the dextrorotatory form.

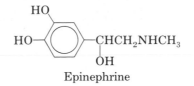

Epinephrine

Starch and cellulose are in a class of compounds called carbohydrates. Both are polymers of glucose (section 6.3B, p. 144) which predominantly exists in a cyclic form. The two differ only in the linkage between the cyclic glucose units (Figure 6.8). The glucose units in starch are connected by α linkages, axial at the glycoside carbon (the carbon with two bonded oxygens), whereas those in cellulose are connected by β linkages, equatorial at the glycoside carbon. This subtle difference (α and β linkages are merely different configurations) is extremely important biologically. The human digestive system contains enzymes (biological catalysts) that can metabolize the α linkages that occur in starch, but none that can metabolize the β linkages in cellulose. Thus we can digest the starch in bread, potatoes, vegetables, and so on, but not the cellulose of grass, paper, wood, and cotton.

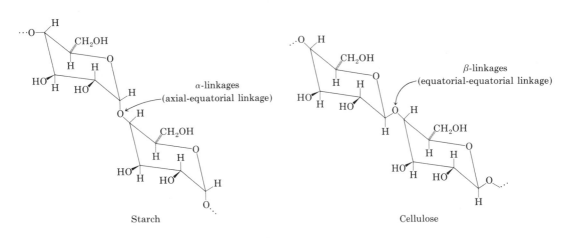

FIGURE 6.8. Starch and Cellulose.

Why does such natural chemical selectivity occur between optical isomers, some of which differ subtly as mirror images? As we have previously stated, enantiomers have identical physical properties except for the rotation of plane-polarized light and identical chemical properties except for reactions with reagents that are themselves chiral. Herein rests nature's ability to be stereoselective. Most biochemical reactions occurring in living organisms are catalyzed by enzymes. Enzymes are proteins, polymers of amino acids, and so have multiple centers of chirality; for example, the enzyme α-amylase, an enzyme found in saliva specifically hydrolyzes starch, but not cellulose. Consider the reaction in which a molecule with one chiral center must bind to a chiral enzyme at three positions.

[PANEL 9] As you can see in Figure 14.9 (p. 363), only one of the two enantiomers can simultaneously bind to all three positions and engage in that particular biochemical process. To interact, the molecule must fit onto the enzyme just as a hand fits into a glove. Since most biochemical processes involve reactions between chiral molecules, most will be highly selective.

END-OF-CHAPTER PROBLEMS

6.8 Chirality: Which of the following objects are chiral and able to exist in enantiomeric forms?

(a) glove	**(b)** nail	**(c)** foot	**(d)** screw
(e) fork	**(f)** spiral staircase	**(g)** pullover sweater	**(h)** scissors
(i) rubber ball	**(j)** pine cone	**(k)** key	**(l)** checkerboard
(m) coiled spring	**(n)** clock	**(o)** ocean wave	**(p)** block
(q) hammer	**(r)** ear	**(s)** golf club	**(t)** umbrella

6.9 Chiral Carbons: Circle the chiral carbons in each of the following molecules. What is the maximum number of optical isomers possible?

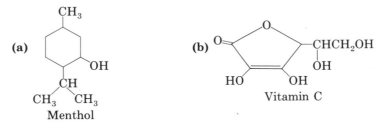

(a) Menthol

(b) Vitamin C

(c) NaO$_2$CCH$_2$CH$_2$CHCO$_2$H
          |
          NH$_2$
Monosodium glutamate

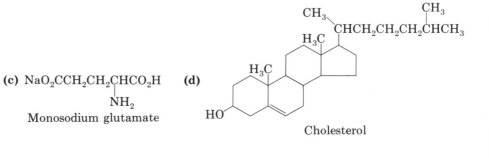

(d) Cholesterol

(e) —CH$_2$CHCH$_3$
              |
              NH$_2$
Amphetamine

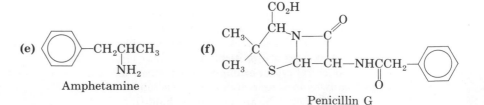

(f) Penicillin G

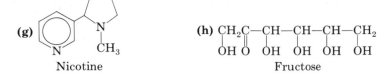

(g)

Nicotine

(h) CH₂C—CH—CH—CH—CH₂
| ‖ | | |
OH O OH OH OH

Fructose

6.10 Optical Isomers: Draw the optical isomers of the following compounds. Label pairs of enantiomers, pairs of diastereomers, and meso structures.

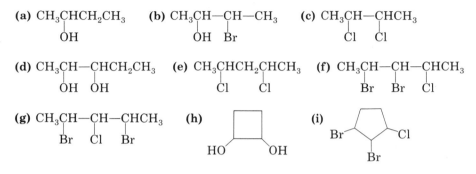

(a) CH₃CHCH₂CH₃
     |
    OH

(b) CH₃CH—CH—CH₃
       | |
      OH Br

(c) CH₃CH—CHCH₃
      | |
     Cl   Cl

(d) CH₃CH—CHCH₂CH₃
      | |
     OH OH

(e) CH₃CHCH₂CHCH₃
     | |
    Cl   Cl

(f) CH₃CH—CH—CHCH₃
      | | |
     Br Br Cl

(g) CH₃CH—CH—CHCH₃
      | | |
     Br Cl Br

(h)

(i)

6.11 Specification of Configuration: Using the designations R, S, Z, and E, specify the configuration of each of the following:

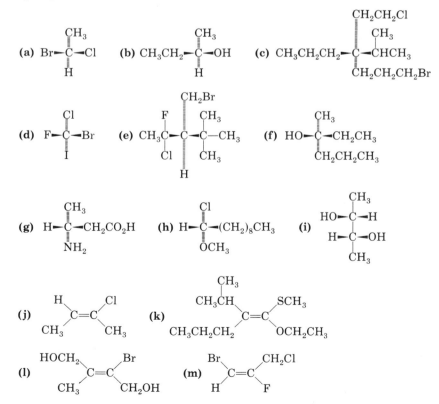

(a) Br—C—Cl
     CH₃ (top)
     H (bottom)

(b) CH₃CH₂—C—OH
     CH₃ (top)
     H (bottom)

(c) CH₃CH₂CH₂—C—CHCH₃
    CH₂CH₂Cl (top)
    CH₃ (top)
    CH₂CH₂CH₂Br (bottom)

(d) F—C—Br
    Cl (top)
    I (bottom)

(e) CH₃C—C—C—CH₃
    F, CH₂Br (top)
    Cl, CH₃ (middle)
    H (bottom)

(f) HO—C—CH₂CH₃
    CH₃ (top)
    CH₂CH₂CH₃ (bottom)

(g) H—C—CH₂CO₂H
    CH₃ (top)
    NH₂ (bottom)

(h) H—C—(CH₂)₈CH₃
    Cl (top)
    OCH₃ (bottom)

(i) HO—C—H, H—C—OH
    CH₃ (top)
    CH₃ (bottom)

(j)
  H, Cl / C=C / CH₃, CH₃

(k)
  CH₃CH (CH₃), SCH₃ / C=C / CH₃CH₂CH₂, OCH₂CH₃

(l)
  HOCH₂, Br / C=C / CH₃, CH₂OH

(m)
  Br, CH₂Cl / C=C / H, F

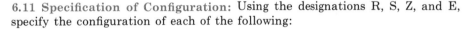

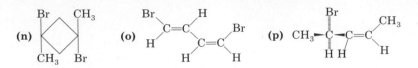

6.12 Specification of Configuration: Draw the following molecules, clearly showing the stereochemistry:

(a) R-2-bromobutane

(b) S-3-methylhexane

(c) R-1-bromo-2-methylbutane

(d) R-2,3-dimethylpentane

(e) Z-2-pentene

(f) E-1,2-dibromo-1-iodo-1-butene

(g) E-1,4-dimethylcyclohexane

(h) Z,Z-2,4-hexadiene

(i) R,Z-1,4-dichloro-1-pentene

6.13 Optical Isomers without Chiral Carbons: Nitrogen and silicon can both act as chiral atoms thus showing optical isomerism. Draw an optically active compound with nitrogen and another with silicon in which the nitrogen and silicon are chiral.

6.14 Optical Isomers without Chiral Atoms: For a compound to rotate plane-polarized light, it must be chiral overall. The presence of chiral atoms, however, is not a necessity for optical activity just as the presence of chiral carbons does not guarantee optical activity (meso compounds, for example). The following compounds are chiral and rotate plane-polarized light even though neither possesses chiral atoms. Among the factors leading to chirality is the fact that one half of the molecule is perpendicular to the other half. Fully explain the geometry of each molecule. Draw a pair of enantiomers in each case and show that they are not superimposable.

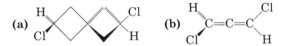

6.15 Stereoisomers: Using three-dimensional drawings, draw the four stereoisomers of 4-bromo-2-pentene.

Chapter 7

Organic Halogen Compounds

STRUCTURE, NOMENCLATURE,
AND PHYSICAL PROPERTIES

7.1   **A. Structure and Properties**

Organic halogen compounds are hydrocarbons in which one hydrogen (or more) has been replaced with a halogen. These compounds can be classified into groups that show similar chemical properties, as summarized in Table 7.1. Alkyl

TABLE 7.1. CLASSES OF ORGANIC HALOGEN COMPOUNDS

Class	General Structure	Examples
Alkyl halides	R—X	CH_3Cl, CH_3CH_2Br
Aryl halides	⬡—X	⬡—Br, CH_3—⬡—I
Vinyl halides	—C=C—X	CH_2=CHCl, CH_3CH=CHBr
Allylic halides	—C=C—C—X	CH_2=CHCH$_2$Br
Benzylic halides	⬡—C—X	⬡—CH$_2$Cl

halides are further described as primary, secondary, or tertiary, depending on the number of alkyl groups connected to the halogenated carbon. If there is one carbon directly connected to the carbon bearing the halogen, the alkyl halide is primary; if there are two it is secondary; and with three it is tertiary.

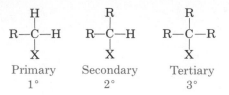

Like most classes of organic compounds, organic halides have boiling points that increase with molecular weight. Thus, chloropropane boils at a higher temperature than chloroethane, which in turn has a higher boiling point than chloromethane. Since the atomic weight of halogens increases in the relation $Cl <$ $Br < I$, the boiling points of particular alkyl halides increase as $R—Cl < R—Br < R—I$. Many organic halides have densities greater than water and most are insoluble in water. Table 7.2 tabulates the boiling points and densities of some representative compounds.

TABLE 7.2. PHYSICAL PROPERTIES OF SOME ORGANIC HALOGEN COMPOUNDS, R—X

Organic Halide	*Chloride*		*Bromide*		*Iodide*	
	BP, °C	*Density, ~20 °C*	*BP, °C*	*Density, ~20 °C*	*BP, °C*	*Density, ~20 °C*
$CH_3—X$	−24	Gas	5	Gas	43	2.28
$CH_3CH_2—X$	12.5	Gas	38	1.44	72	1.93
CH_2X_2	40	1.34	99	2.49	180	3.33
CHX_3	61	1.49	151	2.89	Sublimes	4.01
CX_4	77	1.60	189.5	3.42	Sublimes	4.32
$CH_2=CHX$	−14	Gas	16	Gas	56	
$\langle\bigcirc\rangle—X$	131	1.11	156	1.50	188	1.84

B. IUPAC Nomenclature

In IUPAC nomenclature, halogens are designated by the prefix *halo-* (*fluoro-, chloro-, bromo-, iodo-*): CH_3Cl is chloromethane; and $CH_3CHBrCH_2CH_3$, 2-bromobutane. Nomenclature of these compounds has been previously covered in section 3.2.B.3.

C. Common Nomenclature

Simple alkyl halides are seldom referred to by IUPAC names; they have instead a "salt-type" nomenclature. Just as NaCl is sodium chloride, CH_3Cl is often called methyl chloride. The molecule is divided into an organic part and a halide and named accordingly (see also section 3.3).

The halogen derivatives of methane are most frequently referred to by non-systematic names, as the chlorine derivatives show:

CH_3Cl	CH_2Cl_2	$CHCl_3$	CCl_4
Methyl chloride	Methylene chloride	Chloroform	Carbon tetrachloride

PROBLEMS

7.1. Name the following by the IUPAC system:

(a) $CH_3CHCH_2CH_2CH_3$
    |
    Cl

(b) $BrCH_2CH{=}CHCH_2Br$

(c) 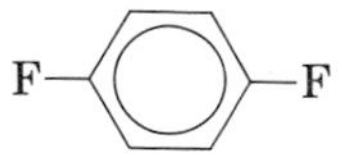 F—⬡—F

7.2. Draw the following compounds:
(a) carbon tetrabromide   **(b)** methylene bromide   **(c)** iodoform
(d) vinyl bromide   **(e)** *p*-nitrobenzylchloride   **(f)** isopropyl iodide

PREPARATIONS OF ORGANIC HALOGEN COMPOUNDS

7.2

A. Free-Radical Halogenation of Alkanes (section 3.6)

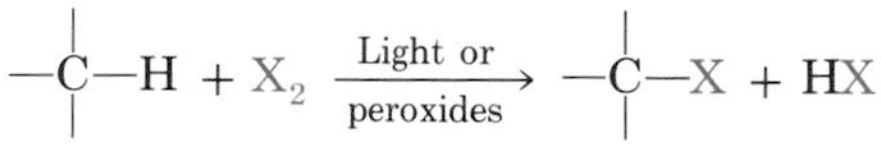

$$-\overset{|}{\underset{|}{C}}-H + X_2 \xrightarrow[\text{peroxides}]{\text{Light or}} -\overset{|}{\underset{|}{C}}-X + HX$$

$$X_2 = Cl_2,\ Br_2$$

B. Addition to Alkenes and Alkynes (section 4.1)

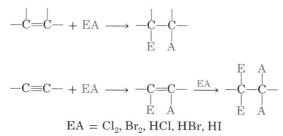

$$EA = Cl_2,\ Br_2,\ HCl,\ HBr,\ HI$$

C. Electrophilic Aromatic Substitution (section 5.6)

$$X = Cl,\ Br$$

D. Conversion of Alcohols to Alkyl Halides (section 8.6.C)

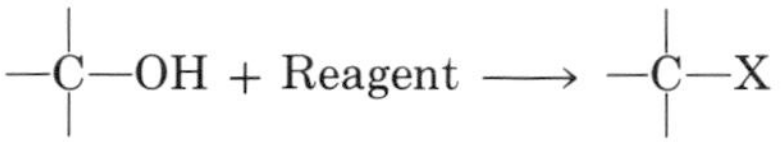

$$-\overset{|}{\underset{|}{C}}-OH + \text{Reagent} \longrightarrow -\overset{|}{\underset{|}{C}}-X$$

$$\text{Reagent} = HCl,\ HBr,\ HI,\ SOCl_2,\ PCl_3,\ PBr_3$$

USES AND OCCURRENCES OF ORGANIC HALOGEN COMPOUNDS

7.3

A. Dry-Cleaning Agents

Dust and dirt cling to body oils, cooking fats and oils, and other organic materials that have been adsorbed onto fabrics. The function of a dry-cleaning agent is to dissolve these fats and oils, thereby loosening the adhering dirt without causing the fabric to swell, shrink, or undergo any other significant alteration.

Chlorinated hydrocarbons have little structural effect on fabrics, are easily removed, and are relatively nonflammable. Trichloroethylene, and tetrachloroethylene especially, are the main present-day dry-cleaning solvents. Trichlorotrifluoroethane is also used but to a lesser extent.

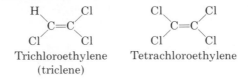

Trichloroethylene   Tetrachloroethylene
(triclene)

B. Addition Polymers

Some of the most important addition polymers (PVC, saran, Teflon, neoprene) are produced from small organic halide monomers (sections 4.2 and 4.3).

C. PCBs and PBBs

A once popular type of plasticizer, the polychlorinated biphenyls (PCB) are also examples of chlorinated hydrocarbons.

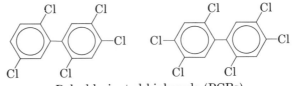

Polychlorinated biphenyls (PCBs)

These have been largely discontinued because they are fairly stable and tend to accumulate in the environment. Like DDT, they enter the food chain and have been found in fish and birds.

In 1973, polybrominated biphenyl (PBB), an industrial flame retardant, was accidentally mixed with livestock feed in Michigan, which led to one of the most widespread and widely publicized cases of chemical contamination in United States history. Because the error was not immediately recognized, farm products from animals exposed to PBB were marketed and reached a large number of Michigan's citizens.

D. Cough Medicines

Chloroform, the handkerchief anesthetic in many old movies, at one time could be found in over 2000 products. More than 80% of its use was as an expectorant and flavoring in cough medicines. On July 8, 1976, the Federal Food and Drug Administration banned the use of chloroform in drug products because of evidence that it was carcinogenic in mice and rats. The Environmental Protection Agency has proposed filtration of municipal water supplies to remove chloroform and other organic halogen compounds. In general, many chlorinated hydrocarbons are becoming suspect as carcinogens. In using these products (paint removers, dry cleaners, liquid drain cleaners, and so on), one should exercise caution.

E. Chlorofluorocarbons—Freons

Freons are small gaseous organic molecules containing carbon, chlorine, fluorine, and sometimes hydrogen. They were developed in the early 1930s by chemists and engineers searching for a new refrigerant to replace the toxic and corrosive ammonia and sulfur dioxide then in use. They became very widely used, especially freon 11 and freon 12, primarily as propellants in aerosol cans for dispensing deodorants, hair sprays, cooking oils, window cleaners, insecticides, paints, shaving cream, and whipped cream, and as refrigerants for freezers, refrigerators, and air conditioning units.

$$CCl_3F \qquad\qquad CCl_2F_2$$
$$\text{Freon 11 (bp 24 °C)} \qquad \text{Freon 12 (bp } -30\text{ °C)}$$

For some time now, freons have been the subject of heated debate that has only recently subsided (not necessarily resolved) due to the orderly phasing out of nonessential use of chlorofluorocarbon propellants. The focus of the controversy lies in the middle of the stratosphere, 15 miles above the earth's surface. Here an ozone shield about 20 miles thick (8–30 miles in altitude) surrounds the earth and absorbs certain levels of ultraviolet radiation. High intensity ultraviolet radiation is deleterious to all forms of life, and it is theorized that surface life on earth did not evolve until after the ozone layer was formed. Even small increases in the amount of UV radiation reaching the earth's surface could cause significant increases in the incidence of skin cancer, growth retardation in some plants, and possibly genetic mutation, interference with plankton growth in the ocean, and climatological changes.

Because of their stability, chlorofluorocarbons are not readily biodegraded or destroyed chemically on the earth's surface. Instead, they slowly diffuse toward the upper atmosphere where they can destroy ozone by a free-radical chain process described in Figure 7.1 (compare to the free-radical chain processes previously discussed in sections 3.6.C, 4.2.A.2, and 4.3.A.2).

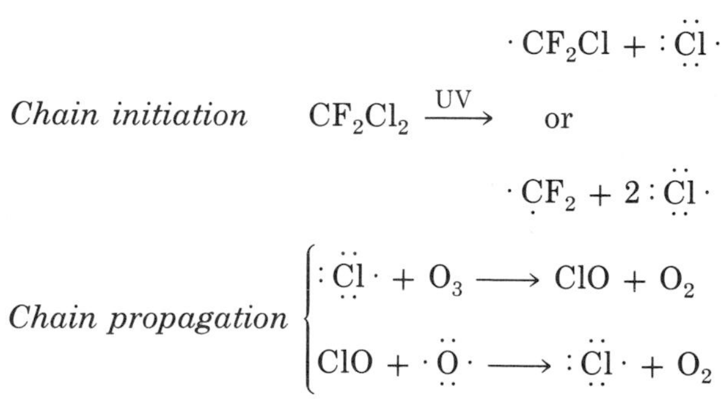

$$\cdot CF_2Cl + :\overset{..}{\underset{..}{Cl}}\cdot$$

Chain initiation $CF_2Cl_2 \xrightarrow{\ UV\ }$ or

$$\cdot CF_2 + 2:\overset{..}{\underset{..}{Cl}}\cdot$$

Chain propagation $\begin{cases} :\overset{..}{\underset{..}{Cl}}\cdot + O_3 \longrightarrow ClO + O_2 \\[6pt] ClO + \cdot\overset{..}{\underset{..}{O}}\cdot \longrightarrow :\overset{..}{\underset{..}{Cl}}\cdot + O_2 \end{cases}$

FIGURE 7.1. Destruction of the ozone layer by a free-radical chain reaction illustrated using freon 12 (CF_2Cl_2).

Ozone molecules in the stratosphere naturally absorb UV radiation and photodissociate. When freon molecules reach the stratosphere, they also photo-

$$O_3 \xrightarrow{\ UV\ } O_2 + O$$

dissociate under the influence of ultraviolet light and form chlorine atoms as illustrated in the chain initiation step for freon 12 (Figure 7.1). Chlorine atoms react with ozone to form chlorine monoxide and oxygen (first propagation step in Figure 7.1), resulting in the depletion of the ozone shield. This would not be too serious were it not for the subsequent reaction of chlorine monoxide with oxygen atoms formed by the natural photodissociation of ozone. This step, the second propagation step, regenerates chlorine atoms that can destroy more ozone. With each chlorine atom formed in chain initiation, thousands of ozone molecules are destroyed.

The earth's ozone shield is also susceptible to damage by nitrogen oxides. These are formed by the combination of nitrogen and oxygen at the high temperatures associated not only with the internal combustion engine but also with supersonic transports flying in the stratosphere, launches of space vehicles, nuclear explosions, and the ever-increasing use of nitrogen fertilizers. Nitrogen oxide causes ozone destruction through a chain reaction similar to that involving chlorine atoms.

$$NO + O_3 \longrightarrow NO_2 + O_2$$
$$NO_2 + O \longrightarrow NO + O_2$$

NUCLEOPHILIC SUBSTITUTION

7.4

A. General Reaction

One of the simplest and most thoroughly studied reactions in organic chemistry is nucleophilic substitution. Although this reaction can take many forms, we will limit our discussion to the reaction of alkyl halides with negative nucleophiles.

Nucleophilic substitution:
$$-\overset{|}{\underset{|}{C}}-\overset{\delta^+ \ \delta^-}{X} + \overset{+}{Na}\overset{-}{Nu} \longrightarrow -\overset{|}{\underset{|}{C}}-Nu + \overset{+}{Na}\overset{-}{X}$$

$$X = Cl, \ Br, \ I \quad \text{and} \quad \overset{-}{Nu} = OH^-, \ SH^-, \ NH_2^-, \ CN^-$$
$$OR^-, \ SR^-, \ NHR^-, \ \overset{-}{C}\equiv CR$$
$$NR_2^-$$

The reaction is a simple substitution reaction. The negative nucleophile Nu^- is attracted to the positive carbon of the polar carbon-halogen bond where it eventually replaces the halide ion. A nucleophile (nucleophile means nucleus loving) essentially is a Lewis base and thus has an unshared electron pair available for bonding. The unshared electron pair on the nucleophile is used in the new carbon-nucleophile bond, and the halide leaves with the electron pair of the carbon-halogen bond. The leaving group is usually less nucleophilic than the nucleophile. The Williamson synthesis of ethers, in which the salt of an alcohol is combined with an alkyl halide, is an example of nucleophilic substitution (in the following example the general anesthetic diethyl ether is prepared).

$$CH_3CH_2Cl + NaOCH_2CH_3 \longrightarrow CH_3CH_2OCH_2CH_3 + NaCl$$

Alkynes are prepared from the sodium salts of terminal alkynes (section 4.6) by nucleophilic substitution as in the preparation of·2-pentyne.

$$CH_3CH_2C\equiv CNa + CH_3Br \longrightarrow CH_3CH_2C\equiv CCH_3 + NaBr$$

PROBLEM

7.3. Before we consider the mechanism of nucleophilic substitution, test your comprehension of the general reaction by writing chemical equations illustrating the reaction of methyl iodide (CH_3I) with each of the following: **(a)** NaOH; **(b)** $NaOCH_2CH_2CH_3$; **(c)** NaSH; **(d)** $NaSCH_3$; **(e)** $NaNH_2$; **(f)** $NaNHCH_2CH_3$; **(g)** $NaN(CH_3)_2$; **(h)** NaCN; **(i)** $NaC\equiv CCH_3$.

B. Reaction Mechanisms, S_N1 and S_N2

How does nucleophilic substitution occur from a mechanistic standpoint? A logical consideration of the question might produce three possibilities.

1. The nucleophile might enter and bond, the halide ion then leaving.
2. The halide ion might leave, followed by the entrance and bonding of the nucleophile.
3. The nucleophile might attack and bond at exactly the same time the halide is leaving.

We can eliminate the first path by realizing that carbon forms no more than four bonds. If the nucleophile bonded to the carbon before the halogen left, this carbon would be pentavalent for that period.

The third alternative is very closely related to the first one. In this mechanism, the formation of the carbon-nucleophile bond and the cleavage of the carbon-halogen bond occur simultaneously. Since the nucleophile enters as the halide leaves, it attacks the carbon from the side opposite to that from which the halide is leaving. This is sterically favorable in that the nucleophile and halide do not hinder each other's movement.

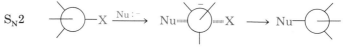

Transition state

This mechanism is called an S_N2 mechanism, which stands for *substitution nucleophilic bimolecular* ("bimolecular" since both reacting species are participating simultaneously in the reaction). Although it appears to be two steps, the mechanism in reality is a concerted, one-step process. The transition state drawing is merely an illustration to conceptualize what is happening.

In the remaining mechanistic possibility (number 2), the halide ion leaves and then the nucleophile enters and bonds. The halogen departs as a negative ion leaving a positive carbon; that is, a carbocation is formed.

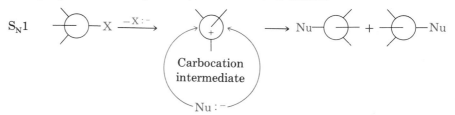

Since the halide has already left, the nucleophile can approach the planar carbocation from either side. Its attack is not restricted, in the way that it was restricted in the S_N2 process. This is a two-step mechanism with a carbocation intermediate of a short but finite lifetime. It is called an S_N1 mechanism, referring to *substitution nucleophilic unimolecular*. "Unimolecular," because in the first, rate-determining step only one of the reacting species is involved, the alkyl halide; the nucleophile doesn't come into play until the second step.

The concept of a reaction intermediate differs from that of a transition state. A transition state is a dynamic ever-changing process, whereas an intermediate is a theoretically isolatable entity, the result of a transition. For example, in the S_N1 mechanism, the process of cleavage of the carbon-halogen bond is the transition state and the carbocation resulting from the complete cleavage is the intermediate. Because an S_N2 reaction occurs in one step, there is no intermediate but only a transition state described by the progressive cleavage of the carbon-halogen bond and the progressive formation of the carbon-nucleophile bond. The transition is over when one bond is completely broken and the other completely formed.

The mechanisms of nucleophilic substitution are similar in concept to a classroom with front and back doors and full of students. Imagine that it is time to change classes. How can the students present be moved out and the new class moved in? As one possibility, the new class could come in and sit down and the other class could then get up and leave. Although a logical suggestion, this would be operationally impossible since the classroom can accommodate only one group of students at a time. For a similar reason, we eliminated our first idea for a nucleophilic substitution mechanism—that in which the nucleophile entered and bonded and then the halide left. If the new class entered from one door and the leaving class left through the other door, there would be a smooth transition. This is analogous to an S_N2 mechanism, in which the nucleophile enters at the same time the halide is leaving but from the opposite side. Finally, if the present class left the room first, then the new class could enter easily using either door. Similarly, in an S_N1 mechanism, if the halide ion leaves completely then the incoming nucleophile can attack the carbocation freely.

PROBLEM

7.4. Write stepwise reaction mechanisms for the following reactions: **(a)** S_N2 reaction between bromomethane and NaOH; **(b)** S_N1 reaction between 2-chloro-2-methylpropane and NaOH solution.

C. Characteristics of S_N1 and S_N2 Reactions

1. Reaction Rates. An S_N2 reaction is a one-step process in which both reacting species are involved and thus the reaction rate depends on the concentrations of both the alkyl halide (RX) and the nucleophile in solution. The rate equals a rate constant k times the concentrations of the two reacting species. Hence the reaction is said to be *bimolecular*.

$$\text{Rate}_{S_N2} = k[\text{RX}][\text{Nu}^-]$$

Increasing the concentration of either species will increase the reaction rate.

An S_N1 mechanism, on the other hand, is a two-step process involving a carbocation intermediate. As one might expect, formation of a carbocation from the neutral alkyl halide is the slow step, which is followed by the very rapid neutralization of the carbocation by the nucleophile. The rate of a chemical reaction depends only on the rate of the slow step and not at all on the faster steps. This is not difficult to comprehend. For example, the rate at which sand falls in an hourglass depends entirely on how long it takes the individual grains of sand to reach and pass through the orifice. It is independent of the time required for them to fall from the orifice to the next chamber. While a particle of sand is falling to the next chamber, other particles are making their way to the orifice, so that both processes are occurring at the same time. The rate will necessarily depend on the slower process. In an S_N1 process, carbocation formation is very slow. Once the ion is formed, it is quickly neutralized. Since carbocation formation involves only the alkyl halide (and not the nucleophile), the reaction rate depends solely on the concentration of the alkyl halide. The reaction is therefore termed *unimolecular.*

$$\text{Rate}_{S_N1} = k[\text{RX}]$$

Doubling the concentration of RX doubles the reaction rate, but doubling, tripling, or even quadrupling the concentration of the nucleophile has no effect whatsoever.

2. Stereochemistry. In an S_N2 reaction, the nucleophile enters as the halide is leaving. Consequently, the nucleophile attacks from the rear to avoid the path of the leaving halide. If the alkyl halide is optically active, inversion of configuration will occur. As the nucleophile begins to bond and the halide leaves, the groups attached to the carbon move from one side to the other much like an umbrella blowing inside out in a strong wind. The substitution product is still optically active but of opposite configuration to the alkyl halide.

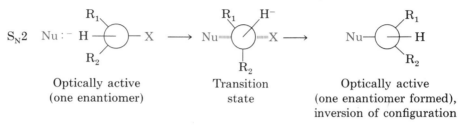

In an S_N1 reaction, the halide leaves, initially forming a planar carbocation that can be neutralized from either side. Reaction of an optically active alkyl halide will result in the formation of an optically inactive racemic mixture. Since the intermediate carbocation can be attacked from either direction, both retention and inversion of configuration will occur and a pair of enantiomers will form in equal amounts.

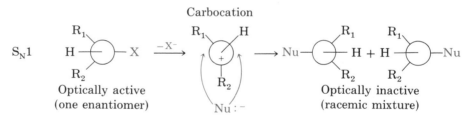

PROBLEM
_____

7.5. Using three-dimensional drawings to clearly show stereochemistry, illustrate the reaction of an optically active form of 2-chlorobutane with sodium hydroxide solution by both an S_N1 and an S_N2 mechanism.

_____

D. Factors Influencing the Reaction Mechanism—S_N2 vs S_N1

We have seen that two mechanisms are possible for nucleophilic substitution. What determines which mechanism will be operative under specific conditions? The following are some of the factors that should be considered.

1. Carbocation Stability. An S_N1 reaction involves an intermediate carbocation; an S_N2 reaction does not. Alkyl halides, which on ionization form stable carbocations, will most likely react by an S_N1 mechanism, whereas those that do not will react by an S_N2 mechanism, for which carbocation formation is unnecessary. We have previously explained the following order of carbocation stability (section 4.1.E.1):

$$\text{\textit{Carbocation stability}} \quad 3° > 2° > 1° > \overset{+}{C}H_3$$

We should expect then that the propensity for an S_N1 mechanism to occur would increase in the direction from primary to tertiary halides, since tertiary halides will form very stable tertiary carbocations.

	R_3CX	R_2CHX	RCH_2X	CH_3X
Alkyl	3°	2°	1°	
halides	S_N1	Mixed	S_N2	S_N2
		S_N1 and S_N2		

2. Steric Effects. In an S_N2 reaction, a nucleophile attacks a saturated carbon and pushes out a halide ion. For a brief period, five groups are coordinated around a single carbon, a relatively crowded condition. In an S_N1 reaction, however, a tetravalent carbon loses a halide ion, forming a trivalent carbocation, a less crowded condition. The bigger the groups around the carbon-halogen bond, the greater the difficulty the nucleophile has in reaching the carbon and displacing the halogen. It follows that an S_N1 reaction, in which steric crowding is minimized, would be preferred. Since alkyl groups are larger than hydrogen atoms, steric crowding increases in the direction from primary to tertiary alkyl halides, and the likelihood of an S_N1 occurring also increases.

Alkyl halides

Crowding increases, S_N1 increases ⟵

3°   2°   1° CH_3X

⟶ Crowding decreases, S_N2 increases

This concept is illustrated in Figure 7.2.

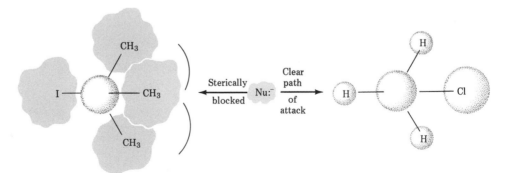

FIGURE 7.2. A nucleophile Nu:⁻ can attack methyl chloride easily; tertiary butyl iodide is sterically very crowded, however, making nucleophilic attack by an S_N2 mechanism difficult.

3. Solvent. An S_N1 reaction proceeds via an ionic mechanism, whereas an S_N2 mechanism is relatively nonionic. Polar solvents (such as water or alcohol) that can aid in forming and stabilizing an intermediate carbocation favor the S_N1 process, whereas nonpolar solvents (such as hexane) favor the S_N2 mechanism.

4. Strength of Nucleophile. Very strong, basic nucleophiles react aggressively, attacking the alkyl halide and pushing out the halide ion. Strong nucleophiles such as $^-OCH_2CH_3$ and NH_2^- promote S_N2 mechanisms. Weak neutral nucleophiles such as water and alcohol are less aggressive on the other hand and wait to be invited in by the carbocation intermediate of the S_N1 mechanism.

ELIMINATION REACTIONS OF ALKYL HALIDES

7.5

A. General Reaction

Alkyl halides react with nucleophiles to form substitution products. But alkyl halides also react with nucleophiles, in particular potassium hydroxide in alcohol, to form alkenes (section 3.7.A, dehydrohalogenation). Hydroxide ion is a nucleophile and can either displace a halide ion to form an alcohol or effect elimination to form an alkene.

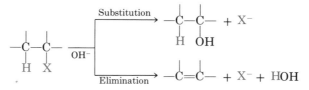

In fact, competition between elimination and substitution is possible when any alkyl halide capable of undergoing elimination is treated with any nucleophile.

B. Elimination Mechanisms—E_1 and E_2

Elimination not only is a competing reaction but is closely related in reaction mechanism to nucleophilic substitution. The reaction can occur by either an E_1 or an E_2 mechanism.

The E_1 mechanism begins exactly like an S_N1—the alkyl halide ionizes to a carbocation. The carbocation, however, instead of being neutralized by the nucleophile ($Nu:^-$), eliminates a proton to form an alkene.

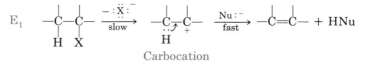

Carbocation

Since the first step is the slow one and only the alkyl halide is involved in this step, the reaction rate depends on the concentration of the alkyl halide.

$$Rate_{E_1} = k[RX]$$

Like an S_N1 reaction, an E_1 reaction is unimolecular—E_1 stands for *elimination unimolecular.*

The E_2 mechanism is a concerted one-step process, as the S_N2 mechanism is. The nucleophile does not directly displace the halogen, however, but attacks and abstracts a hydrogen on an adjacent carbon. The halide ion leaves, simultaneously generating a double bond.

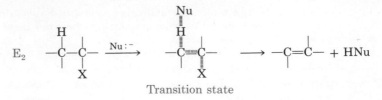

Transition state

Because the reaction occurs in one step, the rate depends on the concentrations of both reacting species.

$$\text{Rate}_{E_2} = k[\text{RX}][\text{Nu}^-]$$

The symbol E_2 stands for *elimination bimolecular*. Figure 7.3 summarizes the E and S_N mechanisms and illustrates their competing nature.

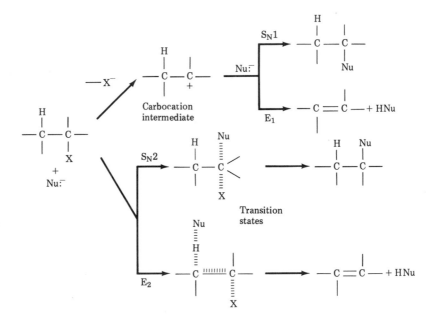

FIGURE 7.3. Competition between S_N1, S_N2, E_1, and E_2 reactions of alkyl halides with nucleophiles.

PROBLEM

7.6. Write the reaction of 2-bromopropane with potassium hydroxide by an E_1 and an E_2 mechanism.

C. Substitution versus Elimination

When alkyl halides are treated with a nucleophile, they can undergo either substitution or elimination. The most important factor in determining which will occur is the stability of the alkene that could be formed by elimination should it predominate. The more substituted an alkene is with alkyl groups, the more stable it is (section 3.8.C). Since a tertiary halide is more substituted with alkyl groups than a primary halide, it will usually form a more highly substituted alkene. Thus, tertiary halides have a greater tendency toward elimination since the alkenes they form are highly stable.

Alkyl halides

Elimination increases ←
3°   2°   1°
Substitution increases →

To illustrate this concept in a practical manner, let us consider the preparation of 2-methoxy-2-methylpropane by the Williamson synthesis of ethers. There are two approaches. One can combine tertiary butyl chloride with sodium methoxide.

$$
\underset{\underset{Cl}{|}}{\overset{\overset{CH_3}{|}}{CH_3CCH_3}} + NaOCH_3 \longrightarrow \underset{\underset{OCH_3}{|}}{\overset{\overset{CH_3}{|}}{CH_3CCH_3}} \quad \text{or} \quad \overset{\overset{CH_3}{|}}{CH_3C{=}CH_2}
$$

Substitution product   Elimination product (predominates)

Since *t*-butyl chloride is a tertiary halide, elimination will preferentially occur and little of the desired product will form. However, if we combine the sodium salt of *t*-butyl alcohol with methyl chloride, which has only one carbon, elimination is impossible and the desired substitution product forms exclusively.

$$
\underset{\underset{CH_3}{|}}{\overset{\overset{CH_3}{|}}{CH_3CONa}} + CH_3Cl \longrightarrow \underset{\underset{CH_3}{|}}{\overset{\overset{CH_3}{|}}{CH_3COCH_3}} + NaCl
$$

2-methoxy-2-methylpropane

PROBLEM

7.7. Write an equation showing the best way to prepare $(CH_3)_2CHOCH_2CH_3$ by the Williamson synthesis.

Panel 10

INSECTICIDES

Insects have been on earth for more than 400 million years. Of the 5 million species fewer than 20% have been identified. In his book *Six-Legged Science,* Brian Hoching estimates the world's insect population to be at least 10^{18} creatures, with a combined body weight 12 times that of the earth's human inhabitants. According to the World Health Organization, insects destroy or consume more than a third of the eatable agricultural products grown by man world-wide. In addition, they damage livestock, timber, and fiber crops and carry deadly diseases like malaria, yellow fever, and encephalitis. The world's human inhabitants have tried to reduce harmful insect activities with many types of biological control, such as crop rotation, selective breeding of insect-resistant plants, radiation-

[PANEL 10] induced sterilization of male insects, bacterial or viral disease infestation of insect populations, insect pheromones (see Panel 12 in Chapter 8), and the use of predator insects. Chemical warfare with insecticides, however, seems to be crucial to saving enough crops to feed the world. The organochlorine insecticides were among the first to be manufactured and utilized on a large scale.

A. Organochlorine Insecticides

In 1874, a German chemist named Othmar Zeidler synthesized dichlorodiphenyltrichloroethane, today known as DDT. Nearly 60 years passed, however,

Dichlorodiphenyltrichloroethane (DDT)

before the practical utility of this substance was discovered by Paul Mueller, a chemist with the Swiss firm J. R. Geigy. The insecticide was used successfully to combat typhus epidemics in Europe and malaria epidemics in the South Pacific, saving many lives. Paul Mueller was awarded a Nobel Prize in 1948.

Because it was an inexpensive and easily produced insecticide, the use of DDT became widespread. But DDT is a relatively stable material and persists in the environment. It is transported globally by air currents and returned to the earth by rainfall. Since DDT is fat-soluble, it is absorbed into the fatty tissue of living organisms and becomes concentrated as higher links are reached in a particular food chain; a predator species absorbs and further concentrates a large portion of the DDT found in its victim. Such concentrations in some birds cause them to lay eggs with thin shells or sometimes just a membrane. These eggs tend to break easily during brooding and as a result some bird populations are sharply reduced. There is also evidence that DDT produces cancerous tumors in laboratory rats and mice. On the basis that it could be carcinogenic for humans, the Environmental Protection Agency, in 1972, placed a nearly total ban on the domestic use of DDT in the United States.

Many other chlorinated hydrocarbons show insecticidal properties (Figure 7.4) and the possible toxicity of these compounds is a general property worth remembering. Aldrin, dieldrin, and heptachlor are effective against soil insects. Toxaphene is a heavily used pesticide against insects attacking livestock, cotton, soybean crops, grains, and vegetables. Chlordane was used principally for household or institutional insect control particularly against ants, termites, and lawn pests. In October, 1974, the Environmental Protection Agency ordered a halt to the sale of dieldrin and aldrin for all but a few uses, and in July, 1975, similar restrictions were put on chlordane and heptachlor. Again, the justification was that these substances caused cancerous tumors in laboratory animals. Organophosphorus and carbamate insecticides become increasingly important as the use of organochlorine insecticides diminishes. But research into insect control must actively continue to develop safe, environmentally compatible insecticides and insect control methods.

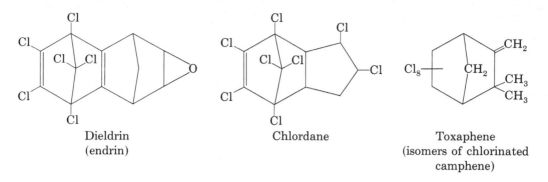

Dieldrin
(endrin)

Chlordane

Toxaphene
(isomers of chlorinated
camphene)

FIGURE 7.4.   Some organochlorine insecticides.

[PANEL 10]   **B. Organophosphorus Insecticides**

Organophosphorus insecticides developed as an outgrowth of World War II research on organophosphorus compounds as nerve gases. Biologically, they act by inhibiting cholinesterases, enzymes found in nerve cells and the brain, which are indispensable to the transmission of impulses in nerves. These insecticides disrupt the nerve impulses, resulting in fatality. Shown here are two common organophosphorus insecticides.

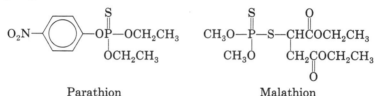

Parathion

Malathion

The organophosphorus compounds have some distinct advantages over the organochlorines. They break down rapidly and do not accumulate in the environment. Likewise, they do not concentrate in organisms. They are more specific than organochlorines, occasionally making it possible to selectively eliminate harmful insects. However, they are more expensive, and since they degrade rapidly, have to be applied more frequently. Also they must be handled with great care as they are more directly toxic to humans.

C. Carbamates

Like organophosphorus insecticides, carbamates are cholinesterase inhibitors. In general, they are safer to handle, more expensive, and effective against some pests that are not susceptible to organophosphorus compounds. Some common examples of carbamate insecticides are shown.

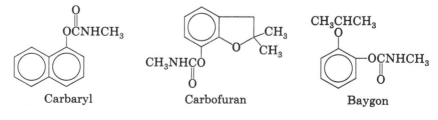

Carbaryl

Carbofuran

Baygon

[PANEL 10] Carbaryl was the first commercial carbamate introduced on a large scale. It is widely used to control insects on economically important crops such as cotton, vegetables, fruits, rice, and sugar cane. Baygon is used to combat mosquitoes, flies, ants, cockroaches, earwigs, spiders, and the like.

D. Pyrethrins

Examples exist in every major class of living organism of species which in nature poison other species. Pyrethrin insecticides are extracts of pyrethrum flowers (*C. cinerariaefolium*). The pyrethrins seem to be gaining in importance as pesticides owing to their rapid action as a contact insecticide and their low mammalian toxicity. Currently, the major use is as a constituent of sprays for flying insects such as flies and mosquitoes.

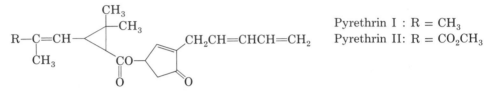

Pyrethrin I : R = CH_3
Pyrethrin II: R = CO_2CH_3

E. Insect Juvenile Hormone

Insect juvenile hormones (naturally secreted by glands in the head of the insect) produced synthetically could become a new generation of chemicals for pest control. To understand their potential benefit, let us trace the development of an insect, which normally begins with an egg and proceeds to larva (immature insect form hatched from egg; quite different from adult form), pupa (intermediate development stage, more similar to adult), and finally adult. Hormones secreted by internal glands control these various stages. A juvenile hormone specific for each kind of insect must be present at the larval stage but absent during the other growth phases. If the hormone is applied to the eggs, the ensuing larvae die, or in some cases the eggs may fail to hatch. If the hormone is present during the pupal stage, metamorphosis into a mature adult will not occur. Applying the juvenile hormone (or a synthetic mimic) to an insect-infested area during the egg or pupal stage of reproduction will prevent the eventual formation of a new generation of mature adults. Such pesticides are very selective, killing only certain insects without endangering other organisms. Following are examples of juvenile hormones for some butterflies and moths (there may be minor variations in structure between species).

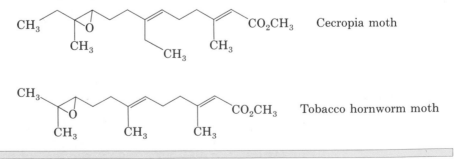

Cecropia moth

Tobacco hornworm moth

END-OF-CHAPTER PROBLEMS

7.8 IUPAC Nomenclature: Name the following compounds:

(a) $CH_3CH_2CHCH_2CH_3$
        $\underset{|}{Br}$

(b) $CH_3C{\equiv}CCH_2CHCH_3$
                    $\underset{|}{I}$

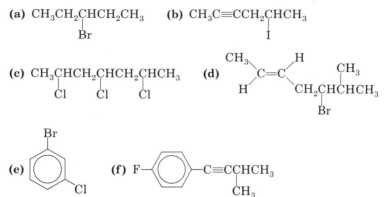

(c) $CH_3CHCH_2CHCH_2CHCH_3$
        $\underset{|}{Cl}$ $\underset{|}{Cl}$ $\underset{|}{Cl}$

(d)

7.9 Nomenclature: Following are some organohalogen compounds that are suspected to be dangerous to human health. Their names periodically appear in newspapers and magazines. Write structures for each.
(a) trichloromethane: chloroform; once in nonprescription cough medicines
(b) 1,2-dibromo-3-chloropropane: DBCP; an agricultural fumigant that diminished sperm count in chemical plant workers
(c) 1,2-dichloroethane: ethylene dichloride; used to make vinyl chloride from which PVC is made, a suspected carcinogen
(d) tetrachloromethane: carbon tetrachloride; prolonged exposure can cause liver and kidney damage or failure
(e) tetrachloroethene and trichloroethylene: dry-cleaning agents suspected of being carcinogens
(f) dichlorodifluoromethane: a freon that could lead to destruction of the ozone shield

7.10 Common Nomenclature: Draw the following compounds:
(a) methyl bromide   (b) methylene chloride
(c) bromoform   (d) carbon tetrafluoride
(e) allyl iodide   (f) vinyl chloride
(g) secondary butyl chloride   (h) isopropyl bromide

7.11 Preparations of Organic Halogen Compounds: Complete the following reactions showing the major organic products:

(a) $CH_4 + 4Cl_2 \xrightarrow{\text{Light}}$ (b) $CH_2{=}CH_2 + Br_2 \longrightarrow$

(c) $HC{\equiv}CH + HCl \longrightarrow$ (d) $CH_3C{\equiv}CH + 2HCl \longrightarrow$

(e) 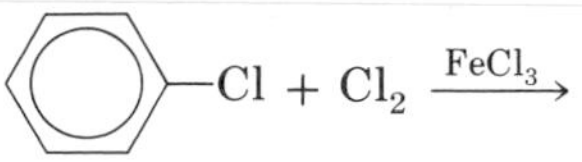 $\text{C}_6\text{H}_5{-}Cl + Cl_2 \xrightarrow{\text{FeCl}_3}$ (f) 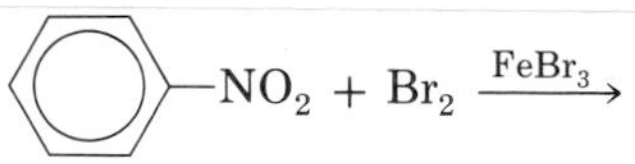 $\text{C}_6\text{H}_5{-}NO_2 + Br_2 \xrightarrow{\text{FeBr}_3}$

7.12 Nucleophilic Substitution: Complete the following reactions showing the major organic products:

(a) $CH_3CHCH_3 + NaOH \longrightarrow$
$|$
Cl

(b) 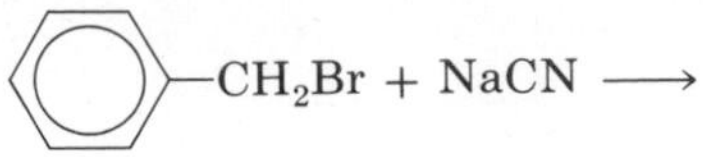 ⬡—$CH_2Br + NaCN \longrightarrow$

(c) $CH_3CH_2I + NaSH \longrightarrow$

(d) $CH_3CH_2CH_2Br + NaN(CH_3)_2 \longrightarrow$

CH_3
$|$
(e) $CH_3CHONa + CH_3CH_2I \longrightarrow$

CH_3
$|$
(f) $CH_3CHCH_2Br + CH_3SNa \longrightarrow$

(g) 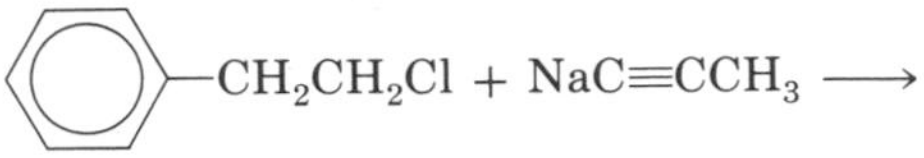 ⬡—$CH_2CH_2Cl + NaC\equiv CCH_3 \longrightarrow$

(h) $CH_3Cl + NaNH_2 \longrightarrow$

7.13 Williamson Synthesis of Ethers: Using as starting materials alkyl halides and sodium alkoxides, prepare the following ethers by the Williamson synthesis:
(a) $CH_3OCH_2CH_2CH_3$ (b) $CH_3CH_2OCH(CH_3)_2$

7.14 Nucleophilic Substitution in Preparing Alkynes: Prepare the following alkynes from ethyne (acetylene) and alkyl halides as organic starting materials (see sections 4.6 and 7.4.A):

(a) $CH_3CH_2C\equiv CH$ (b) $CH_3C\equiv CCH_2CH_3$ (c) 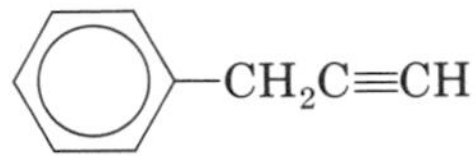 ⬡—$CH_2C\equiv CH$

7.15 Elimination Reactions: Complete the following reactions showing the major organic products:

(a) $CH_3CHCH_3 + KOH \longrightarrow$
$|$
Cl

(b) $CH_3CHCH_2CH_2CH_3 \xrightarrow{KOH}$
$|$
Br

CH_3
$|$
(c) $CH_3CHCHCH_2CH_3 + KOH \longrightarrow$
$|$
I

(d) $CH_3CH_2CHBr_2 + 2KOH \longrightarrow$

7.16 Reaction Mechanisms: Using the following reagents, write mechanisms for
(a) S_N1, (b) S_N2, (c) E_1, and (d) E_2 reactions.   For the S_N1 and S_N2 reactions, show stereochemistry.

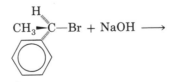

7.17 S_N1, S_N2 Mechanisms: Make a chart comparing an S_N1 and an S_N2 reaction of an alkyl halide and a nucleophile with regard to the following:
(a) rate expression
(b) reaction intermediates
(c) stereochemistry
(d) relative rates of reaction of 1°, 2°, and 3° halides
(e) effect of increasing concentration of nucleophile

(f) effect of increasing concentration of alkyl halide
(g) effect of an ionic or polar solvent
(h) effect of a nonpolar solvent
(i) effect of bulky groups around reaction center
(j) strength of nucleophile

7.18 Nucleophilic Substitution: The following compounds are responsible for the odor and flavor of garlic. Write a nucleophilic substitution reaction showing the preparation of each.
(a) $CH_2{=}CHCH_2SH$ **(b)** $CH_2{=}CHCH_2SCH_2CH{=}CH_2$

7.19 Structure and Use: Using words or a simple general structural unit, describe the following commercial compounds:
(a) dry-cleaning agents   **(b)** PCBs
(c) freons   **(d)** organochlorine insecticides
(e) organophosphorus insecticides   **(f)** carbamate insecticides

7.20 Geometric Isomerism: There are eight possible geometrical isomers of the insecticide benzene hexachloride (BHC) (1,2,3,4,5,6-hexachlorocyclohexane). Draw them.

7.21 E_1, E_2 Stereochemistry: E_2 eliminations occur by antielimination, in which the abstracted hydrogen and leaving halogen on the adjacent carbon are as far apart as possible (by C—C bond rotation). Because in E_1 reactions a carbocation is formed, such stereospecificity is not observed. Write the product or products of an E_2 elimination and an E_1 elimination of the following compound (shown in sawhorse diagram and Newman projection):

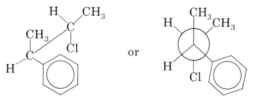

Chapter 8

Alcohols, Phenols, and Ethers

STRUCTURE AND NOMENCLATURE

8.1   Alcohols, phenols, and ethers can be thought of as derivatives of water in which one or both hydrogens are replaced with organic groups. Alcohols have only one hydrogen replaced by an alkyl group, while ethers have both replaced, and in phenols, one hydrogen is replaced by an aromatic ring (Table 8.1). Alcohols are further classified as primary, secondary, and tertiary according to the number of alkyl groups directly bonded to the alcohol carbon.

$$
\begin{array}{ccc}
\text{H} & \text{H} & \text{R} \\
| & | & | \\
\text{R}-\text{C}-\text{H} & \text{R}-\text{C}-\text{R} & \text{R}-\text{C}-\text{R} \\
| & | & | \\
\text{OH} & \text{OH} & \text{OH} \\
1° & 2° & 3° \\
\text{Primary} & \text{Secondary} & \text{Tertiary}
\end{array}
$$

TABLE 8.1.   WATER, ALCOHOLS, PHENOLS AND ETHERS

Water	H—O—H		
Alcohols	R—O—H	CH_3OH	CH_3CH_2OH
Phenols	Ar—OH	(phenol with OH)	(naphthol with OH)
Ethers	R—O—R	$CH_3CH_2OCH_2CH_3$	

A. IUPAC Nomenclature of Alcohols

1. Saturated Alcohols. The names of most simple organic compounds are based on the name of the longest continuous chain of carbon atoms. To name alcohols, the *-e* of the name of the parent hydrocarbon is replaced by *-ol*, the suffix ending for alcohols. The position of the alcohol function on the hydrocarbon chain is described by a number. The chain is numbered so as to give the alcohol group the lowest possible number.

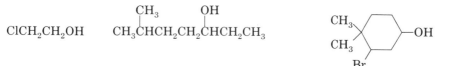

$$CH_3OH \qquad CH_3CH_2OH \qquad CH_3CH_2CH_2OH \qquad CH_3\overset{\overset{\displaystyle OH}{|}}{C}HCH_3$$

Methanol   Ethanol   1-Propanol   2-Propanol

Functional groups named by suffixes, such as *-ol* for alcohols, always get the lowest possible number over other groups on the chain named by prefixes even if a higher combination of numbers results. If the alcohol function is on carbon-1, its number can be omitted and understood to be one.

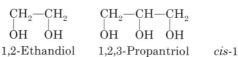

ClCH₂CH₂OH $CH_3\overset{\overset{\displaystyle CH_3}{|}}{C}HCH_2CH_2\overset{\overset{\displaystyle OH}{|}}{C}HCH_2CH_3$

2-Chloroethanol   6-Methyl-3-heptanol   3-Bromo-4,4-dimethylcyclohexanol
                            (not 2-methyl-5-heptanol)

Compounds containing two or more alcohol groups are named as *diols*, *triols*, etc.

$$\begin{matrix} CH_2{-}CH_2 \\ | \qquad | \\ OH \quad\; OH \end{matrix} \qquad \begin{matrix} CH_2{-}CH{-}CH_2 \\ | \qquad | \qquad | \\ OH \quad OH \quad OH \end{matrix}$$

1,2-Ethandiol   1,2,3-Propantriol   *cis*-1,3-Cyclopentandiol

PROBLEM

8.1. Name by the IUPAC system of nomenclature,

(a) CH₃CH₂CH₂CH₂OH   **(b)** $CH_3\overset{\overset{}{}}{C}HCH_2CH_3$ **(c)** $CH_3\overset{\overset{\displaystyle CH_3}{|}}{C}CH_2\overset{\overset{\displaystyle CH_3}{|}}{C}CH_2CH_3$
                                               OH   CH₃ OH

(d) HO—⬡—OH
                Br

2. Unsaturated Alcohols. To name unsaturated alcohols or alcohols in general,

1. use the Greek word for the number of carbons in the longest continuous chain.
2. follow this by the suffix *-an* if the chain is saturated, *-en* if it contains a carbon-carbon double bond, and *-yn* if it contains a carbon-carbon triple bond.
3. next put in the suffix *-ol* to designate the alcohol function.
4. number the chain, giving the lowest possible number to the alcohol group. Incorporate the appropriate numbers in the suffix as shown in the example below.
5. complete the name by naming all other groups with prefixes.

For example, let us name the following compound:

$$\underset{\text{OH}}{\overset{\overset{\overset{\displaystyle\text{CH}_3}{|}}{}}{\underset{6}{\text{CH}_2}=\underset{5}{\text{CH}}\underset{4}{\text{CH}_2}\underset{3}{\text{CH}}\underset{2}{\text{CH}}\underset{1}{\text{CH}_3}}}$$

1. The longest chain is five carbons, *hex*.
2. There is a carbon-carbon double bond, *hexen*.
3. The alcohol function is designated by the suffix *-ol*, *hexenol*.
4. Number the chain to give the alcohol group the lowest possible number. Incorporate these numbers in the suffix, 5-hexen-3-ol. The first number, 5, refers to the position of the double bond; the second, 3, locates the alcohol group.
5. Name all other substituents (2-methyl) with prefixes. The complete name is

2-methyl-5-hexen-3-ol

PROBLEM

8.2. Name by the IUPAC system of nomenclature,

(a) $\underset{\underset{\text{CH}_3}{|}}{\text{CH}_3\text{C}}=\text{CH}\underset{\underset{\text{OH}}{|}}{\text{CH}}\text{CH}_3$ **(b)** $\text{CH}_3\text{CH}_2\text{CH}_2\text{C}\equiv\text{CCH}_2\text{OH}$

(c) $\text{CH}_2=\text{CH}\underset{\underset{\text{OH}}{|}}{\text{CH}}\text{CH}=\text{CH}_2$

B. IUPAC Nomenclature of Ethers

To name an ether, first find the longest continuous chain of carbon atoms. The substituents attached to this chain can be pictured as alkyl groups containing an oxygen. For this reason, they are referred to as *alkoxy groups*. Just as CH_3- is a methyl group, $\text{CH}_3\text{O}-$ is a methoxy group.

CH_3CH_2-	ethyl	$\text{CH}_3\text{CH}_2\text{O}-$	ethoxy
$\text{CH}_3\text{CH}_2\text{CH}_2-$	propyl	$\text{CH}_3\text{CH}_2\text{CH}_2\text{O}-$	propoxy

These groups are named as prefixes and their positions designated by a number.

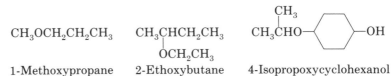

1-Methoxypropane
2-Ethoxybutane
4-Isopropoxycyclohexanol

PROBLEM

8.3. Name by the IUPAC system of nomenclature,

(a) $\text{CH}_3\text{CH}_2\text{CH}_2\text{OCH}_2(\text{CH}_2)_5\text{CH}_3$ **(b)** $(\text{CH}_3\text{O})_2\text{CH}_2$ **(c)** $\text{CH}_3\text{CH}_2\text{OCH}_2\text{CH}_2\text{OH}$

C. IUPAC Nomenclature of Phenols

Phenols are named by dropping the *e* from the parent aromatic hydrocarbon and adding the suffix *-ol*. Although it is the IUPAC method, phenols are seldom named in this manner. Some of the common names (such as phenol and naphthol) have been accepted into the IUPAC system (section 5.4).

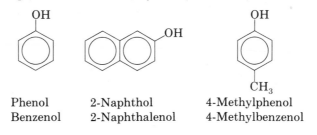

Phenol	2-Naphthol	4-Methylphenol
Benzenol	2-Naphthalenol	4-Methylbenzenol

PROBLEM

8.4. Name by the IUPAC system of nomenclature,

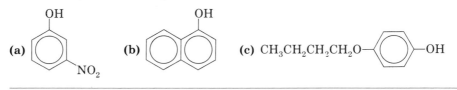

(a)   (b)   (c) $CH_3CH_2CH_2CH_2O$—⟨ ⟩—OH

D. Common Nomenclature of Alcohols and Ethers

Alcohols and ethers are frequently referred to by common names. In everyday terminology, the alkyl group or groups connected to the oxygen are named first, followed by the class of compound, alcohol or ether.

$$CH_3CH_2OH \qquad CH_3CHCH_3 \qquad CH_3CH_2OCH_3 \qquad CH_3CHOCHCH_3$$
$$ \underset{OH}{|} \overset{CH_3\;\;CH_3}{|\quad\;\;|}$$

Ethyl	Isopropyl	Ethyl methyl	Diisopropyl
alcohol	alcohol	ether	ether

PROBLEM

8.5. Draw the following compounds: **(a)** tertiary butyl alcohol; **(b)** pentyl alcohol; **(c)** diethylether; **(d)** ethyl cyclopentyl ether; **(e)** *meta*-nitrophenol; **(f)** 1-naphthol.

PREPARATIONS OF ALCOHOLS AND ETHERS

8.2

A. Hydration of Alkenes

1. Addition of Water to Alkenes (section 4.1.B.4)

$$-\overset{|}{C}=\overset{|}{C}- + H_2O \xrightarrow{H^+} -\overset{|}{\underset{OH}{C}}-\overset{|}{\underset{H}{C}}-$$

2. Hydroboration (section 4.4)

$$6 -\overset{|}{C}=\overset{|}{C}- + B_2H_6 \longrightarrow 2 \left(-\overset{|}{\underset{H}{C}}-\overset{|}{C}-\right)_3 B \xrightarrow[OH^-]{H_2O_2,} 6 -\overset{|}{\underset{H}{C}}-\overset{|}{\underset{OH}{C}}-$$

B. Nucleophilic Substitution (sections 7.4.A, 7.5.C)

$$-\underset{|}{\overset{|}{C}}-X + NaNu \longrightarrow -\underset{|}{\overset{|}{C}}-Nu + NaX$$

X = Cl, Br, I   Nu = OH for alcohol synthesis
                 = OR for ether synthesis (Williamson synthesis)

C. Reduction of Aldehydes and Ketones

1. Catalytic Hydrogenation (section 9.5.C)

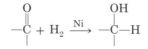

2. Reduction by Lithium Aluminum Hydride (section 9.5.D)

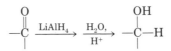

3. Grignard Synthesis of Alcohols (section 9.5.E)

$$R-X + Mg \xrightarrow{Ether} RMgX \qquad (Grignard\ reagent)$$

$$-\underset{|}{\overset{O}{\overset{\|}{C}}} + RMgX \longrightarrow -\underset{|}{\overset{OMgX}{\overset{|}{C}}}-R \xrightarrow[H^+]{H_2O,} -\underset{|}{\overset{OH}{\overset{|}{C}}}-R + MgXOH$$

PHYSICAL PROPERTIES—HYDROGEN BONDING

8.3   **A. Boiling Points**

The melting points and especially the boiling points of alcohols and ethers generally increase with increasing molecular weight within a homologous series (Table 8.2) as they do in other classes of organic compounds (section 3.4). However, alcohols exhibit unusually high boiling points. For example, compare the boiling points of the following pairs of compounds, each pair having approximately the same molecular weight:

	H_2O	CH_4	CH_3OH	CH_3CH_3	CH_3CH_2OH	$CH_3CH_2CH_3$
mol wt	18	16	32	30	46	44
bp °C	100	−164	65	−89	78.5	−42

Why does water boil higher than methane by 264°, methanol higher than ethane by 154°, and ethanol higher than propane by 120°? The answer cannot be higher molecular weight because the molecular weights of each pair are almost equivalent. In fact, water boils higher than propane by 142° even though propane has a molecular weight 2.5 times that of water. To explain these facts, let us review boiling on a molecular scale (section 3.4.A). Recall that in the liquid phase, the molecules of a compound are in continuous motion, but attractive forces between the molecules restrict their movement within a certain volume (Figure 3.1). In the gas phase, the molecules are in constant, random motion with no intermolecu-

TABLE 8.2. PHYSICAL PROPERTIES OF ALCOHOLS, PHENOLS, AND ETHERS

Compound	Molecular Weight	Melting Point, °C	Boiling Point, °C
Alcohols			
CH_3OH	32	− 94	65
CH_3CH_2OH	46	−117	78.5
$CH_3CH_2CH_2OH$	60	−127	97
Ethers			
CH_3OCH_3	46	−139	− 23
$CH_3OCH_2CH_3$	60	—	11
$CH_3O(CH_2)_2CH_3$	74	—	39
Phenols			
⬡—OH	94	43	182

lar attractive forces. Each molecule is independent of the others. To convert a liquid to a gas, one must provide enough energy to overcome the intermolecular forces of attraction. The stronger the attractions between molecules, the greater the energy required to break them and the higher the boiling point.

Now apply these principles to the example. Methane, ethane, and propane are composed of nonpolar carbon-carbon and carbon-hydrogen bonds. As such, the existing attractions between molecules are very weak. In water, methanol, and ethanol, however, there are very polar oxygen-hydrogen bonds ($^\delta{-}O—H^{\delta+}$). Because of this polarization, electrostatic attractions exist between these molecules (specifically between the hydrogen of one molecule and oxygen of another) that are not possible in nonpolar compounds. Further, due to the minute size of hydrogen, close intermolecular association is possible, providing maximum attractions (see Figure 8.1). This phenomenon is called *hydrogen-bonding* and occurs in molecules where hydrogen is bonded to a strongly electronegative element such as nitrogen, oxygen, or fluorine.

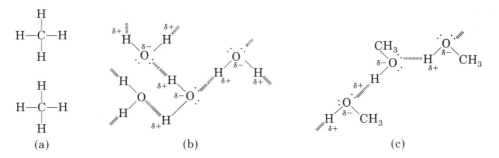

(a)   (b)   (c)

FIGURE 8.1. (a) Methane is a nonpolar compound with only weak intermolecular attractions. Consequently, it has a very low boiling point. (b) Water has strong attractions between molecules owing to its capacity to hydrogen-bond and thus has a relatively high boiling point. (c) Methanol and other alcohols can hydrogen-bond much like water and as a result have relatively high boiling points.

Since ethers have no hydrogen on the oxygen, hydrogen-bonding does not occur, and they have considerably lower boiling points than alcohols of identical molecular weights.

	CH_3CH_2OH	CH_3OCH_3	$CH_3CH_2CH_2OH$	$CH_3OCH_2CH_3$
mol wt	46	46	60	60
bp °C	78.5	−23	97	11

The greater the capacity for hydrogen-bonding, the higher the boiling point as illustrated by the following compounds of similar molecular weights (analyze why the capacity for hydrogen-bonding increases in these examples):

	$CH_3CH_2CH_2CH_3$	$CH_3CH_2CH_2OH$	$CH_3\overset{\displaystyle O}{\overset{\displaystyle \|}{C}}OH$	$HOCH_2CH_2OH$
mol wt	58	60	60	62
bp °C	−0.5	97	118	198

B. Viscosity

The viscosity (thickness, resistance to flow) of a liquid is influenced by hydrogen-bonding. Compare, for example, the hydrocarbon hexane, a gasoline component, with glycerol, a triol.

Hexane

mol wt = 86

bp   = 69 °C $CH_3CH_2CH_2CH_2CH_2CH_3$

Glycerol

mol wt = 92 $\underset{\text{OH OH OH}}{CH_2 CH CH_2}$

bp   = 290 °C

Glycerol is a thick syrupy liquid (very unlike gasoline) which is often used as a lubricant in laboratories.

C. Solubility

Alcohols of low molecular weight are water-soluble owing to their ability to hydrogen-bond with water (Figure 8.2). As the molecular weight increases, however, water-solubility decreases. The rule "like dissolves like" applies here. An alcohol has a waterlike portion (—OH) and a hydrocarbonlike portion (the alkyl group). As the molecular weight of an alcohol increases, the proportion of hydrocarbon to hydroxy increases. The alcohol becomes more like an alkane, less like water, and less soluble in water.

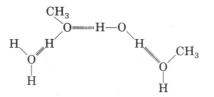

FIGURE 8.2. Solubility of methanol in water. Note the hydrogen-bonding between methanol and water.

PROBLEMS

8.6. Ethylene glycol (section 8.4.B.1) is a good radiator antifreeze because it has a high boiling point and is soluble in water in all proportions. These properties are due to hydrogen-bonding. Draw illustrations of ethylene glycol hydrogen-bonding with itself and with water in solution.

8.7. Which has the higher boiling point, butanoic acid ($CH_3CH_2CH_2\overset{\overset{O}{\|}}{C}OH$, odor of rancid butter) or ethyl acetate ($CH_3\overset{\overset{O}{\|}}{C}OCH_2CH_3$, solvent in fingernail polish remover)? Explain.

USES OF ALCOHOLS AND ETHERS

8.4

A. Alcohols

1. Methyl Alcohol. Methyl alcohol, methanol, is sometimes called wood alcohol because in the early part of this century it was produced by the destructive distillation of wood (such as beech, birch, hickory, maple, and oak). Today, it is synthesized by the reduction of carbon monoxide with hydrogen gas and a catalyst. It is also available from coal liquefaction by much the same process (see Panel 3 in Chapter 3). Methanol is converted into formaldehyde or used to synthesize other chemicals. Recently, it has received attention as a possible gasoline or gasoline blending agent and as a material which can be converted into conventional fuels synthetically.

Unlike ethyl alcohol, which can be ingested to a certain limit, methanol is highly toxic and its consumption can lead to blindness and possibly death.

2. Disinfectant Action of Ethyl and Isopropyl Alcohol. Ethyl and isopropyl alcohols are effective antiseptics, because they can either destroy microorganisms or at least inhibit their metabolic processes and reproductive capacities. To carry out such activities, ethyl alcohol is used in mouth washes (5%–30% concentration), aerosol disinfectants, and medicines. A 70% solution of ethyl alcohol in water is more effective than pure alcohol because it reduces the surface tension of bacteria cells better than pure ethanol, allowing more efficient penetration.

Isopropyl alcohol, common rubbing alcohol sold in drug stores, is an even more effective antiseptic than ethyl alcohol. Ethyl alcohol is the alcohol in alcoholic beverages.

Panel 11

BEVERAGE ALCOHOL

Ethyl alcohol is commonly referred to as *grain alcohol* or *beverage alcohol* because it is produced by the fermentation of the natural sugars (or sugar from

[PANEL 11] hydrolyzed starches) found in grapes and grains.* Enzymatic fermentation of sucrose is described by the following equations:

$$C_{12}H_{22}O_{11} + H_2O \xrightarrow{\text{Sucrase}} 2C_6H_{12}O_6$$

Sucrose   Glucose

$$C_6H_{12}O_6 \xrightarrow{\text{Zymase}} 2CH_3CH_2OH + 2CO_2$$

Glucose   Ethanol

A solution containing no more than 14% alcohol can be produced by fermentation because yeast cells die or cease reproduction at this concentration due to the aforementioned antiseptic action of alcohol. Ethyl alcohol is produced commercially by the hydration of ethylene.

$$CH_2{=}CH_2 + H_2O \xrightarrow{H_2SO_4} CH_3CH_2OH$$

The fermentation of agricultural wastes to produce ethanol for use in gasoline, called gasohol, is gaining in economic importance.

Alcoholic beverages can be divided into three categories—beers, wines, and spirits. All involve fermentation of sugars. In addition, beverages can acquire other alcohols, carboxylic acids, esters, aldehydes, and ketones through fermentation or absorption. These are called *congeners* and contribute to the taste, aroma, and color.

Beer is the fermentation product of barley and hops. The alcohol content varies from 2%–12%, although most beers are about 4%–5%. Typical American beer is 90% water, 5% carbohydrate, and 3.5% alcohol, and the remainder consists of CO_2, protein, and minerals.

Wines are fermented from the juice of grapes and are classified as natural or sparkling wines and fortified or aromatic wines. Natural table and beverage wines and also sparkling wines like champagne contain less than 14% alcohol. Sparkling wines are also highly carbonated. Fortified wines like sherry, and aromatic wines like vermouth contain about 15%–23% alcohol. Since fermentation cannot naturally produce this concentration, these beverages must be fortified by alcohol from other sources.

Distilled spirits include rums, whiskeys, gins, and vodkas. In the production of these beverages, a carbohydrate mash is allowed to ferment and is then distilled to produce beverages with a 40%–50% alcohol content. Congeners are also collected in the process, giving the beverages their characteristic flavors. Whiskey is made from a fermented mash of at least 51% corn, with the remainder largely barley. Rye whiskeys arise from mashes of at least 51% rye grain. Rum is distilled from the fermented juice of sugar cane or molasses. Gin and vodka are alcohol-water mixtures derived from grain mashes. Vodka is unflavored, but during distillation to produce gin, the alcohol-water vapors are passed over flavoring agents such as juniper berries or in the case of sloe gin, sloe berries (a type of wild plum). The alcohol content of distilled spirits is expressed in terms of "proof,"

---

*For many, fermentation refers to the natural production of alcohol. Yet fermentation is a much more universal process, encompassing any chemical change caused by living microorganisms acting on organic materials. It is one of the oldest chemical processes used by man. In addition to making alcoholic beverages, fermentation is responsible for the aging of meat and cheese, and also the production of bread, foods, animal feeds, drugs, antibiotics, hormones, and other materials. In 1857, when Louis Pasteur proved that alcoholic fermentation is caused by living cells (yeast), the ancient art of fermentation graduated from the realm of magic to the world of scientific understanding.

[PANEL 11] which originated from an old method of testing whiskey by pouring it on gunpowder. If the gunpowder ignited, this was "proof" that the beverage did not contain too much water. Alcohol concentration in terms of proof is double the percentage of alcohol by volume (a 100-proof vodka is 50% alcohol, 50% water by volume). It is illegal to drive with a blood alcohol concentration of 0.10% or higher. Individuals with 0.3% are visibly intoxicated; those with 0.4% are anesthetized and incapable of voluntary action. The concentration 0.5%–1% leads to coma and death.

About 80% of adult men and 67% of adult women in the United States drink. Approximately 4.5% of the adult population suffers from alcoholism. Intoxication is a factor in over half the fatal traffic accidents. Not only can an alcoholic cause mental anguish to his or her family but 30%–50% of the offspring of alcoholic mothers suffer from alcoholic infant syndrome, which includes retarded physical development, flat face, narrow eyes, smaller brain, learning disabilities, and coordination problems. Recent evidence suggests that even a single heavy drinking episode by a nonalcoholic mother during pregnancy can sometimes produce the infant alcohol syndrome. Alcoholics on the average have a 10–15-year decreased life expectancy.

Although ethyl alcohol is inexpensive to produce, it is heavily taxed when used in beverages. To avoid this tax when ethanol is used for other commercial or industrial purposes, manufacturers must denature the alcohol to make it unfit for ingestion. Denaturing agents include substances such as methyl and isopropyl alcohols.

B. Polyhydric Alcohols

Polyhydric alcohols are alcohols with more than one hydroxyl group per molecule. Two of the most important examples are ethylene glycol and glycerol.

$$\underset{\substack{|\\OH}}{CH_2}\underset{\substack{|\\OH}}{CH_2} \qquad \underset{\substack{|\\OH}}{CH_2}\underset{\substack{|\\OH}}{CH}\underset{\substack{|\\OH}}{CH_2}$$

Ethylene glycol   Glycerol
(1,2-Ethanediol)   (1,2,3-Propanetriol)

1. Ethylene Glycol. The principal commercial use of ethylene glycol is as an antifreeze in automobile radiators. Its unique properties make it especially suitable for this purpose. (See problem 8.6.) It has a high boiling point (198 °C) and will not readily boil out of a hot radiator. It is soluble in water in all proportions. And it is noncorrosive. Other applications of ethylene glycol include its use as a hydraulic brake fluid and in the production of such polymers as Dacron.

2. Glycerol. Glycerol is a sweet, syrupy liquid obtained as a by-product of soap manufacture and through synthesis from propene. It is used commercially as a humectant to preserve moistness in tobacco, cosmetics, and the like. A *humectant* is an agent that attracts and retains moisture. Glycerol is particularly effective because of its capacity to hydrogen-bond with water. Another important application of glycerol occurs in the manufacture of polymers and in cellulose polymers as softening agents.

Glycerol can be converted into nitroglycerin by treatment with concentrated nitric and sulfuric acids.

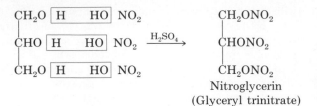

Nitroglycerin
(Glyceryl trinitrate)

The powerfully explosive character of nitroglycerin arises from its rapid conversion, sometimes merely on minor shock, from a liquid occupying a relatively small volume to a large volume of hot expanding gases. Four moles of nitroglycerin, occupying a volume of just over half a liter, decompose to 29 moles of hot gases (one mole of a gas occupies 22.4 liters at STP) that probably expand to at least 10,000–20,000 times the original volume.

$$4\begin{matrix} CH_2ONO_2 \\ | \\ CHONO_2 \\ | \\ CH_2ONO_2 \end{matrix} \longrightarrow 6N_2 + O_2 + 12CO_2 + 10H_2O$$

Although nitroglycerin is a very powerful explosive, its shock sensitivity makes it extremely dangerous to use. In 1886, Alfred Nobel discovered that this undesirable property could be mitigated by mixing nitroglycerin with diatomaceous earth and sawdust. The resulting material, dynamite, made Nobel a very wealthy man. In his will, he specified that the income from the investment of his fortune be applied to the establishment of cash prizes in various disciplines. Today Nobel Prizes are awarded in physics, chemistry, physiology or medicine, literature, economics, and peace.

Many nitro compounds and nitrates are used as explosives, as is evident from Figure 8.3. Nitroglycerin is also used medicinally for people with heart trouble to dilate blood vessels and arteries.

C. Ethyl Ether, $CH_3CH_2OCH_2CH_3$

Diethyl ether has been used since 1846 as a general anesthetic. A general anesthetic acts on the brain and produces unconsciousness as well as an insensitivity to pain. Diethyl ether is also used as a solvent for lipid materials.

PHENOLS

8.5

A. Medicinal Applications

In terms of its medicinal use, phenol has four properties worth noting:

Phenol

1. ability to act as antiseptic, disinfectant
2. ability to act as a local anesthetic
3. skin irritancy
4. toxicity when ingested

The first two properties have positive medicinal value. In all medicinal formulations, however, properties 3 and 4 must also be considered. To prevent skin irrita-

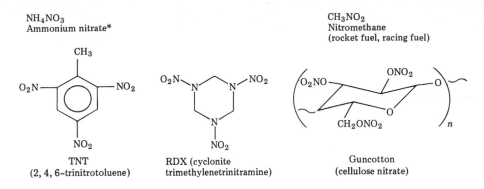

FIGURE 8.3. Some nitro and nitrate explosives.* Ammonium nitrate is a fertilizer and an industrial explosive when mixed with petroleum. Explosions of ships loading ammonium nitrate in Texas City on April 16, 1947, killed 576 people and constituted one of the world's worst chemical disasters.

tion or possible ingestion or absorption through the skin of large amounts, phenol is found only in very small quantities in over-the-counter medications.

Since phenol has medicinal properties, it is reasonable to suspect that other phenols may have similar properties. Many related structures are much more effective for certain uses than phenol itself. Because of these antiseptic and anesthetic activities, phenols are found in a variety of commercial products including soaps, deodorants, disinfectant sprays and ointments, first aid sprays, gargles, lozenges, muscle rubs, and home disinfectant sprays. Figure 8.4 summarizes some of these compounds and their uses. Note the phenol units in each molecule.

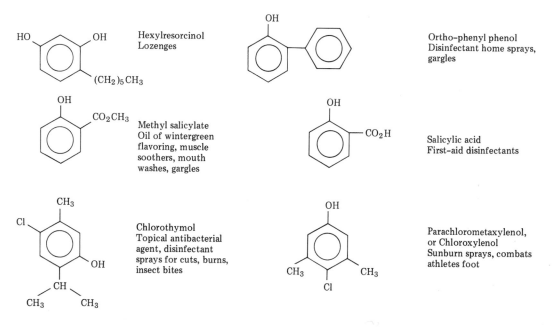

FIGURE 8.4. Some commercially important phenols.

B. Photographic Developers

Black and white film and photographic paper are produced by spreading a thin layer of colloidal silver bromide dispersed in gelatin on transparent plastic film or paper. On exposure to light, some silver ions in each grain are converted to metallic silver. Photographic developers reduce silver ion to silver metal. The grains already containing silver nuclei are reduced to silver much more rapidly than those containing few or no silver nuclei and these areas develop large silver deposits that appear dark after developing. Photographic developers must themselves be easily oxidized (since they must reduce Ag^+). Hydroquinone and *p*-methylaminophenol have this property and are popular black and white developers.

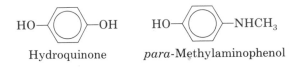

Hydroquinone   *para*-Methylaminophenol

C. Antioxidants

Antioxidants are added to most foods and cosmetics that contain fat and oil to retard rancidification and to gasoline to prevent polymerization and gum formation. An antioxidant is a compound that is more susceptible to oxidation than the material it is protecting and can terminate oxidative chain reactions. We have just seen that the ease of oxidation of phenols makes some of them good photographic developers. For the same reason phenols are effective antioxidants. Following are some common phenol-type antioxidants and preservatives:

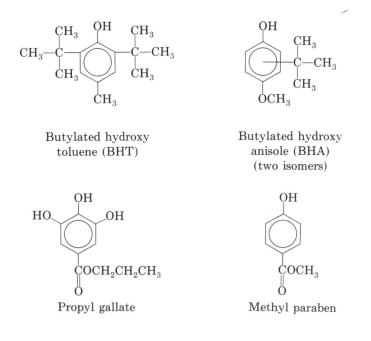

Butylated hydroxy
toluene (BHT)

Butylated hydroxy
anisole (BHA)
(two isomers)

Propyl gallate

Methyl paraben

D. Tetrahydrocannabinol

Tetrahydrocannabinol (THC) is both a phenol and a phenolic ether that is found in marijuana, a preparation made from the leaves, seeds, small stems, and flowers of the weed *Cannabis sativa.*

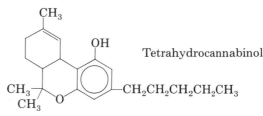

Tetrahydrocannabinol

Next to alcohol, marijuana is probably the most-used intoxicating drug in the United States. Effects of smoking marijuana include euphoria, increased appetite, a feeling of mind expansion, and accelerated pulse rate. In a medicinal application, marijuana has been found to relieve pressure in cases of glaucoma and alleviate the nausea accompanying cancer therapy.

REACTIONS OF ALCOHOLS, PHENOLS, AND ETHERS

8.6   What would you predict as the most likely sites for chemical reaction in alcohols, phenols, and ethers? One of the most obvious considerations would be the polar bonds in the molecules. All three classes of compounds have polar carbon-oxygen bonds, and alcohols and phenols also have polar oxygen-hydrogen bonds.

$$\overset{\delta+\ \ \delta-}{\text{C—O}} \qquad \overset{\delta-\ \ \delta+}{\text{O—H}} \qquad -\overset{..}{\underset{..}{\text{O}}}-$$

Polar bonds   Lone pair of electrons

In addition, oxygen has two lone pairs of electrons, and is therefore a Lewis base. Chemical reagents are attracted to this area of alcohols, phenols, and ethers, and most reactions will involve the C—O bond or the O—H bond, or both. Of course, ethers, without an O—H bond, cannot undergo reactions involving that bond. Phenols are not as reactive as alcohols and ethers at the C—O bond, since resonance between the benzene ring and a lone pair of electrons on oxygen strengthens this bond (Figure 8.5).

A. Salts of Alcohols and Phenols

1. Acidity of Phenols. The characteristic property that differentiates phenols from alcohols is acidity. Phenols are weakly acidic and can be neutralized by sodium hydroxide. Alcohols do not react with sodium hydroxide.

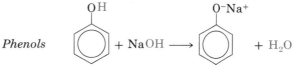

Alcohols $ROH + NaOH \longrightarrow$ no reaction

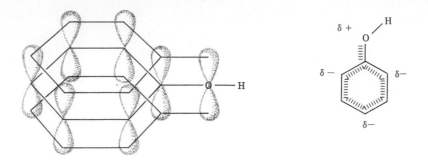

FIGURE 8.5. Resonance in phenols. A lone pair of electrons occupying a p orbital on the oxygen overlaps the π cloud of the benzene ring. This strengthens the C—O bond, which is therefore more difficult to break. In addition, this resonance makes the ring more negative and polarizes the O—H bond.

Why this difference in acidity? In phenols the hydroxy group is directly attached to an aromatic ring, allowing resonance interaction between a lone pair of electrons on oxygen and the aromatic π cloud (Figure 8.5). The ring becomes more negative, the hydroxy group more positive. Because of this resonance, the O—H bond is more polar in phenols than in alcohols and consequently more acidic.

Besides examining the reactants, we should also look at the products in order to fully explain phenol's acidity. The salt of a phenol is a negative phenoxide ion, whereas the salt of an alcohol would be a negative alkoxide ion.

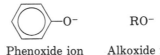

Phenoxide ion   Alkoxide ion

Again, due to resonance, the phenoxide ion is more stable than the alkoxide ion. The negative charge is spread throughout the benzene ring, and thereby effectively dispersed (Figure 8.6). In the alkoxide ion, however, no resonance is possible and the negative charge is concentrated on a single atom.

In summary, we have said that the acidity of phenols is due to increased polarization of the O—H bond and delocalization of the negative charge in the phenoxide ion, both by resonance. In each case, the aromatic ring becomes more negative. We have previously seen an effect of this charge transferral from the hydroxy (an activating group) to the phenyl group in electrophilic aromatic substitution (section 5.6.D). To only monobrominate benzene, pure bromine, a catalyst, and possibly heat are necessary. The electrophile Br^+ attacks the ring, replacing a hydrogen ion. In the bromination of phenol, however, tribromination occurs instantaneously even if a dilute water solution of bromine is used and no catalyst. This is due to the increased negative character of the aromatic ring (Figures 8.5 and 8.6).

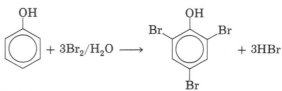

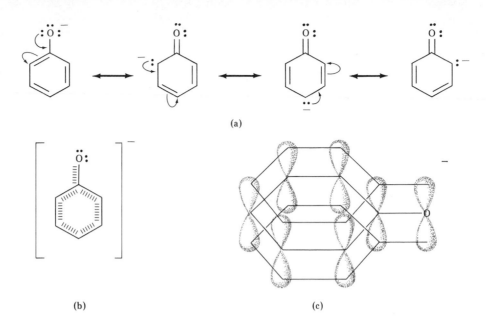

(a)

(b)   (c)

FIGURE 8.6. Resonance stabilization of the phenoxide ion. (a) Resonance forms. (b) Resonance hybrid. (c) Bonding picture.

PROBLEM

8.8. Write an equation illustrating the reaction between *ortho*-phenylphenol (Figure 8.4), an antiseptic in throat gargles and home disinfectant sprays, and sodium hydroxide.

2. Reaction of Alcohols with Sodium Metal. Salts of alcohols are useful in organic synthesis (for example the Williamson synthesis of ethers). They are prepared by treating alcohols with sodium metal, forming sodium alkoxides.

$$2ROH + 2Na \longrightarrow 2RONa + H_2$$

The reaction can be understood as an oxidation-reduction. Sodium releases its outer-shell electron to the alcohol, forming a hydrogen atom and an alkoxide ion. The combination of two hydrogen atoms produces hydrogen gas.

$$
\begin{array}{ll}
RO\!:\!H \quad Na & RO\!:^- \quad Na^+ \quad H\cdot \\
RO\!:\!H \quad Na & RO\!:^- \quad Na^+ \quad H\cdot
\end{array}
$$

The reaction is simply an extension of the reaction of active metals, such as sodium, with water to produce a metal hydroxide and hydrogen gas.

$$2Na + 2H_2O \longrightarrow 2NaOH + H_2$$

8.9. Write an equation illustrating the reaction (if any) of ethanol (beverage alcohol) with **(a)** sodium hydroxide and **(b)** sodium metal.

B. Dehydration of Alcohols

Alcohols dehydrate under acidic conditions by elimination (section 3.7.B):

$$-\overset{|}{\underset{OH}{C}}-\overset{|}{\underset{H}{C}}- \xrightarrow{H_2SO_4} -\overset{|}{C}=\overset{|}{C}- + H_2O$$

The mechanism involves initial protonation of the basic oxygen, followed by carbocation formation, and subsequent proton elimination.

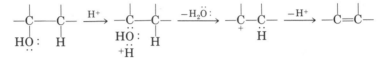

8.10. Write equations illustrating the dehydration, using sulfuric acid, of **(a)** 2-propanol and **(b)** 3-methyl-2-pentanol.

C. Reactions of Alcohols and Ethers with Hydrogen Halides

1. Alcohols. Alcohols react with hydrogen halides to produce alkyl halides by a substitution reaction.

$$ROH + HX \longrightarrow RX + H_2O$$

HX = HCl, HBr, HI

The reaction with HCl (catalyzed by $ZnCl_2$) is the basis of the Lucas test for distinguishing among low-molecular-weight alcohols. Although the alcohols are soluble in the Lucas reagent, the alkyl halide formed is not. As it forms, the solution becomes cloudy with droplets of alkyl halide. Primary alcohols require several hours of reaction time, even with heat. Secondary alcohols react in 5 to 15 minutes when heated, and tertiary alcohols react almost instantaneously at room temperature.

This reaction and its mechanism are analogous to dehydration of alcohols (section 3.7.B) and nucleophilic substitution (section 7.4), both of which we have already studied. Let us consider the mechanism for the reaction of a tertiary alcohol with a hydrogen halide, HX. Because alcohols have a lone pair of electrons, they are Lewis bases and react with the hydrogen ion of HX, a Lewis acid.

$$\underset{:\overset{..}{O}H}{\overset{R}{\overset{|}{R C R}}} + H^+ \longrightarrow \underset{+:\overset{..}{O}H}{\underset{\overset{|}{H}}{\overset{R}{\overset{|}{R C R}}}} \quad \text{(Oxonium ion)}$$

The oxonium ion formed loses a water molecule to generate a carbocation. Note that the formation of an oxonium ion and its conversion to a carbocation are also the first two steps in the acid-catalyzed dehydration of alcohols. However, when concentrated hydrogen halide is used, the carbocation is neutralized by the large concentration of halide ion; elimination of a hydrogen ion to form an alkene does not occur as readily. The result is an alkyl halide (compare this mechanism with that in section 8.6.B). The mechanism is S_N1.

$$
\begin{array}{ccccc}
\text{R} & & \text{R} & & \text{R} \\
\text{RCR} & \longrightarrow & \overset{+}{\text{RCR}} & \overset{:\ddot{\text{X}}:^-}{\longrightarrow} & \text{RCR} \\
| & & \text{Carbocation} & & | \\
+:\ddot{\text{O}}\text{H} & & & & :\ddot{\text{X}}: \\
\ddot{\text{H}} & & & &
\end{array}
$$

Secondary and tertiary alcohols react by an S_N1 mechanism as shown. But as in nucleophilic substitution reactions involving primary alkyl halides, primary alcohols react by an S_N2 mechanism, thus avoiding the less stable primary carbocation. Protonation still initiates the reaction, but then the halide ion enters and pushes out the water molecule.

$$
\begin{array}{ccccccc}
\text{H}\quad\text{H} & & \text{H}\quad\text{H} & & {}^{\delta-}\text{H}\quad\text{H}\ {}^{\delta+} & & \\
\diagdown\diagup & & \diagdown\diagup & & \diagdown\quad\diagup & & \\
\text{C}-\ddot{\text{O}}\text{H} & \overset{\text{H}^+}{\longrightarrow} & \text{C}-\overset{+}{\ddot{\text{O}}}\text{H} & \overset{:\ddot{\text{X}}:^-}{\longrightarrow} & :\ddot{\text{X}}\text{------C------}\ddot{\text{O}}\text{H} & \longrightarrow & \text{RCH}_2\ddot{\text{X}}: + \text{H}_2\ddot{\text{O}}: \\
| & & |\quad\ddot{\text{H}} & & |\quad\ddot{\text{H}} & & \\
\text{R} & & \text{R} & & \text{R} & &
\end{array}
$$

PROBLEMS

8.11. To compare similarities and differences, write reaction mechanisms side-by-side for the reaction of 2-methyl-2-propanol (*t*-butyl alcohol) with **(a)** H_2SO_4 (dehydration) and **(b)** HBr (substitution).

8.12. Write an S_N2 mechanism for the reaction of methanol with HBr.

---

2. Ethers. Ethers react with hydrogen halides in a manner similar to that of alcohols. This would be expected since the O—H bond of the alcohols is not involved in that reaction. Rather, the reaction involves protonation of the alcohol oxygen followed by cleavage of the carbon-oxygen bond. Ethers are capable of both these actions. As a consequence, ethers react with 1 mole of hydrogen halide to produce 1 mole of alkyl halide and 1 mole of alcohol.

$$\text{R—O—R} + \text{HX} \longrightarrow \text{RX} + \text{ROH}$$
HX = HCl, HBr, HI

If a second mole of hydrogen halide is available, the alcohol will likewise be halogenated. In the presence of excess hydrogen halide, therefore, ethers produce 2 moles of alkyl halide.

$$\text{ROR} + 2\text{HX} \longrightarrow 2\text{RX} + \text{H}_2\text{O}$$
HX = HCl, HBr, HI

The cleavage of ethers by hydrogen halide can occur by an S_N1 or an S_N2 mechanism just as the halogenation of alcohols did. If the carbon involved in the cleavage is primary, an S_N2 process predominates; if the carbon is secondary or tertiary, the mechanism is S_N1. Figure 8.7 summarizes the mechanisms of ether cleavage and compares them with alcohol halogenation.

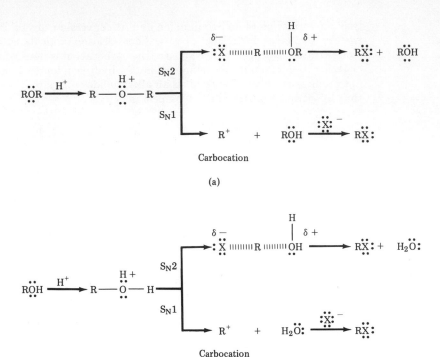

FIGURE 8.7. Comparison of mechanisms of the reactions of (a) ethers and (b) alcohols with hydrogen halides.

PROBLEM

8.13. Write equations illustrating the reaction of each of the following with excess HBr: **(a)** CH_3CH_2OH; **(b)** $CH_3CH_2OCH_2CH_3$; and **(c)** $CH_3OCH_2CH_3$.

D. Conversion of Alcohols to Alkyl Halides

In addition to treating alcohols with hydrogen halides, alkyl halides can be prepared by reaction with thionyl chloride, $SOCl_2$, or phosphorus trihalides, PX_3.

$$-\overset{|}{\underset{|}{C}}-OH + SOCl_2 \longrightarrow -\overset{|}{\underset{|}{C}}-Cl + HCl + SO_2$$

$$3 -\overset{|}{\underset{|}{C}}-OH + PX_3 \longrightarrow 3 -\overset{|}{\underset{|}{C}}-X + H_3PO_3$$

E. Oxidation of Alcohols

Primary and secondary alcohols are oxidized by $KMnO_4$, $K_2Cr_2O_7$, or CrO_3. Oxidation can be visualized as the successive insertions of oxygen into a carbon-hydrogen bond followed by elimination of hydrogen from carbon-oxygen bonds.

Consider, for example, the complete oxidation of methane (natural gas) to carbon dioxide using this concept.

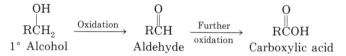

$$CH_4 \longrightarrow CH_3OH \longrightarrow HC\!=\!O \longrightarrow HOC\!=\!O \longrightarrow O\!=\!C\!=\!O$$

Very simply, in organic chemistry, oxidation involves the formation of new carbon-oxygen bonds, whereas reduction involves the formation of new carbon-hydrogen bonds.

Primary alcohols are oxidized to aldehydes, which are, in turn, easily oxidized to carboxylic acids.

$$RCH_2 \xrightarrow{\text{Oxidation}} RCH \xrightarrow[\text{oxidation}]{\text{Further}} RCOH$$

1° Alcohol   Aldehyde   Carboxylic acid

Secondary alcohols oxidize to ketones.

$$RCHR \xrightarrow{\text{Oxidation}} RCR$$

2° Alcohol   Ketone

Tertiary alcohols usually do not oxidize under mild conditions. The dichromate oxidation of ethyl alcohol is used to measure blood alcohol levels in those suspected of drunk driving.

PROBLEM

8.14. Write reaction equations illustrating the oxidation of the four isomeric alcohols with the formula $C_4H_{10}O$.

EPOXIDES

8.7

Three-membered ring cyclic ethers are called *epoxides* or *oxiranes*. The most commercially important epoxide is ethylene oxide, which is used in the petrochemical industry as an intermediate in the production of antifreeze, synthetic fibers, resins, paints, adhesives, films, cosmetics, and synthetic detergents. Ethylene oxide is prepared from ethylene.

$$CH_2\!\!-\!\!CH_2 \qquad \text{Ethylene oxide}$$
$$\diagdown O \diagup$$

A. Reactions of Ethylene Oxide

The importance of ethylene oxide as an industrial chemical lies in its propensity toward ring-opening reactions. Like cyclopropane (section 2.8.A), ethylene oxide suffers from acute angle strain because of the distortion of normal bond angles from 109° to approximately 60°. This strain is relieved by cleavage of a polar carbon-oxygen bond under either acidic or basic conditions. For example, over half the ethylene oxide produced commercially is hydrolyzed to ethylene glycol, which is used as antifreeze, in brake fluids, and in the manufacture of polyester fibers. The reaction mechanism is very similar to that of acid cleavage of ethers to alkyl halides (section 8.6.C.2). First, the epoxide oxygen is protonated. Nucleophilic attack of a water molecule and loss of a hydrogen ion follow.

$$CH_2-CH_2 \xrightarrow{H^+} CH_2-CH_2 \xrightarrow{H_2O} CH_2-CH_2 \xrightarrow{-H^+} CH_2-CH_2$$

The use of alcohols instead of water to effect the ring opening produces ether-alcohol compounds commercially known as cellosolves such as methyl cellosolve, which is added to jet fuels to prevent formation of ice crystals.

$$CH_3OH + CH_2-CH_2 \xrightarrow{H^+} CH_3OCH_2CH_2OH$$
Methyl cellosolve

Reaction with ammonia produces ethanolamine, which is used to remove hydrogen sulfide and carbon dioxide from natural gas.

$$NH_3 + CH_2-CH_2 \longrightarrow H_2NCH_2CH_2OH$$
Ethanolamine

B. Epoxy Resins

Epoxy resins are manufactured from epichlorohydrin and bisphenol A. The involved (though not complex) process includes reactions already studied: acidity

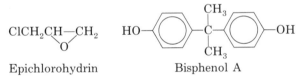

Epichlorohydrin   Bisphenol A

of phenols, ring opening of epoxides, and nucleophilic substitution on alkyl halides. The structure of an epoxy resin follows. Treatment of the developing polymer with a triamine causes crosslinking between polymer chains and gives the resin added strength.

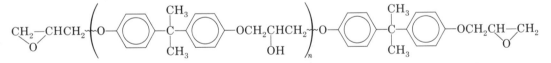

Epoxy resins have tremendous adhesive properties and are used extensively to bind glass, porcelain, metal, and wood. These resins, because of their inertness, hardness, and flexibility make excellent protective coatings. Fiberglass boat hulls, for example, have a metal frame coated with a thick layer of spun glass trapped in a set epoxy resin.

SULFUR ANALOGUES OF ALCOHOLS AND ETHERS

8.8   Since sulfur is directly below oxygen in group VI of the periodic table, there are sulfur counterparts of alcohols and ethers. The sulfur analogues of alcohols are called mercaptans, thiols, or alkyl hydrogen sulfides, and the sulfur analogues of ethers are thioethers, or sulfides.

$$CH_3CH_2CH_2CH_2SH \qquad CH_3SCH_2CH_2CH_3$$
Butanethiol   Methyl propyl sulfide

Thiols and sulfides are especially noted for their strong, often unpleasant odors, as is evident from these examples.

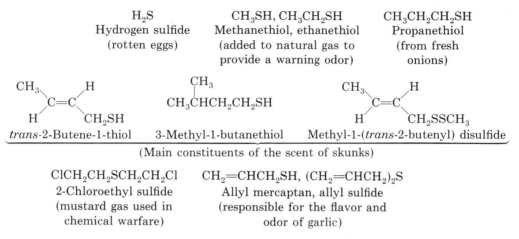

H_2S CH_3SH, CH_3CH_2SH $CH_3CH_2CH_2SH$

Hydrogen sulfide   Methanethiol, ethanethiol   Propanethiol

(rotten eggs)   (added to natural gas to   (from fresh

provide a warning odor)   onions)

trans-2-Butene-1-thiol   3-Methyl-1-butanethiol   Methyl-1-(*trans*-2-butenyl) disulfide

(Main constituents of the scent of skunks)

$$ClCH_2CH_2SCH_2CH_2Cl \qquad CH_2{=}CHCH_2SH, \ (CH_2{=}CHCH_2)_2S$$

2-Chloroethyl sulfide   Allyl mercaptan, allyl sulfide

(mustard gas used in   (responsible for the flavor and

chemical warfare)   odor of garlic)

The amino acid cysteine is a thiol, and cystine, another amino acid, is a disulfide. They can be interconverted by oxidation and reduction.

$$2HOCCHCH_2SH \xrightleftharpoons[\text{Reduction}]{\text{Oxidation}} HOCCHCH_2S{-}SCH_2CHCOH$$

Cysteine   Cystine

The disulfide unit in cystine is important in determining the shapes of protein molecules. The cleavage and recombination of the disulfide units of cystine in hair is the basis of hair permanents.

Enzymes possessing the thiol group react with heavy metal ions such as those of mercury and lead. This can precipitate or deactivate the enzyme and is partly the basis of mercury and lead poisoning.

Panel 12

INSECT PHEROMONES

Writing, talking, painting, the use of telephones, radio, and television—these are the means of human communication. We know from their behavior that insects and other animals also communicate among themselves. They, and perhaps even humans, do so on a chemical level. A *pheromone* is a chemical substance that, when secreted by an individual of a species, can elicit a certain type of behavior in other members. Alarm pheromones warn of danger; aggregating or recruiting pheromones direct others to a food source; primer pheromones regulate caste systems in social insects; and sex pheromones attract the opposite sex and elicit sexual behavior. Table 8.3 gives the structures of some representative pheromones.

TABLE 8.3. INSECT PHEROMONES

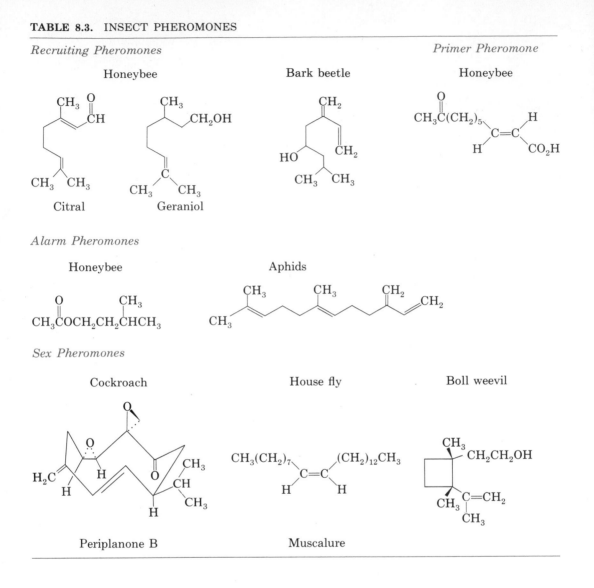

Recruiting Pheromones   *Primer Pheromone*

Honeybee   Bark beetle   Honeybee

Citral   Geraniol

Alarm Pheromones

Honeybee   Aphids

Sex Pheromones

Cockroach   House fly   Boll weevil

Periplanone B   Muscalure

[PANEL 12]   Sex pheromones (sometimes called sex attractants) can be quite useful in insect control. They can be used as bait in traps to attract large numbers of insects, which can then be efficiently destroyed with chemical insecticides. Alternatively, sex pheromones sprayed in the air can so severely confuse a male insect that it becomes impossible to find a female and mate. Extremely small amounts of these pheromones elicit such behavior in a large number of insects. For example, one hundredth of a gram of periplanone B, the American cockroach sex excitant, can excite 100 billion cockroaches weighing a total of approximately 10,000 tons.

END-OF-CHAPTER PROBLEMS

8.15 Isomerism and Nomenclature: For the molecular formula $C_5H_{12}O$: **(a)** draw all alcohols; **(b)** classify the alcohols as 1°, 2°, or 3°; **(c)** name the alcohols by the IUPAC system; **(d)** draw all ethers; **(e)** name the ethers by the IUPAC system.

8.16 IUPAC Nomenclature: Name the following compounds by the IUPAC system of nomenclature:

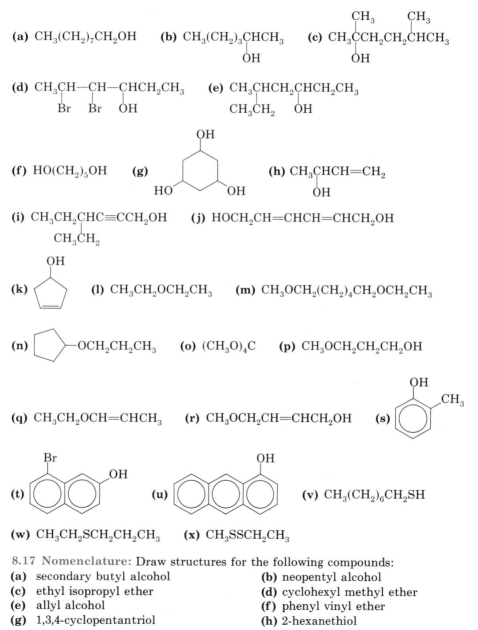

(a) $CH_3(CH_2)_7CH_2OH$

(b) $CH_3(CH_2)_3CHCH_3$
                     OH

(c) $CH_3\underset{\underset{OH}{|}}{C}CH_2CH_2\underset{\underset{CH_3}{}}{C}HCH_3$ (with CH_3 groups)

(d) $CH_3CH\!-\!CH\!-\!CHCH_2CH_3$
          Br Br OH

(e) $CH_3CHCH_2CHCH_2CH_3$
         CH_3CH_2 OH

(f) $HO(CH_2)_5OH$

(g)

(h) $CH_3CHCH\!=\!CH_2$
         OH

(i) $CH_3CH_2CHC\!\equiv\!CCH_2OH$
     CH_3CH_2

(j) $HOCH_2CH\!=\!CHCH\!=\!CHCH_2OH$

(k)

(l) $CH_3CH_2OCH_2CH_3$

(m) $CH_3OCH_2(CH_2)_4CH_2OCH_2CH_3$

(n) $-OCH_2CH_2CH_3$

(o) $(CH_3O)_4C$

(p) $CH_3OCH_2CH_2CH_2OH$

(q) $CH_3CH_2OCH\!=\!CHCH_3$

(r) $CH_3OCH_2CH\!=\!CHCH_2OH$

(s)

(t)

(u)

(v) $CH_3(CH_2)_6CH_2SH$

(w) $CH_3CH_2SCH_2CH_2CH_3$

(x) $CH_3SSCH_2CH_3$

8.17 Nomenclature: Draw structures for the following compounds:

(a) secondary butyl alcohol
(b) neopentyl alcohol
(c) ethyl isopropyl ether
(d) cyclohexyl methyl ether
(e) allyl alcohol
(f) phenyl vinyl ether
(g) 1,3,4-cyclopentantriol
(h) 2-hexanethiol

(i)   ethyl propyl disulfide
(k)   *p*-methoxyphenol
(m) 2-methoxybutanol

(j)   ethyl propyl sulfide
(l)   2-ethyl-4-isopropylcyclohexanol
(n)   5-hydroxy-3-pentenoic acid

8.18 Physical Properties: For each of the following sets of compounds, arrange the members in order of increasing boiling point:

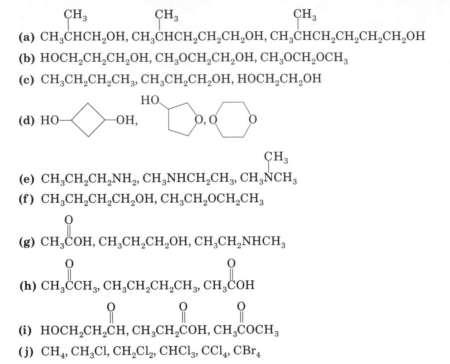

(a) $CH_3\overset{\underset{\displaystyle CH_3}{|}}{C}HCH_2OH$, $CH_3\overset{\underset{\displaystyle CH_3}{|}}{C}HCH_2CH_2CH_2OH$, $CH_3\overset{\underset{\displaystyle CH_3}{|}}{C}HCH_2CH_2CH_2CH_2OH$

(b) $HOCH_2CH_2CH_2OH$, $CH_3OCH_2CH_2OH$, $CH_3OCH_2OCH_3$

(c) $CH_3CH_2CH_2CH_3$, $CH_3CH_2CH_2OH$, $HOCH_2CH_2OH$

(d) $HO\text{—}\square\text{—}OH$, (HO-cyclopentane, O), (O-cyclohexane, O)

(e) $CH_3CH_2CH_2NH_2$, $CH_3NHCH_2CH_3$, $CH_3\overset{\underset{\displaystyle CH_3}{|}}{N}CH_3$

(f) $CH_3CH_2CH_2CH_2OH$, $CH_3CH_2OCH_2CH_3$

(g) $CH_3\overset{\overset{\displaystyle O}{||}}{C}OH$, $CH_3CH_2CH_2OH$, $CH_3CH_2NHCH_3$

(h) $CH_3\overset{\overset{\displaystyle O}{||}}{C}CH_3$, $CH_3CH_2CH_2CH_3$, $CH_3\overset{\overset{\displaystyle O}{||}}{C}OH$

(i) $HOCH_2CH_2\overset{\overset{\displaystyle O}{||}}{C}H$, $CH_3CH_2\overset{\overset{\displaystyle O}{||}}{C}OH$, $CH_3\overset{\overset{\displaystyle O}{||}}{C}OCH_3$

(j) CH_4, CH_3Cl, CH_2Cl_2, $CHCl_3$, CCl_4, CBr_4

8.19 Physical Properties: For each of the following pairs of compounds, compare the boiling points of the members. Why is the ortho isomer the lower-boiling structure in each case, compared with the other isomer? Explain fully. Use drawings if necessary.

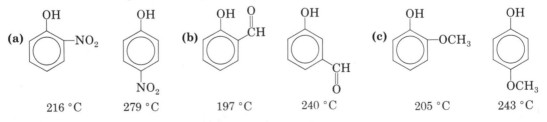

(a)   216 °C   279 °C   (b) 197 °C   240 °C   (c) 205 °C   243 °C

8.20 Water Solubility: Sucrose ($C_{12}H_{22}O_{11}$), table sugar, dissolves to the extent of 200 g per 100 ml of water. How can one account for this tremendous solubility?

8.21 Preparations of Alcohols: Complete the following reactions showing the major organic products:

(a) $CH_3CH=CHCH_3 + H_2O \xrightarrow{H^+}$

(b) $\overset{\overset{\displaystyle CH_3}{|}}{CH_3CH_2C}=CHCH_3 + H_2O \xrightarrow{H^+}$

(c) $CH_3CH_2CH_2CH=CH_2 \xrightarrow{B_2H_6} \xrightarrow[OH^-]{H_2O_2,}$

(d) $CH_3CH_2CH_2Br + NaOH \longrightarrow$

8.22 Preparations of Alcohols: Write reaction equations illustrating the preparations of **(a)** 2-methyl-1-propanol and **(b)** 2-methyl-2-propanol from methylpropene.

8.23 Williamson Synthesis of Ethers: There are two ways to prepare 2-ethoxypropane by the Williamson synthesis. Write a reaction equation for each.

8.24 Williamson Synthesis of Ethers: Write equations showing the best way for preparing each of the following ethers by the Williamson synthesis:

(a) $\overset{\overset{\displaystyle CH_3}{|}}{CH_3CH}OCH_3$

(b) $CH_3CH_2CH_2O\overset{\overset{\displaystyle CH_3}{|}}{\underset{\underset{\displaystyle CH_3}{|}}{C}}CH_3$

(c) 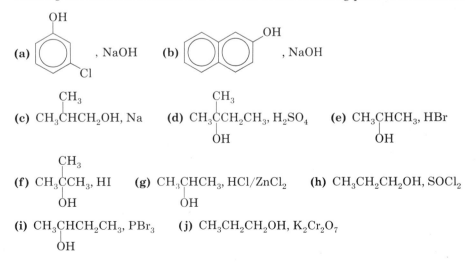 —$CH_2OCH_2CH_2CH_3$

8.25 Williamson Synthesis of Ethers: Show how 2-ethoxy-2-methylpropane can be prepared by the Williamson synthesis using alcohols as starting materials.

8.26 Reactions of Alcohols: Write equations illustrating the reaction of

(I) $CH_3CH_2CH_2CH_2OH$ (II) $CH_3\overset{\underset{\displaystyle OH}{|}}{CH}CH_2CH_3$ (III) $CH_3\overset{\overset{\displaystyle CH_3}{|}}{\underset{\underset{\displaystyle OH}{|}}{C}}CH_3$

with the following reagents:
(a) Na   **(b)** H_2SO_4 **(c)** $HCl/ZnCl_2$ **(d)** $Na_2Cr_2O_7/H^+$

8.27 Reactions of Alcohols, Phenols, Ethers, and Epoxides: Write equations showing the reaction between the members of the following pairs of substances:

(a) [structure: phenol with Cl substituent], NaOH   **(b)** [structure: naphthalenol with OH], NaOH

(c) $CH_3\overset{\overset{\displaystyle CH_3}{|}}{CH}CH_2OH$, Na   **(d)** $CH_3\overset{\overset{\displaystyle CH_3}{|}}{\underset{\underset{\displaystyle OH}{|}}{C}}CH_2CH_3$, H_2SO_4 **(e)** $CH_3\overset{\underset{\displaystyle OH}{|}}{CH}CH_3$, HBr

(f) $CH_3\overset{\overset{\displaystyle CH_3}{|}}{\underset{\underset{\displaystyle OH}{|}}{C}}CH_3$, HI   **(g)** $CH_3\overset{\underset{\displaystyle OH}{|}}{CH}CH_3$, $HCl/ZnCl_2$ **(h)** $CH_3CH_2CH_2OH$, $SOCl_2$

(i) $CH_3\overset{\underset{\displaystyle OH}{|}}{CH}CH_2CH_3$, PBr_3 **(j)** $CH_3CH_2CH_2OH$, $K_2Cr_2O_7$

(k) CH_3CHCH_3, $K_2Cr_2O_7$
　　　　|
　　　　OH

(l) $CH_3OCH_2CHCH_3$, 2HBr
　　　　　　　　　|
　　　　　　　　　CH_3

　　　　　　　　CH_3
　　　　　　　　|
(m) $CH_3CH_2OCH_2CH_3$, HCl　　**(n)** $CH_3CHOCH_2CH_3$, 2HI

(o) $CH_3CH{-\!-}CHCH_3 + H_2O/H^+$　　**(p)** $CH_2{-\!-}CH_2 + CH_3CH_2OH/H^+$
　　　　　　＼O／　　　　　　　　　　　　　　　＼O／

(q) $CH_2{-\!-}CH_2 + CH_3NH$　　**(r)** $CH_2{-\!-}CH_2$, HBr
　　　＼O／　　　　　　　|　　　　　　　＼O／
　　　　　　　　　　　　CH_3

8.28 Reaction Mechanisms: Write step-by-step reaction mechanisms for each of the following reactions:

(a) $CH_3CHCH_3 + H_2SO_4 \longrightarrow$　　**(b)** $CH_3CH_2OH + HBr \longrightarrow$
　　　　|
　　　　OH

　　　　CH_3
　　　　|
(c) $CH_3CCH_3 + HBr \longrightarrow$　　**(d)** $CH_2{-\!-}CH_2 + CH_3OH \xrightarrow{\ H^+\ }$
　　　　|　　　　　　　　　　　　　　　＼O／
　　　　OH

8.29 Reaction Mechanisms: Write the reaction mechanism and products of the reaction between 2-methoxy-2-methylpropane (methyl *t*-butyl ether) and 2 moles of HBr. (*Hint:* See Figure 8.7; both an S_N1 and S_N2 mechanism should be included in the correct answer. Why?)

8.30 Consumer Chemistry: Some time when you're in a grocery store or drug store do some of the following activities:
(a) Examine the labels of several mouthwashes, aerosol disinfectant cleaners, extracts (such as vanilla extract), and any other products that may contain ethyl alcohol (often merely called alcohol). Record the percentage of alcohol in each.
(b) Find the phenolic compounds listed in Figure 8.4 and possibly some other phenols by reading labels on commercial products likely to contain phenols. Suggest a reason for including the phenol in the product.
(c) Examine the labels of some packaged foods, cosmetics, and other products that may contain antioxidants such as those described in section 8.5.C.

8.31 Qualitative Analysis: Suggest and explain a chemical method (preferably a simple test tube reaction) for distinguishing between the members of the following sets of compounds. Tell what you would do and see.

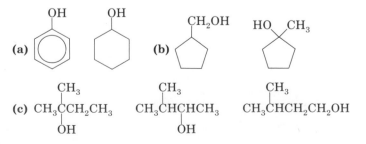

8.32 Epoxide Chemistry: Ammonia reacts with three molecules of ethylene oxide to form triethanolamine, used as an intermediate in the manufacture of detergents, waxes, polishes, herbicides, toilet goods, and cement additives. Write a structure for triethanolamine and rationalize its formation.

8.33 Acidity of Phenols: Ortho, meta and para nitrophenols are more acidic than phenol itself. In addition, both the ortho and para isomers are more acidic than the meta. Explain these facts. (*Hint:* Refer to sections 8.6.A.1 and 5.6.D and Figures 8.5 and 8.6.)

Chapter 9

Aldehydes and Ketones

STRUCTURE OF ALDEHYDES AND KETONES

9.1   Aldehydes and ketones are structurally very similar; both have a carbon-oxygen double bond often called a *carbonyl group*. Like any double bond, the carbonyl group is composed of a σ bond and a π bond, but unlike carbon-carbon double bonds it is polar due to oxygen's greater electronegativity (Figure 9.1).

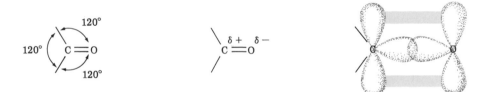

FIGURE 9.1. Structure of the carbonyl group.

Aldehydes and ketones differ in that aldehydes have at least one hydrogen atom bonded to the carbonyl group, whereas in ketones the carbonyl group has carbons bonded on each side.

Table 9.1 gives physical properties for some representative aldehydes and ketones.

Aldehydes and ketones are quite prevalent in nature. They appear as natural fragrances and flavorings. In addition, carbonyl groups and their derivatives

TABLE 9.1. ALDEHYDES AND KETONES

Structure	Melting Point, °C	Boiling Point, °C
Aldehydes		
HCHO	− 92	− 21
CH_3CHO	−121	20
CH_3CH_2CHO	− 81	49
⟨◯⟩—CH (with O double bond)	− 26	178
Ketones		
CH_3COCH_3	− 94	56
$CH_3COCH_2CH_3$	− 86	80

are the main structural features of carbohydrates and appear as functional groups in other natural compounds including dyes, vitamins, and hormones.

NOMENCLATURE OF ALDEHYDES AND KETONES

9.2

A. IUPAC Nomenclature of Aldehydes and Ketones

1. Simple Aldehydes and Ketones. As with other organic compounds, the names of aldehydes and ketones are based on the name of the longest continuous chain of carbon atoms. To name aldehydes, the -*e* of the parent hydrocarbon is replaced by -*al* (the suffix for aldehydes), and by -*one* to name ketones (-*one* is the suffix ending for ketones). The chain is numbered so as to give the functional group the lowest possible number.

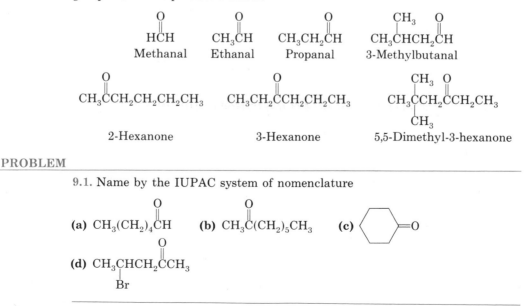

PROBLEM

9.1. Name by the IUPAC system of nomenclature

(a) $CH_3(CH_2)_4CH$ (with O double bond)   **(b)** $CH_3C(CH_2)_5CH_3$ (with O double bond)   **(c)** ⬡=O

(d) $CH_3CHCH_2CCH_3$ (with O double bond, and Br substituent)

2. Polyfunctional Aldehydes and Ketones.   Compounds with two aldehyde or two ketone groups are named *dials* and *diones,* respectively.   But what about a compound that possesses both an aldehyde and a ketone group or maybe even an alcohol group?   In these cases, one group is named using the normal suffix and the rest are named by prefixes.   The group highest in the following table takes the suffix, and the chain is numbered to give it the lowest possible number.

Functional Group	Suffix	Prefix
Aldehyde	-al	carboxaldo-
Ketone	-one	keto-
Alcohol	-ol	hydroxy-

Rationalize the following examples using these principles:

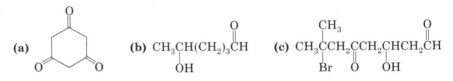

$CH_3CCH_2CCH_3$	CH_3CCH_2CH	$CH_3CHCH_2CCH_3$
2,4-Pentandione	3-Ketobutanal	4-Hydroxy-2-pentanone

PROBLEM
_____

9.2. Name by the IUPAC system of nomenclature

(a)   (b) $CH_3CH(CH_2)_3CH$ (c) $CH_3CCH_2CCH_2CHCH_2CH$

3. Unsaturated and Polyfunctional Aldehydes and Ketones.   Let us apply the nomenclature we have learned to compounds containing several important structural features.   The following procedure is useful in naming more complex molecules:

1. Determine and name the longest continuous chain of carbons.
2. Designate carbon-carbon single bonds by *-an,* double bonds by *-en,* and triple bonds by *-yn.*
3. Name the most important functional group (aldehyde > ketone > alcohol) with a suffix.
4. Number the carbon chain giving preference to the functional group named by a suffix.   Identify, in the name, the position of this group and any carbon-carbon double or triple bonds.
5. Name (and number) all other groups with prefixes.

Example 9.1: Name citral (lemon flavor and odor).

$$\underset{8}{CH_3}\underset{7}{C}=\underset{6}{CH}\underset{5}{CH_2}\underset{4}{CH_2}\underset{3}{C}=\underset{2}{CH}\underset{1}{CH}$$

with CH₃ and CH₃, O substituents

1. There are eight carbons in the longest chain, *oct.*
2. There are two carbon-carbon double bonds, *octadien.*
3. The aldehyde is the only functional group and is named with a suffix, *octadienal.*

4. The chain is numbered from the aldehyde group. The double bonds at carbons 2 and 6 are identified. Because the aldehyde is at carbon-1, designating it with a number is optional: 2,6-octadienal.
5. The methyls are named with prefixes. The complete name is

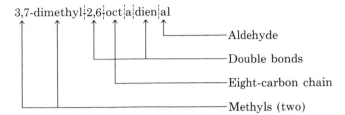

Example 9.2: Name

$$HC\equiv CCHCH_2CCH_2CH_2$$

with positions 7 6 5 4 3 2 1 and substituents OH, O, Br

1. Seven-carbon chain, *hept*.
2. One carbon-carbon triple bond, *heptyn*.
3. The ketone takes preference over the alcohol and gets the suffix, *heptynone*.
4. The carbon chain is numbered to give the lowest number to the ketone group. The positions of the groups taking suffix designations (triple bond and ketone) are indicated, 6-heptyn-3-one.
5. All other groups are named with prefixes. The complete name is

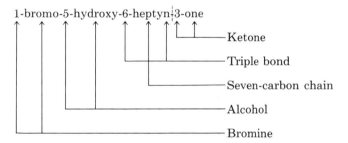

PROBLEM

9.3. Name by the IUPAC system of nomenclature

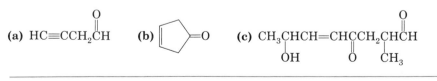

(a) $HC\equiv CCH_2\overset{\displaystyle O}{\overset{\|}{C}}H$ (b) [cyclopentane] $=O$ (c) $CH_3CHCH=CHCCH_2CHCH$ with substituents OH, O, CH₃

B. Common Nomenclature

The use of trivial names for aldehydes, particularly simple ones, is very prevalent. As we will see in Chapter 11, the common names of aldehydes are related to those of carboxylic acids.

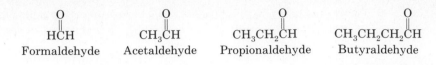

$$\overset{O}{\overset{\|}{HCH}} \qquad \overset{O}{\overset{\|}{CH_3CH}} \qquad \overset{O}{\overset{\|}{CH_3CH_2CH}} \qquad \overset{O}{\overset{\|}{CH_3CH_2CH_2CH}}$$

Formaldehyde　　Acetaldehyde　　Propionaldehyde　　Butyraldehyde

The simplest ketone, 2-propanone, is commonly called *acetone* (solvent in finger-nail polish remover).　The common names of other ketones are derived by naming the alkyl groups attached to the carbonyl.

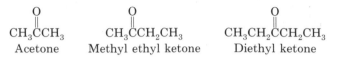

$$\overset{O}{\overset{\|}{CH_3CCH_3}} \qquad \overset{O}{\overset{\|}{CH_3CCH_2CH_3}} \qquad \overset{O}{\overset{\|}{CH_3CH_2CCH_2CH_3}}$$

Acetone　　Methyl ethyl ketone　　Diethyl ketone

PROBLEM

9.4. Write structures for the following compounds:
(a) isobutyraldehyde; **(b)** 2-chloropropionaldehyde; **(c)** methyl propyl ketone; **(d)** methyl phenyl ketone.

SOME PREPARATIONS OF ALDEHYDES AND KETONES

9.3　　A. Hydration of Alkynes (section 4.1.C.2)

$$-C\equiv C- + H_2O \xrightarrow[HgSO_4]{H_2SO_4,} -\overset{\overset{\displaystyle H}{|}}{C}-\overset{\overset{\displaystyle O}{\|}}{\underset{\underset{\displaystyle H}{|}}{C}}-$$

B. Ozonolysis of Alkenes (section 4.5.B)

$$-\overset{|}{C}=\overset{|}{C}- \xrightarrow[H_2O]{O_3 \quad Zn,} -\overset{|}{C}=O + O=\overset{|}{C}-$$

C. Friedel Crafts Reaction (section 5.6.A.2)

$$\bigcirc + R\overset{O}{\overset{\|}{C}}Cl \xrightarrow{AlCl_3} \bigcirc-\overset{O}{\overset{\|}{C}}R + HCl$$

D. Oxidation of Alcohols (section 8.6.E)

$$-\overset{\overset{\displaystyle OH}{|}}{\underset{\underset{\displaystyle |}{}}{C}}-H \xrightarrow[\text{agent}]{\text{Oxidizing}} -\overset{O}{\overset{\|}{C}}$$

REACTIONS OF ALDEHYDES AND KETONES— OXIDATION OF ALDEHYDES

9.4　　Aldehydes and ketones are structurally similar and consequently they show similar chemical properties.　They do differ significantly in one chemical property—susceptibility to oxidation. Aldehydes are easily oxidized under mild conditions; ketones are not.

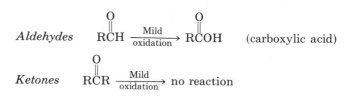

Aldehydes $R\overset{O}{\overset{\|}{C}}H \xrightarrow[\text{oxidation}]{\text{Mild}} R\overset{O}{\overset{\|}{C}}OH$ (carboxylic acid)

Ketones $R\overset{O}{\overset{\|}{C}}R \xrightarrow[\text{oxidation}]{\text{Mild}}$ no reaction

This difference in reactivity is the basis of the following diagnostic tests for distinguishing between aldehydes and ketones.

A. Tollens' "Silver Mirror" Test

Aldehydes can be distinguished from ketones using Tollens' reagent, which is a solution of silver nitrate in ammonium hydroxide (actually $Ag(NH_3)_2OH$). As the aldehyde is oxidized to the salt of a carboxylic acid, silver ion (Ag^+) is reduced to metallic silver. Ketones give no reaction.

$$R\overset{O}{\overset{\|}{C}}H + 2Ag(NH_3)_2{}^+ + 3OH^- \longrightarrow R\overset{O}{\overset{\|}{C}}O^- + 2Ag\downarrow + 4NH_3 + 2H_2O$$

If the reaction is allowed to proceed slowly in a clean test tube, metallic silver is deposited on the glass walls, creating a smooth reflective surface; hence the name *silver mirror* test. If the silver is plated on smooth, flat glass and a backing put over the silver layer to protect it from scratching, a mirror of good quality can be made. One looks through the glass to the silver layer.

B. Benedict's or Fehling's Test

Benedict's and Fehling's reagents consist of a basic solution of copper(II) ion complexed with citrate or tartrate ions, respectively. As the reaction proceeds, the aldehyde is oxidized to the salt of the carboxylic acid. In the process, the deep blue copper(II) ion complex is reduced to brick red copper(I) oxide. Ketones generally give no reaction.

$$R\overset{O}{\overset{\|}{C}}H + 2Cu^{2+} \text{ (complex)} + 5OH^- \longrightarrow R\overset{O}{\overset{\|}{C}}O^- + Cu_2O\downarrow + 3H_2O$$

This test is used clinically to detect glucose in the urine, a condition characteristic of disorders such as diabetes, in which the body is unable to metabolize glucose normally. Glucose is an aldehyde and gives a positive result.

$$\underset{\substack{| \quad |\\ OH \ \ OH\\ \text{Glucose}}}{CH_2(CH)_4\overset{O}{\overset{\|}{C}}H} + Cu^{2+} \text{ (complex)} \xrightarrow{OH^-} \underset{\substack{| \quad |\\ OH \ \ OH}}{CH_2(CH)_4\overset{O}{\overset{\|}{C}}O^-} + Cu_2O\downarrow$$

The average person has 60–100 mg of glucose per 100 ml of blood. If the concentration reaches the renal threshold level, 150–180 mg per 100 ml of blood, the kidneys spill glucose into the urine.*

---

*See Panel 14, Chapter 10, for a discussion of diabetes.

PROBLEM

9.5. Write equations showing the reactions of the isomers propanal and 2-propanone with **(a)** Tollens' reagent and **(b)** Benedict's reagent.

REACTIONS OF ALDEHYDES AND KETONES—ADDITION

9.5
A. General Considerations

1. General Reaction. Addition is the characteristic chemical reaction of most compounds possessing a multiple bond. Recall that addition is a very simple process in which atoms or groups add to adjacent atoms of a multiple bond. For example, we saw in Chapter 4 (section 4.1) that alkenes add a variety of reagents such as hydrogen, halogens, hydrogen halides, and water as summarized in the following equation:

$$Alkenes \qquad -\overset{|}{\underset{|}{C}}=\overset{|}{\underset{|}{C}}- + EA \longrightarrow -\overset{|}{\underset{\underset{E}{|}}{C}}-\overset{|}{\underset{\underset{A}{|}}{C}}- \qquad EA = H_2, X_2, HX, H_2O$$

Aldehydes and ketones possess a carbon-oxygen double bond and, as we might expect, addition is their most characteristic chemical reaction also. Unlike the double bond in alkenes, the carbonyl group has a permanent polarity. Consequently, unsymmetrical reagents (H—Nu) always add so that the positive portion bonds to the negative oxygen and the negative portion to the positive carbon.

$$Aldehydes\ and\ Ketones \qquad \overset{O^{\delta-}}{\underset{|}{\overset{||}{\underset{|}{C}}{}^{\delta+}}} + \overset{\delta+}{H}-\overset{\delta-}{Nu} \longrightarrow \overset{O-H}{\underset{|}{\overset{|}{\underset{|}{C}}-Nu}}$$

Although aldehydes and ketones add a variety of reagents, the reactions are generally not as simple as those of alkenes. This is because the product of straight addition is frequently unstable and either exists in equilibrium with the starting materials or reacts further to form a more stable substance. For example, the product of addition of water or hydrogen halide to a carbonyl compound usually comprises only a small portion of the equilibrium mixture between it and the starting materials. Even when it is formed in significant amounts, seldom can it be isolated from the reaction mixture.

$$-\overset{O}{\underset{|}{\overset{||}{C}}} + H_2O \underset{\text{Elimination}}{\overset{\text{Addition}}{\rightleftharpoons}} -\overset{OH}{\underset{|}{\overset{|}{C}}-OH} \qquad\qquad -\overset{O}{\underset{|}{\overset{||}{C}}} + HX \underset{\text{Elimination}}{\overset{\text{Addition}}{\rightleftharpoons}} -\overset{OH}{\underset{|}{\overset{|}{C}}-X}$$

Compounds of this type in which a carbon possesses an —OH or —NH group and one or more —OH, —OR, —NH$_2$, or —X (halogen) groups, are usually unstable and readily undergo elimination.

In subsequent sections, we will examine specific examples of aldehyde and ketone addition reactions. As you study the reactions, note that in each case the first step of the reaction is simple addition. After you have studied each reaction, consult the summary in Table 9.2 (page 223) so you can compare the similarities between the various reactions.

2. Mechanism of Nucleophilic Addition. Owing to the positive character of the carbonyl carbon, aldehydes and ketones generally react by a nucleophilic addition mechanism. In this type of mechanism, a nucleophile (Lewis base) is attracted to and bonds to the partially positive carbonyl carbon. The reaction can be initiated by either an acid or base.

In base-initiated nucleophilic addition, the nucleophile attacks the carbonyl carbon first and provides both electrons for the new carbon-nucleophile bond. The π electrons of the carbonyl are displaced to the oxygen, forming an anion. Abstraction of a hydrogen ion from HNu (or from neutralization with acid) by the negative oxygen completes the addition process.

$$\textit{Base-Initiated Addition} \qquad \overset{\ddot{O}}{\underset{|}{\overset{\|}{-C}}}\;:\overset{-}{Nu} \longrightarrow \overset{:\ddot{O}:^-}{\underset{|}{-C}}:Nu \xrightarrow{H:Nu} \overset{:\ddot{O}:H}{\underset{|}{-C}}:Nu + :\overset{-}{Nu}$$

In acid-initiated nucleophilic addition, a hydrogen ion bonds to the partially negative carbonyl oxygen; a carbocation results. The formation of a carbocation enhances the attraction of the nucleophile to the carbonyl carbon. Note that the reaction is acid-catalyzed: a hydrogen ion initiates the process and is returned in the final step.

$$\textit{Acid-Initiated Addition} \qquad \overset{\ddot{O}}{\underset{|}{\overset{\|}{-C}}} \xrightarrow{H^+} \overset{:\ddot{O}:H}{\underset{|}{-C^+}} \xrightarrow{H:\ddot{N}u} \overset{:\ddot{O}:H}{\underset{\underset{H}{|}}{-\overset{+}{C}:\ddot{N}u}} \xrightarrow{-H^+} \overset{:\ddot{O}:H}{\underset{|}{-C}}:Nu$$

Most of the reactions described below proceed by one of the previously outlined mechanisms.

B. Addition of Hydrogen Cyanide

1. General Reaction. Hydrogen cyanide adds to aldehydes and ketones to form a class of compounds known as *cyanohydrins* or *hydroxy nitriles*. In adding, the hydrogen ion bonds to the negative oxygen and the negative cyanide to the positive carbonyl carbon.

$$\overset{O}{\underset{|}{\overset{\|}{-C}}} + \overset{\delta^+ \;\; \delta^-}{HCN} \longrightarrow \overset{OH}{\underset{|}{-C}}-CN$$

Because the cyanide group is easily hydrolyzed to a carboxylic acid, cyanohydrins are useful intermediates in organic synthesis. The reaction is useful in preparing biological molecules such as hydroxy acids and carbohydrates. A variation of the reaction can lead to amino acids. For example, the hydroxy acid lactic acid (one enantiomer is found in sore muscles, the other in sour milk) can be prepared from ethanal by addition of HCN followed by hydrolysis.

$$\overset{O}{\underset{}{\overset{\|}{CH_3CH}}} + HCN \xrightarrow[HCN]{\substack{\text{Addition} \\ \text{of}}} \overset{OH}{\underset{\underset{H}{|}}{CH_3C}}-CN \xrightarrow[\substack{\text{(hydrolysis} \\ \text{of nitrile)}}]{H_2O/H^+} \overset{OH}{\underset{\underset{H}{|}}{CH_3C}}-\overset{O}{\underset{}{\overset{\|}{C}}OH}$$

Since hydrogen cyanide is extremely toxic (cyanide ion binds to blood hemoglobin and respiratory cytochromes in preference to oxygen), this reaction must be performed carefully in a fume hood.

2. Reaction Mechanism. Since hydrogen cyanide is a weak acid and poor nucleophile, HCN addition to aldehydes and ketones is often performed by mixing the aldehyde or ketone with a sodium cyanide solution followed by neutralization with acid. Under these conditions, the cyanide ion adds first and the reaction is base initiated.

$$\underset{\displaystyle|}{\overset{\displaystyle :\ddot{O}}{\underset{\displaystyle|}{\overset{\displaystyle \|}{-C}}}} \quad :\bar{C}{\equiv}N: \longrightarrow \underset{\displaystyle|}{\overset{\displaystyle :\ddot{O}:^-}{-\overset{\displaystyle|}{C}:C{\equiv}N}} \xrightarrow{\text{HCN}} \underset{\displaystyle|}{\overset{\displaystyle :\ddot{O}:H}{-\overset{\displaystyle|}{C}:C{\equiv}N:}} + :\bar{C}{\equiv}N:$$

PROBLEM

9.6. **(a)** Write reaction equations illustrating the addition of HCN to 2-butanone and to benzaldehyde. **(b)** Write a reaction mechanism for the addition of hydrogen cyanide to ethanal.

C. Reduction to Alcohols: Catalytic Hydrogenation

Addition of hydrogen to aldehydes and ketones, catalytically and under pressure, results in the formation of primary and secondary alcohols, respectively.

$$\textit{Aldehydes} \quad \overset{\displaystyle O}{\overset{\displaystyle \|}{R C H}} + H_2 \xrightarrow[\text{pressure}]{\text{Ni,}} \underset{\displaystyle H}{\overset{\displaystyle OH}{\underset{\displaystyle|}{R\overset{\displaystyle|}{C}H}}} \quad 1° \text{ alcohol}$$

$$\textit{Ketones} \quad \overset{\displaystyle O}{\overset{\displaystyle \|}{R C R}} + H_2 \xrightarrow[\text{pressure}]{\text{Ni,}} \underset{\displaystyle H}{\overset{\displaystyle OH}{\underset{\displaystyle|}{R\overset{\displaystyle|}{C}R}}} \quad 2° \text{ alcohol}$$

The reaction and its mechanism are analogous to that of addition of hydrogen to alkenes (section 4.1.B.1). It does not involve nucleophilic addition as do other reactions of carbonyl compounds.

PROBLEM

9.7. Write reaction equations illustrating the conversion of propanal and 2-propanone to primary and secondary alcohols, respectively, by hydrogenation. Why can't tertiary alcohols be prepared in this manner?

D. Reduction to Alcohols with Lithium Aluminum Hydride

1. General Reaction. A second and often more convenient method for the reduction of aldehydes and ketones to alcohols involves the use of metal hydrides such as lithium aluminum hydride ($LiAlH_4$) or sodium borohydride ($NaBH_4$). The procedure involves treating a carbonyl compound with lithium aluminum hydride in ether followed by hydrolysis in water or dilute acid.

$$\underset{\displaystyle|}{\overset{\displaystyle O}{\overset{\displaystyle \|}{-C}}} \xrightarrow{\text{LiAlH}_4} \xrightarrow[\text{H}^+]{\text{H}_2\text{O,}} \underset{\displaystyle|}{\overset{\displaystyle O-H}{-\overset{\displaystyle|}{C}-H}}$$

As with catalytic hydrogenation, primary and secondary alcohols can be prepared by this reaction.

2. Reaction Mechanism. When this reaction is examined closely, it proves to be an example of nucleophilic addition. Structurally, lithium aluminum hydride has an aluminum in the 3+ oxidation state with four bonded hydride ions (negative hydrogen ions, $H:^-$).

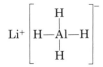

Being negative, the hydride ions are attracted to the positive carbonyl carbon and provide the electrons for a new carbon-hydrogen bond (remember, reduction involves the formation of new carbon-hydrogen bonds). The reaction is base initiated. One mole of $LiAlH_4$ actually reduces 4 moles of carbonyl and produces 4 moles of alkoxide ion (salt of alcohols). Treatment with water neutralizes the metal alkoxides to alcohols (Figure 9.2).

$$4R\overset{O}{\underset{R}{C}} + LiAlH_4 \longrightarrow \left(R\overset{O^-}{\underset{R}{CH}}\right)_4 Li^+Al^{+3} \xrightarrow[H^+]{H_2O,} 4R\overset{OH}{\underset{R}{CH}} + \begin{array}{c} Li^+ \\ Al^{3+} \end{array} \quad salts$$

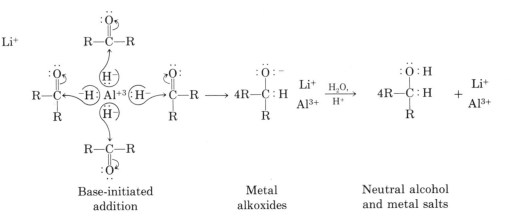

Base-initiated addition   Metal alkoxides   Neutral alcohol and metal salts

FIGURE 9.2. Mechanism of reduction with lithium aluminum hydride.

PROBLEM

9.8. (a) Write reaction equations showing the reduction of propanal and 2-propanone to primary and secondary alcohols, respectively, with lithium aluminum hydride. **(b)** Write a detailed reaction mechanism for the reduction of ethanal with lithium aluminum hydride.

3. Biological Reductions.   Organisms exist by the chemical conversion of foods such as carbohydrates to more elementary compounds like lactic acid or the ultimate combustion products, CO_2 and H_2O.   In the process, energy is released and used to power the thousands of other reactions going on at the same time throughout the organism.   Complete combustion involves O_2:

$$C_{12}H_{22}O_{11} \quad + 12O_2 \longrightarrow 12CO_2 + 11H_2O$$
Carbohydrate

Both anaerobic (without O_2) and aerobic metabolism (that involving O_2) consist of a series of oxidations and reductions along with other reactions.   Several of these steps involve the compound nicotinamide adenine dinucleotide (NAD⁺) and an enzyme.   During the reaction of NAD⁺ with some metabolic intermediate, generally termed MH₂, hydrogen, H_2, leaves the metabolite and is transferred to NAD⁺ in the form of a hydride ion, H : ⁻, forming NADH and leaving H⁺ as a by-product (Figure 9.3).

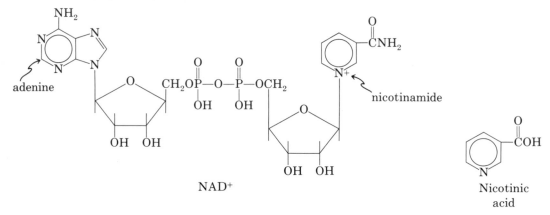

For example ethanol can be oxidized to acetaldehyde in the liver and kidneys as well as by yeast cells.

$$CH_3CH_2OH + NAD^+ \rightleftharpoons CH_3\overset{\overset{O}{\|}}{C}H + NADH + H^+$$

The reaction can be reversed by another type of compound that can extract the hydride ion (oxidize the NADH).   NAD⁺ and NADP⁺ (nicotinamide adenine dinucleotide phosphate) act with enzymes in more than 250 different redox reactions.   They are known as coenzymes and aid these biological catalysts in com-

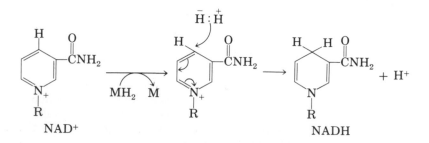

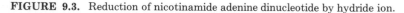

FIGURE 9.3.   Reduction of nicotinamide adenine dinucleotide by hydride ion.

pleting complex reactions efficiently and at a speed vital for continued existence. Nicotinic acid is also known as niacin, part of the water-soluble B-vitamin-complex group.

E. Grignard Addition—Preparation of Alcohols

1. General Reaction. One of the most versatile preparations of alcohols was developed by the French chemist Victor Grignard (1871–1935). His efforts won him a Nobel Prize in 1912. The Grignard reagent is prepared by treating an organic halide with magnesium metal in dry ether. The magnesium metal reacts slowly, forming a solution of the Grignard reagent.

$$RX + Mg \xrightarrow{\text{Ether}} RMgX$$

Grignard
reagent

R = alkyl or aromatic
X = Cl, Br, I

In reacting with aldehydes and ketones, the Grignard reagent adds to the carbon-oxygen double bond, with the negative alkyl group attacking the carbonyl carbon and the positive magnesium going to the negative oxygen. The resulting alkoxide is then hydrolyzed to an alcohol. Primary, secondary, and tertiary alcohols can be prepared in this fashion. Reaction with formaldehyde results in primary alcohols. With any other aldehyde, secondary alcohols are formed, and ketones are used to synthesize tertiary alcohols.

Formaldehyde
$$\underset{\displaystyle}{\overset{O}{\overset{\|}{H C H}}} + RMgX \longrightarrow \underset{R}{\overset{OMgX}{H C H}} \xrightarrow[H^+]{H_2O,} \underset{R}{\overset{OH}{H C H}} \quad 1° \text{ alcohols}$$

Aldehydes
$$\underset{\displaystyle}{\overset{O}{\overset{\|}{R_1 C H}}} + RMgX \longrightarrow \underset{R}{\overset{OMgX}{R_1 C H}} \xrightarrow[H^+]{H_2O,} \underset{R}{\overset{OH}{R_1 C H}} \quad 2° \text{ alcohols}$$

Ketones
$$\underset{\displaystyle}{\overset{O}{\overset{\|}{R_1 C R_2}}} + RMgX \longrightarrow \underset{R}{\overset{OMgX}{R_1 C R_2}} \xrightarrow[H^+]{H_2O,} \underset{R}{\overset{OH}{R_1 C R_2}} \quad 3° \text{ alcohols}$$

2. Reaction Mechanism. The reaction of Grignard reagents with carbonyl compounds is an example of base-initiated nucleophilic addition. Analyzing the charge distribution in the Grignard reagent, we find that since the magnesium is positive, the organic portion of the reagent must be negative and therefore a very powerful nucleophile.

$$\overset{-}{R}: \overset{2+}{Mg} : \overset{..}{\underset{..}{X}} :$$

When a Grignard reagent is mixed with an aldehyde or ketone, the negative alkyl group quickly attacks the positive carbonyl carbon and provides the two electrons needed for the new carbon-carbon bond. The π electrons are displaced to the

oxygen, forming the alcohol salt that is then neutralized to an alcohol with water
and acid.

$$\underset{\substack{\ddots\\ \delta+ \\ -\overset{|}{\underset{|}{C}}}}{\overset{\overset{\ddot{O}}{\underset{\delta-}{\parallel}}}{}}\ \overset{2+}{\underset{\bar{}}{R}:\ \overset{\bar{}}{MgX}}\ \longrightarrow\ \overset{:\overset{..}{O}:^{-}\ \overset{+}{MgX}}{-\overset{|}{\underset{|}{C}}:R}\ \xrightarrow{H_2O}\ \overset{:\overset{..}{O}:H}{-\overset{|}{\underset{|}{C}}:R}\ +\ MgXOH$$

Note that the alkyl group of a Grignard reagent essentially acts as a carbanion. It is for this reason that Grignard reactions must be performed in scrupu-
lously dry ether. Even traces of moisture can neutralize the reagent.

$$-\overset{|}{\underset{|}{C}}:^{-}\ \overset{+}{MgX}\ +\ \overset{\delta+}{H}-\overset{\delta-}{OH}\ \longrightarrow\ -\overset{|}{\underset{|}{C}}:H\ +\ MgXOH$$

PROBLEMS

9.9. (a) Write a reaction equation illustrating the preparation of a Grignard reagent from chloromethane. **(b)** Write reaction equations showing the preparation
of alcohols using the Grignard reagent methyl magnesium chloride and each of the
following carbonyl compounds: formaldehyde (methanal), propanal, and 2-pro-
panone.

9.10. Write a reaction mechanism for the reaction of methyl magnesium chloride
with formaldehyde, followed by hydrolysis.

3. Grignard Synthesis of Alcohols. How can a Grignard synthesis be
planned? First, one must recognize that during the reaction the Grignard reagent
always provides one alkyl group to the final alcohol product, and the others, if
any, must come from the carbonyl compound. So to make a primary alcohol,
one chooses formaldehyde as the carbonyl compound because it possesses no
alkyl groups—the one alkyl group is provided by the Grignard reagent. For a
secondary alcohol, an aldehyde provides one alkyl group and the Grignard reagent
the other.

Let us illustrate a specific problem, using the tertiary alcohol 2-phenyl-2-
butanol.

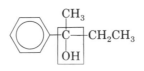

First identify the alcohol function (boxed in the formula as shown) and then
realize that two of the attached alkyl groups come from a carbonyl compound (a
ketone) and the third from a Grignard reagent. One could start with a carbonyl
compound possessing a methyl and ethyl group and a Grignard reagent with a
phenyl group.

Method 1

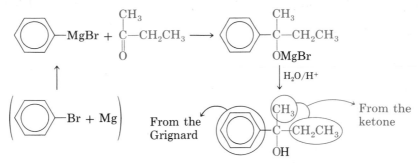

Two other combinations are possible. The ketone could be used to provide the phenyl and methyl groups and the Grignard to provide the ethyl group.

Method 2

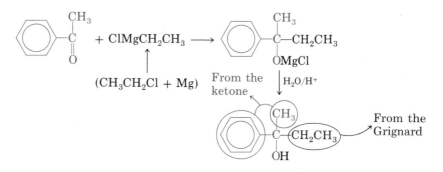

Finally, the phenyl and ethyl groups could be provided by the ketone and the methyl by the Grignard.

Method 3

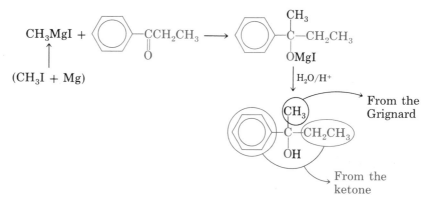

PROBLEM

9.11. Using the Grignard synthesis of alcohols, illustrate two methods for preparing 2-butanol.

4. Other Reactions of the Grignard Reagent and Organometallic Compounds. Grignard reagents are extremely versatile and react with a variety of other types of compounds. Reaction with ethylene oxide (section 8.7) converts an alkyl halide to a primary alcohol with a chain two carbons longer than the original halide.

$$RMgX + CH_2\!\!-\!\!CH_2 \longrightarrow RCH_2CH_2OMgX \xrightarrow[H^+]{H_2O,} RCH_2CH_2OH$$

Pouring a prepared Grignard reagent over crushed dry ice (carbon dioxide) followed by neutralization with weak acid produces carboxylic acids.

$$RMgX + O\!\!=\!\!C\!\!=\!\!O \longrightarrow R\overset{\text{O}}{\overset{\|}{C}}OMgX \xrightarrow[H^+]{H_2O} R\overset{\text{O}}{\overset{\|}{C}}OH$$

Some other organometallic compounds are alkyllithium reagents (RLi) and sodium acetylides, $RC\!\!\equiv\!\!CNa$. Acetylides (section 4.6) react with aldehydes and ketones as do Grignards to produce alcohols.

$$RC\!\!\equiv\!\!C\!:\!\overset{-}{}\overset{+}{Na} + \underset{R''}{\overset{\text{O}}{\overset{\|}{C}}R'} \longrightarrow \underset{R''}{\overset{\text{ONa}}{\overset{|}{RC\!\!\equiv\!\!C\!\!-\!\!C}}R'} \xrightarrow[H^+]{H_2O,} \underset{R''}{\overset{\text{OH}}{\overset{|}{RC\!\!\equiv\!\!C\!\!-\!\!C}}R'}$$

F. Alcohol Addition—Acetal Formation

1. General Reaction. Alcohols add to aldehydes and ketones to form hemiacetals (or hemiketals), which can condense with a second molecule of alcohol to produce acetals (or ketals). Note that the first step, hemiacetal formation, simply involves addition of the polar O—H group of the alcohol to the polar C=O of the aldehyde or ketone.

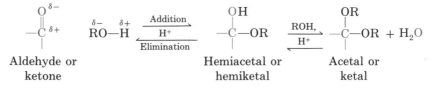

Aldehyde or ketone   Hemiacetal or hemiketal   Acetal or ketal

The hemiacetal is in equilibrium with the starting carbonyl compound, but the acetal can be isolated in a stable state if the water by-product is removed during its formation. The following example shows the formation of the methyl hemiacetal and dimethyl acetal of cyclopentanone:

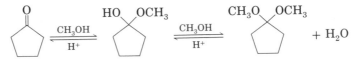

Carbohydrates usually exist in hemiacetal or acetal forms. The most prevalent example is glucose. Glucose in the open-chain form possesses both aldehyde and alcohol functions. In nature, glucose exists predominantly in a cyclic hemiacetal form, which arises by addition of the alcohol function on carbon-5 to the carbonyl.

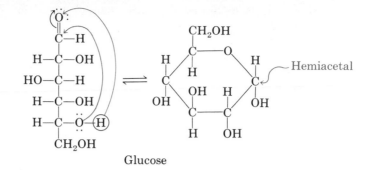

Glucose

2. Reaction Mechanism.   Hemiacetal formation occurs by an acid-initiated nucleophilic substitution mechanism.   The entire process is an equilibrium.

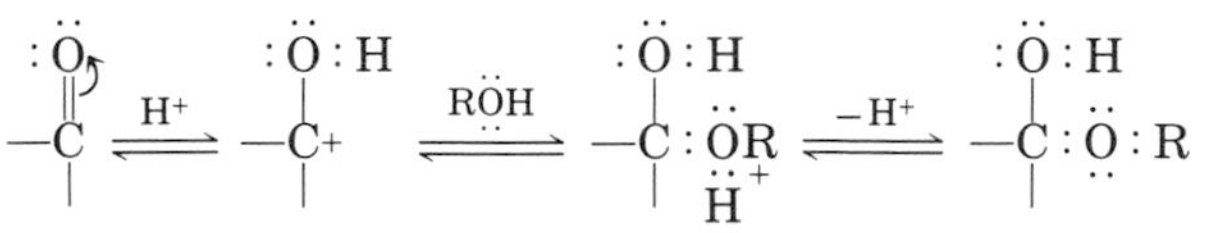

Reaction with a second mole of alcohol yields an acetal essentially by an intermolecular dehydration.

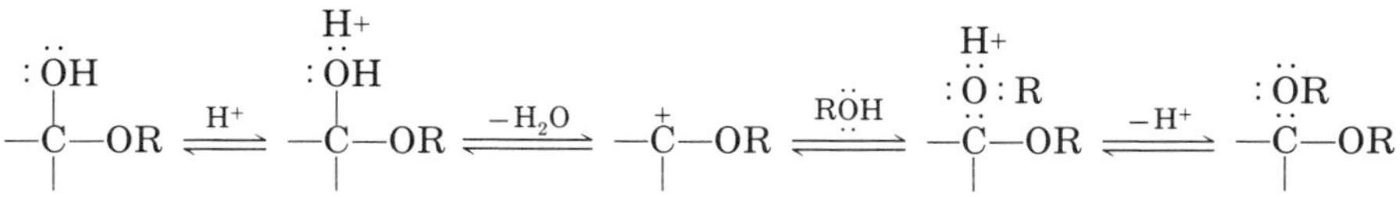

PROBLEM

9.12. Write structures for **(a)** the hemiacetal and **(b)** acetal that is formed in the reaction of benzaldehyde with ethanol.

G. Addition of Ammonia Derivatives

Primary amines (RNH_2) add to aldehydes and ketones in a way analogous to the way alcohols add to form hemiacetals.   The nucleophilic addition reaction involves addition of a polar nitrogen-hydrogen bond from the amine to the polar carbon-oxygen double bond of the carbonyl.   The initial addition product is not stable and a molecule of water is eliminated between the carbon and nitrogen to form a double bond.   The product is called an *imine.*

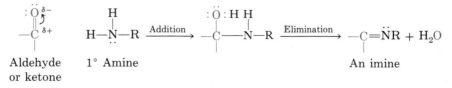

Aldehyde   1° Amine   An imine
or ketone

Aldehydes and ketones react with a variety of ammonia derivatives to form crystalline derivatives that can be used to characterize the compound.   Note that in each of the following reactions, the carbonyl reacts with an —NH_2 group.   A carbon-nitrogen double bond forms in place of the original carbon-oxygen double bond.

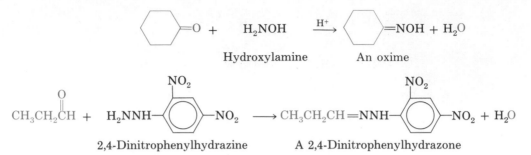

Imine formation is important biochemically since many enzymes use an —NH_2 group of an amino acid to react with and bind a carbonyl substrate to the enzyme. In the rods of the eye, for example, 11-*cis*-retinal combines with a large protein molecule, opsin, through an imine function to form rhodopsin, which is operative in converting light impulses into nerve impulses (see Panel 1, Chapter 2).

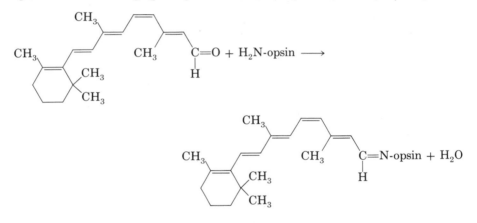

PROBLEM

9.13. Write equations showing reactions between the following substances: **(a)** benzaldehyde and 2,4-dinitrophenylhydrazine; **(b)** 2-pentanone and hydroxylamine.

REACTIONS INVOLVING α-HYDROGENS

9.6

A. Acidity of α-Hydrogens

Hydrogens on a carbon directly attached to a carbonyl group are referred to as *alpha*-(α-)hydrogens.

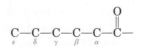

TABLE 9.2.   ADDITION REACTIONS OF ALDEHYDES AND KETONES

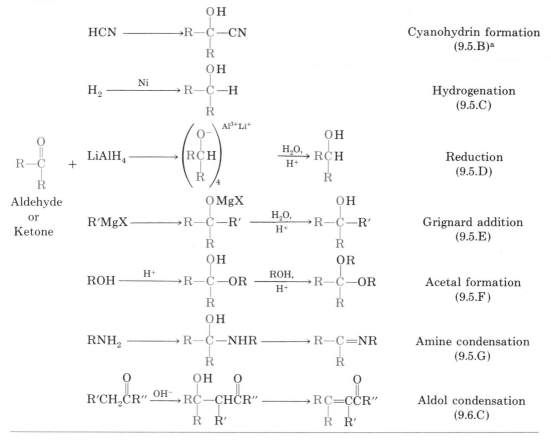

NOTE: The first step of all of these reactions is simple addition to the carbon-oxygen double bond.
[a] Section number

Due to the electron-withdrawing power of the carbon-oxygen double bond, the
α-carbon-hydrogen bonds are polar and weakly acidic. Strong bases like sodium
hydroxide can abstract the hydrogens, forming a carbanion.

As shown in Figure 9.4, this carbanion is resonance-stabilized: the negative charge
is delocalized between the α-carbon and the carbonyl oxygen. The greater the
charge dispersal, that is, the greater the number of atoms involved in accommo-
dating the destabilizing character of the negative charge, the greater the stability
of the carbanion. This phenomenon of delocalization can be compared to everyday experiences. A given quantity of heat, for instance, applied to a specific area
of the body could cause a severe burn; if the same amount of heat were dispersed
over the entire body, it would produce only gentle warming. A strong wind blowing all day is less destructive than a tornado touching down for a few seconds.

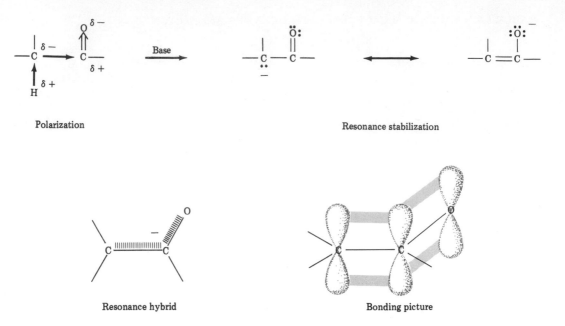

Polarization

Resonance stabilization

Resonance hybrid

Bonding picture

FIGURE 9.4. α-Hydrogens are acidic due to polarization of the carbon-hydrogen bond by the carbonyl group and resonance stabilization of the carbanion (enolate ion).

α-Hydrogens are acidic, then, because the carbon-hydrogen bond is polarized by the adjacent carbonyl function and the resulting carbanion is resonance-stabilized. The carbanion is usually referred to as an *enolate ion* because it is the salt of the enol formed when the carbanion is neutralized by acid.

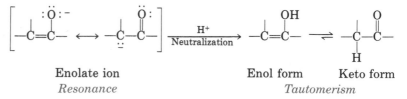

Enolate ion   Enol form   Keto form
Resonance   *Tautomerism*

Enols are relatively unstable, and although they exist in equilibrium with the free aldehyde or ketone (keto form), the equilibrium usually favors the keto form. This type of interconversion is called *keto-enol tautomerism.* Tautomerism differs from resonance in that it involves the movement of both electrons and atoms (a hydrogen in this case), whereas in drawing resonance structures one varies the positions of electrons only.

B. Haloform Reaction

1. General Reaction. α-Hydrogen acidity is responsible for a significant part of the chemistry of carbonyl compounds. We consider only two reactions in this group, the haloform reaction and the aldol condensation.

In the haloform reaction, methyl aldehydes and ketones when treated with a halogen in basic solution are converted to a haloform and the salt of a carboxylic acid.

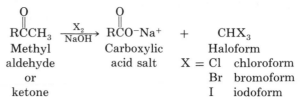

The methyl group becomes a haloform and the remainder of the molecule forms a carboxylic acid salt. When this reaction is performed with iodine, it is known as the iodoform test; it is used as a diagnostic tool for identifying compounds with

$$\overset{O}{\overset{\|}{}}$$

the general structure $RCCH_3$, that is, methyl aldehydes and ketones. A positive test is indicated by the formation of the yellow precipitate iodoform, which has a medicinal odor.

2. Reaction Mechanism. In the haloform reaction, the α-hydrogens of a methyl aldehyde or ketone are initially replaced by halogens. Each substitution is a two-step process, commencing with the abstraction of an acidic α-hydrogen by the hydroxide base; from this reaction, the resonance-stabilized enolate carbanion is formed. The carbanion displaces a halide from a halogen molecule, completing the substitution.

$$\underset{\underset{H}{|}}{\overset{O}{\overset{\|}{RC}}CH_2} + OH^- \longrightarrow \overset{O}{\overset{\|}{RC}}\overset{..}{CH_2} \xrightarrow{X_2} \underset{\underset{X}{|}}{\overset{O}{\overset{\|}{RC}}CH_2} + X^-$$

The acidity of the remaining α-hydrogens is enhanced by the electronegativity of the halogen, and they subsequently undergo abstraction and replacement with increasing facility.

$$\underset{\underset{X}{|}}{\overset{O}{\overset{\|}{RC}}CH_2} \xrightarrow{OH^-} \underset{\underset{X}{|}}{\overset{O}{\overset{\|}{RC}}CH:^-} \xrightarrow{X_2} \underset{\underset{X}{|}}{\overset{O}{\overset{\|}{RC}}CHX} \xrightarrow{OH^-} \underset{\underset{X}{|}}{\overset{O}{\overset{\|}{RC}}\overset{..}{C}{}^-{-}X} \xrightarrow{X_2} \underset{\underset{X}{|}}{\overset{O}{\overset{\|}{RC}}\overset{X}{\overset{|}{C}}{-}X}$$

The trihalogenated aldehyde or ketone is unstable to base and is attacked at the carbonyl carbon, with the displacement of a trihalomethyl anion. This carbanion is neutralized by the newly formed carboxylic acid.

$$\underset{\underset{X}{|}}{\overset{O}{\overset{\|}{RC}}\overset{X}{\overset{|}{C}}{-}X} \xrightarrow{OH^-} R{-}\underset{\underset{OH}{|}}{\overset{O^-}{\overset{|}{C}}}{-}\overset{X}{\underset{X}{\overset{|}{C}}}{-}X \longrightarrow \overset{O}{\overset{\|}{RC}}OH + {}^-{:}\overset{X}{\underset{X}{\overset{|}{C}}}{-}X \longrightarrow \overset{O}{\overset{\|}{RC}}O^- + H\overset{X}{\underset{X}{\overset{|}{C}}}{-}X$$

PROBLEM

9.14. (a) Write an equation illustrating the iodoform reaction using a 2-pentanone. **(b)** Will 3-pentanone give a positive iodoform test? **(c)** Only one aldehyde gives a positive iodoform test. Draw its structure.

C. The Aldol Condensation

1. General Reaction. Under conditions of basic catalysis, a carbon-hydrogen bond alpha to the carbonyl group of an aldehyde or ketone can be made to add to the carbon-oxygen double bond of another molecule. Since the product formed when an aldehyde is subjected to these conditions has both an aldehyde and alcohol function, the reaction is called the *aldol condensation.*

$$2RCH_2\overset{\overset{\displaystyle O}{\|}}{C}H \xrightarrow{\;OH^-\;} RCH_2\overset{\overset{\displaystyle OH}{|}}{C}H\overset{}{C}H\overset{\overset{\displaystyle O}{\|}}{C}H$$
$$\underset{\displaystyle R}{|}$$

An aldol

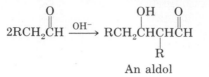

Aldols are easily dehydrated, because the resulting double bond is conjugated with the carbonyl group, which creates an extended system of overlapping p orbitals, that is, a resonance-stabilized structure.

$$RCH_2\overset{\overset{\displaystyle OH}{|}}{C}H\overset{}{-}\overset{\overset{\displaystyle H}{|}}{C}\overset{}{-}\overset{\overset{\displaystyle O}{\|}}{C}H \xrightarrow{\;Dehydration\;} RCH_2CH{=}\overset{\overset{\displaystyle O}{\|}}{C}CH$$
$$\underset{\displaystyle R}{|} \qquad\qquad \underset{\displaystyle R}{|}$$

In some cases, the aldol dehydrates spontaneously on formation or during acid neutralization of the reaction mixture, and its isolation becomes impossible. In summary, the aldol condensation involves addition of an α-carbon-hydrogen bond to the carbonyl group. The resulting aldol can sometimes be isolated, but it often dehydrates. The overall process resembles the reaction of aldehydes and ketones with ammonia derivatives (section 9.5.G and Table 9.2).

$$2CH_3\overset{\overset{\displaystyle O}{\|}}{C}H \xrightarrow{\;OH^-\;} CH_3\overset{\overset{\displaystyle OH}{|}}{C}HCH_2\overset{\overset{\displaystyle O}{\|}}{C}H \xrightarrow{\;H^+\;} CH_3CH{=}CH\overset{\overset{\displaystyle O}{\|}}{C}H$$

$$\left(CH_3\overset{\overset{\displaystyle O}{\|}}{\underset{\displaystyle H}{C}}\overset{\frown}{\underset{}{}}\overset{\displaystyle H}{\underset{}{}}\overset{\displaystyle O}{CH_2}\overset{\overset{\displaystyle O}{\|}}{C}H\right)$$

$$2CH_3CH_2\overset{\overset{\displaystyle O}{\|}}{C}H \xrightarrow{\;OH^-\;} CH_3CH_2\overset{\overset{\displaystyle OH}{|}}{C}H\overset{}{C}H\overset{\overset{\displaystyle O}{\|}}{C}H \xrightarrow{\;H^+\;} CH_3CH_2CH{=}\overset{\overset{\displaystyle O}{\|}}{C}CH$$
$$\underset{\displaystyle CH_3}{|} \qquad\qquad\qquad \underset{\displaystyle CH_3}{|}$$

$$\left(CH_3CH_2\overset{\overset{\displaystyle O}{\|}}{\underset{\displaystyle H}{C}}\overset{\frown}{\underset{}{}}\overset{\displaystyle H}{\underset{}{}}\overset{\displaystyle O}{\underset{\displaystyle CH_3}{C}H}\overset{\overset{\displaystyle O}{\|}}{C}H\right)$$

2. Mechanism of the Aldol Condensation. The aldol condensation depends on the acidity of α-hydrogens in aldehydes and ketones. Let us consider the base-catalyzed condensation of acetaldehyde. To initiate the reaction, a hydroxide base abstracts an α-hydrogen, generating the resonance-stabilized enolate anion.

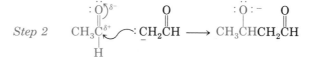

$$\text{Step 1} \quad CH_3\overset{\displaystyle O}{\overset{\|}{C}}H + OH^- \longrightarrow \underset{\cdot\cdot}{C}H_2{-}\overset{\displaystyle O}{\overset{\|}{C}}H + H_2O$$

Once the carbanion is formed, it attacks the positive carbonyl carbon of another acetaldehyde molecule by a nucleophilic addition mechanism in an effort to neutralize itself. As the carbanion bonds, the π electrons of the carbonyl group are transferred completely to the oxygen, forming an alkoxide ion.

$$\text{Step 2} \quad CH_3\overset{\delta^+}{\underset{H}{C}}\overset{:\ddot{O}^{\delta-}}{\|} \curvearrowleft :\underset{-}{C}H_2\overset{\displaystyle O}{\overset{\|}{C}}H \longrightarrow CH_3\overset{:\ddot{O}:^-}{C}HCH_2\overset{\displaystyle O}{\overset{\|}{C}}H$$

The alkoxide ion is neutralized by a water molecule, and the catalyst regenerated in the process.

$$\text{Step 3} \quad CH_3\overset{:\ddot{O}:^-}{C}HCH_2\overset{\displaystyle O}{\overset{\|}{C}}H + H_2\ddot{O}: \longrightarrow CH_3\overset{:\ddot{O}H}{C}HCH_2\overset{\displaystyle O}{\overset{\|}{C}}H + :\ddot{O}H^-$$

PROBLEM

9.15. Write reaction equations illustrating the aldol condensation using butanal. Include both the aldol and dehydration product.

3. Crossed Aldol Condensations. Aldol condensations between two different carbonyl compounds can be performed successfully as long as one of the reactants has no α-hydrogens. For example, by mixing benzaldehyde (which has no α-hydrogens) and a base and slowly adding acetaldehyde a drop at a time (to prevent its condensing with itself), one can synthesize cinnamaldehyde, the primary component of cinnamon oil.

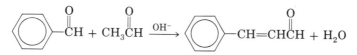

4. Aldol Additions in Nature. One step in the synthesis of glucose by green plants involves a crossed aldol addition (without subsequent dehydration) between the monophosphates of glyceraldehyde and dihydroxyacetone to produce fructose 1,6-diphosphate. The reaction is catalyzed by the enzyme aldolase.

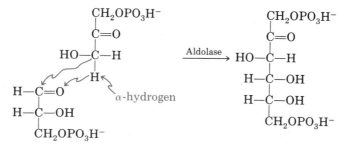

Panel 13

FORMALDEHYDE POLYMERS

Formaldehyde is a gas with a pungent, suffocating odor, very irritating to mucous membranes. It has such a great propensity toward polymerization, that it can polymerize spontaneously and explosively. In fact, it is usually sold and transported as low-molecular-weight polymers, all of which are easily depolymerized to free formaldehyde.

Formaldehyde can undergo a base-initiated nucleophilic addition polymerization to form the high-molecular-weight polymers known as Delrin and Celcon. End groups (such as $R = CH_3$, $CH_3\overset{\overset{\displaystyle O}{\|}}{C}$) are tacked on to the end of the polymer to prevent it from unraveling. These thermoplastics are very tough and resilient materials with high tensile strength and resistance to stress cracking. They are used in a variety of engineering applications such as in gears, bearings, pump parts, business machinery, and instrument housings.

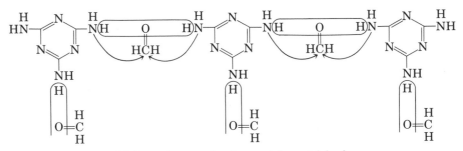

Formaldehyde also forms three-dimensional, highly crosslinked thermosetting copolymers with phenol, urea, and melamine. Urea-formaldehyde resins are used as adhesives in the manufacture of plywood and particle board as well as in a variety of other applications. Formaldehyde condenses with melamine (as illustrated using lasso chemistry) to produce the familiar Melmac of Melmac plastic dishes.

Melmac from melamine and formaldehyde

The first manufactured polymer of commercial importance was a formaldehyde-phenol resin developed by Leo Baekeland in 1907. At the time, Baekeland was interested in synthesizing an artificial shellac. Shellac had been collected from the Indian female Lac insect, which attaches itself to a tree at birth and spends its entire life sucking sap. After ingesting the sap, the insect secretes a sticky resin, which accumulates until the insect is completely surrounded and trapped, except for a few openings for breathing purposes and for newborn insects to crawl through. After producing a new generation of insects, the females die and are collected, ground, and processed into shellac. To produce one pound of shellac requires six months of labor by 150,000 Lac insects.

[PANEL 13]   Baekeland, in his research toward discovering a synthetic shellac, found in chemistry journals that in 1871 Adolf von Baeyer, a German chemist, had produced a hard, hornlike, insoluble, gray material on heating phenol (a disinfectant) and formaldehyde (biological preservative, methanal, HCHO). To Baeyer and other chemists of his time, such substances were troublesome and annoying; not only were the scientists unable to analyze, identify, and classify the materials but it was often impossible to clean the products out of their laboratory equipment. Baekeland, however, recognized these properties as having potential utility and dedicated himself to developing this material into a useful polymer. In 1907 he successfully controlled the reaction to produce a thermosetting plastic, which he called Bakelite.

Although the chemistry of Bakelite formation is complex, it involves reactions we have studied—addition and aromatic substitution. Bakelite can be formed by heating phenol and formaldehyde under either acidic or basic conditions. Under acidic conditions, a hydrogen ion adds to the negative oxygen of formaldehyde, this reaction producing a carbocation, which attacks the π cloud of the phenol ring. Electrophilic aromatic substitution, specifically a modification of the Friedel-Crafts reaction, occurs.

$$\overset{\delta-}{\underset{\delta+}{H\overset{O}{\overset{\|}{C}}H}} + H^+ \longrightarrow H-\overset{OH}{\underset{+}{C}}-H \qquad (\overset{+}{C}H_2OH)$$

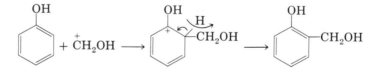

Overall, addition has occurred on formaldehyde's carbon-oxygen double bond, and substitution has occurred on the benzene ring—an alcohol (CH_2OH) is the result. The alcohol is protonated by the acid (see section 3.7.B.2, dehydration of alcohols), a molecule of water is expelled, and the resulting carbocation attacks another phenol ring. (Once again the Friedel-Crafts reaction is evident.)

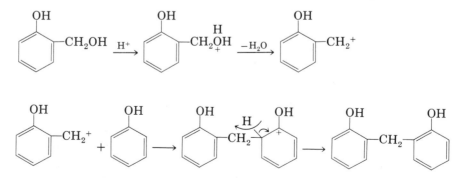

The hydroxy group is both a strong activating group and an ortho, para director (remember that phenol readily tribrominates at o,p positions). Thus, the reactions outlined above can occur at all ortho and para positions, producing a gigantic polymer that expands in three dimensions.

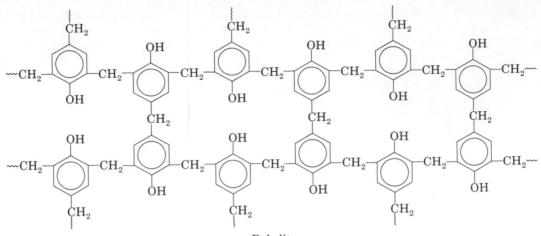

Bakelite
(a three-dimensional thermosetting plastic)

[PANEL 13]   Phenolic resins are widely used. Molded products are common, such as handles on electrical and cooking utensils, electrical plates and switches, appliances, and some business machines. One of the largest single uses is as a bonding adhesive in plywood and particle board.

Baekeland was already an established chemist with an excellent reputation when he embarked on his investigation into phenolic resins. While in his thirties, he invented Velox, the first photographic paper that could be exposed with artificial light. George Eastman, who developed the Kodak camera and later a large manufacturing company, invited Baekeland to his plant to discuss purchase of the Bakelite patent. Baekeland had decided that he would ask for $50,000 and compromise on no less than $25,000. Eastman's initial offer was $1 million.

END-OF-CHAPTER PROBLEMS

9.16 IUPAC Nomenclature: Name by the IUPAC system of nomenclature

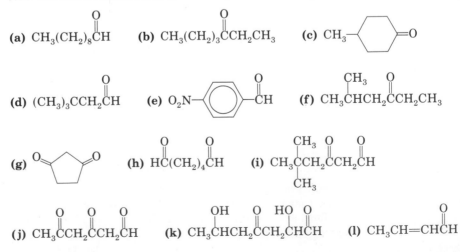

(m) $HC{\equiv}CCH_2\overset{\overset{O}{\|}}{C}CH_3$ **(n)** $CH_2{=}CH\overset{\overset{O}{\|}}{C}CH{=}CH\overset{\overset{O}{\|}}{C}H$

(o)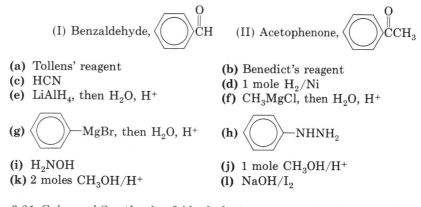

9.17 IUPAC Nomenclature: Draw structures for each of the following compounds: **(a)** 3-heptanone; **(b)** octanal; **(c)** 5-ketohexanal; **(d)** 3,7-dihydroxy-5-ketoheptanal; **(e)** 3-cyclopentenone; **(f)** 1,1,1,5,5,5-hexabromo-2,4-pentandione; **(g)** 4-keto-7-bromo-7-ethyl-9-hydroxy-2,5-nonadiynal; **(h)** m-methylbenzaldehyde; **(i)** 1-phenyl-2-butanone.

9.18 Common Nomenclature: Draw structures for each of the following compounds: **(a)** butyl ethyl ketone; **(b)** acetone; **(c)** formaldehyde; **(d)** chloroacetaldehyde; **(e)** dipropyl ketone; **(f)** diphenyl ketone

9.19 Preparations of Aldehydes and Ketones: Complete the following reactions illustrating some preparations of aldehydes and ketones:

(a) $CH_3C{\equiv}CCH_3 + H_2O \xrightarrow[HgSO_4]{H_2SO_4,}$

(b) $CH_3\overset{\overset{CH_3}{|}}{C}{=}CHCH_3 \xrightarrow[H_2O]{O_3 \quad Zn,}$

(c) [benzene ring] $+ CH_3CH_2CH_2\overset{\overset{O}{\|}}{C}Cl \xrightarrow{AlCl_3}$

(d) $CH_3CH_2\overset{\overset{OH}{|}}{C}HCH_2CH_3 \xrightarrow{Na_2Cr_2O_7}$

9.20 Reactions of Aldehydes and Ketones: Write equations illustrating the reaction (if any) of the following two compounds with each of the reagents listed:

(I) Benzaldehyde, [benzene ring]$\overset{\overset{O}{\|}}{C}H$ (II) Acetophenone, [benzene ring]$\overset{\overset{O}{\|}}{C}CH_3$

(a) Tollens' reagent   **(b)** Benedict's reagent
(c) HCN   **(d)** 1 mole H_2/Ni
(e) $LiAlH_4$, then H_2O, H^+ **(f)** CH_3MgCl, then H_2O, H^+

(g) [benzene ring]$-MgBr$, then H_2O, H^+ **(h)** [benzene ring]$-NHNH_2$

(i) H_2NOH **(j)** 1 mole CH_3OH/H^+
(k) 2 moles CH_3OH/H^+ **(l)** $NaOH/I_2$

9.21 Grignard Synthesis of Alcohols: Prepare the following alcohols in all the possible ways using the Grignard synthesis:

(a) $CH_3CH_2CH_2CH_2OH$ **(b)** $CH_3\overset{\overset{}{}}{\underset{\underset{OH}{|}}{C}}HCH_3$ **(c)** $CH_3CH_2\overset{\overset{CH_3}{|}}{\underset{\underset{OH}{|}}{C}}CH_2CH_2CH_3$

9.22 Organometallic Chemistry: Complete the following reactions showing the major organic products:

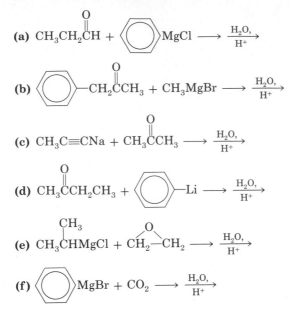

(a) $CH_3CH_2\overset{\overset{\displaystyle O}{\|}}{C}H$ + ⬡MgCl ⟶ $\overset{\text{H}_2\text{O,}}{\underset{\text{H}^+}{\longrightarrow}}$

(b) ⬡—$CH_2\overset{\overset{\displaystyle O}{\|}}{C}CH_3$ + CH_3MgBr ⟶ $\overset{\text{H}_2\text{O,}}{\underset{\text{H}^+}{\longrightarrow}}$

(c) $CH_3C{\equiv}CNa$ + $CH_3\overset{\overset{\displaystyle O}{\|}}{C}CH_3$ ⟶ $\overset{\text{H}_2\text{O,}}{\underset{\text{H}^+}{\longrightarrow}}$

(d) $CH_3\overset{\overset{\displaystyle O}{\|}}{C}CH_2CH_3$ + ⬡—Li ⟶ $\overset{\text{H}_2\text{O,}}{\underset{\text{H}^+}{\longrightarrow}}$

(e) $CH_3\overset{\overset{\displaystyle CH_3}{|}}{C}HMgCl$ + $\overset{\overset{\displaystyle O}{\diagup\;\diagdown}}{CH_2{-}CH_2}$ ⟶ $\overset{\text{H}_2\text{O,}}{\underset{\text{H}^+}{\longrightarrow}}$

(f) ⬡MgBr + CO_2 ⟶ $\overset{\text{H}_2\text{O,}}{\underset{\text{H}^+}{\longrightarrow}}$

9.23 Aldol Condensation: Write reaction equations illustrating the aldol condensation of the following compounds using sodium hydroxide as the base. Show the aldol initially formed and the unsaturated aldehyde or ketone produced by dehydration.

(a) $CH_3CH_2CH_2CH_2\overset{\overset{\displaystyle O}{\|}}{C}H$ (b) $CH_3\overset{\overset{\displaystyle CH_3}{|}}{C}HCH_2\overset{\overset{\displaystyle O}{\|}}{C}H$ (c) ⬡$\overset{\overset{\displaystyle O}{\|}}{C}CH_3$

9.24 Crossed Aldol Condensation: Write equations showing the preparation of benzalacetophenone by the aldol condensation of benzaldehyde and acetophenone (methyl phenyl ketone, see problem 9.20). Show both the initial aldol and then the dehydration product.

9.25 Aldol Condensation: Show how the following compounds could be prepared by the aldol condensation:

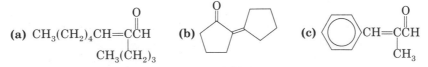

(a) $CH_3(CH_2)_4CH{=}\overset{\overset{\displaystyle O}{\|}}{\underset{\underset{\displaystyle CH_3(CH_2)_3}{|}}{C}}H$ (b)   (c) ⬡$CH{=}\overset{\overset{\displaystyle O}{\|}}{\underset{\underset{\displaystyle CH_3}{|}}{C}}CH$

9.26 Acetal Formation: Write the acetal or ketal that would result from reaction of the following compounds: **(a)** propanal and 2 moles ethanol; **(b)** propanone and 2 moles methanol; **(c)** cyclohexanone and 1 mole 1,2-ethanediol.

9.27 Synthesis of Familiar Compounds: Write equations showing the preparation of the following compounds by the steps described:

(a) rubbing alcohol by the hydrogenation of acetone

(b) formaldehyde by the oxidation of methanol

(c) chloral hydrate (knockout drops, sleep producer) by treatment of acetalde-hyde with 3 mol chlorine and sodium hydroxide (the carbon-carbon bond cleavage in the haloform reaction does not occur in this synthesis) followed by addition of water to the aldehyde function. (This is one of the few cases in which water forms a stable addition product with a carbonyl compound.)

(d) insect repellent 6-12 by aldol condensation (without dehydration) of butyral-dehyde followed by complete hydrogenation

(e) acetic acid (vinegar taste and odor in dilute water solutions) by oxidation of acetaldehyde

(f) rose oil by reaction of phenylmagnesium bromide with ethylene oxide followed by acid hydrolysis

9.28 Preparations of Alcohols: Show in as many ways as possible how 2-penta-nol could be prepared from carbonyl compounds by the Grignard synthesis and by reduction with H_2/Ni and $LiAlH_4$.

9.29 Keto-Enol Tautomerism: **(a)** Most keto-enol equilibrium mixtures consist primarily of the keto form. However, the enol of 2,4-cyclohexadienone is consid-erably more stable than the ketone and exists to the exclusion of the ketone. Draw the enol form and explain its unusual stability. **(b)** The enamine, $CH_2=CH-NHCH_3$, is part of a tautomeric mixture of which the other tautomer is the more stable component. Draw the other tautomer.

9.30 Reaction Mechanisms: Write (1) products and (2) reaction mechanisms for the reaction of propanal with the following reagents:

(a) CH_3MgCl, then H_2O, H^+ (b) NaCN, then H^+

(c) $LiAlH_4$, then H_2O (d) H_2NOH/H^+

(e) NaOH (aldol condensation)   (f) $2CH_3OH/H^+$

9.31 Reaction Mechanisms: Write stepwise reaction mechanisms for the three reactions involving organometallic compounds shown in section 9.5.E.4.

9.32 Acidity of α-Hydrogens: There are three distinct types of hydrogens in the following molecule. Arrange them in order of increasing acidity. Explain your order.

$$CH_3CH_2\overset{\overset{\displaystyle O}{\|}}{C}CH_2\overset{\overset{\displaystyle O}{\|}}{C}CH_2CH_3$$

9.33 Aldol-type Condensations: Aldehydes and ketones can engage in aldol-type condensations with other molecules that have acidic hydrogens. Show the base-catalyzed aldol condensation product of benzaldehyde with the following compounds:

(a) CH_3NO_2 (b) CH_3CN (c) $CH_3O\overset{\overset{\displaystyle O}{\|}}{C}CH_2\overset{\overset{\displaystyle O}{\|}}{C}OCH_3$

9.34 Organic Qualitative Analysis: Show how you could chemically distin-guish between the compounds in the following sets. Tell what you would do and see.

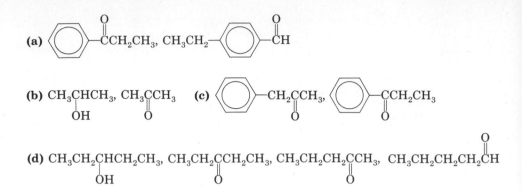

(a)

(b) CH_3CHCH_3, CH_3CCH_3 **(c)**

(d) $CH_3CH_2CHCH_2CH_3$, $CH_3CH_2CCH_2CH_3$, $CH_3CH_2CH_2CCH_3$, $CH_3CH_2CH_2CH_2CH$

9.35 Carbohydrate Chemistry: Below are the structures of sucrose (cane sugar, table sugar) and lactose (5% of human milk and cow's milk). **(a)** Identify any acetal or hemiacetal linkages. **(b)** Which of these sugars would react with Benedict's reagent?

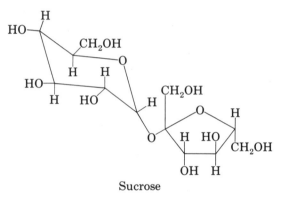

Sucrose

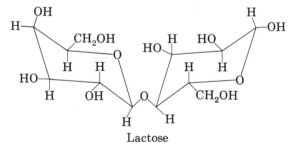

Lactose

9.36 Organic Qualitative Analysis: An unknown compound has the formula C_7H_{14}. Ozonolysis (O_3, Zn/H_2O) gives two substances, both of which react with 2,4-dinitrophenylhydrazine to give a solid derivative. One of the ozonolysis products gives both a positive Tollens' and a positive iodoform test, whereas the other is negative to both of these tests. What is the structure of the unknown?

Chapter 10

Carbohydrates

The term *carbohydrate* is an understandable misnomer. This class of compounds is of the general chemical formula $C_n(H_2O)_m$, and appears, without structural analysis, to contain hydrates of carbon. However, the structures of the simplest carbohydrates—glyceraldehyde and dihydroxyacetone—illustrate their actual chemical nature, that of polyhydroxyaldehydes (aldoses), polyhydroxyketones (ketoses), and their derivatives.

$$
\begin{array}{cc}
\text{HC}=\text{O} & \text{H}_2\text{COH} \\
\text{HCOH} & \text{C}=\text{O} \\
\text{H}_2\text{COH} & \text{H}_2\text{COH} \\
\text{C}_3\text{H}_6\text{O}_3 & \text{C}_3\text{H}_6\text{O}_3 \\
\text{Glyceraldehyde} & \text{Dihydroxyacetone}
\end{array}
$$

Another common name for carbohydrate is *saccharide*. This term is derived from several languages and means "sugar." Indeed the most familiar saccharides are sweet, including table sugar, or sucrose, fructose (honey), and glucose (grape sugar).

MONOSACCHARIDES

10.1   The monosaccharides are single carbohydrate units. Many are found naturally, and characteristically they contain a carbonyl functional group ($C=O$) and also alcohol groups ($C-OH$).

A. Nomenclature

In general, saccharides are given the suffix *-ose*. Three-carboned monosaccharides are called *tri*oses. Those containing four carbons are *tetr*oses, five carbons *pent*oses, six carbons *hex*oses, and seven carbons *hept*oses. Monosaccharides containing the carbonyl group as a ketone are given the prefix *keto*-; those with an aldehyde have the prefix *aldo*-. A six-carboned ketose, for example, is a ketohexose; a five-carboned aldose is an aldopentose. Individual monosaccharides also have common names, such as ribose, arabinose, glucose, and sorbose, which are used most often.

B. Structure

You have seen many of the structural aspects of carbohydrates previously in simpler compounds. Since carbohydrates are larger molecules and contain more than one type of functional group, their overall structures will reflect the effects of this multifunctionality.

1. Optical Isomerism. The smallest aldose, glyceraldehyde, has one chiral carbon atom, which gives rise to two optical isomers.

$$
\begin{array}{ll}
\text{H—C=O} & \qquad\qquad \text{H—C=O} \\
\quad| & \qquad\qquad\qquad | \\
\text{H—C—OH} \quad \text{D-Glyceraldehyde} & \text{HO—C—H} \quad \text{L-Glyceraldehyde} \\
\quad| & \qquad\qquad\qquad | \\
\text{CH}_2\text{OH} & \qquad\qquad \text{CH}_2\text{OH}
\end{array}
$$

D-Glyceraldehyde is the compound used as a reference for the designation of carbohydrate structure. That is, any carbohydrate molecule with its bottom chiral carbon in the same configuration as D-glyceraldehyde is named as a D-sugar. Figure 10.1 illustrates the structures of the D-aldoses containing from three to six carbon atoms. In each case note the configuration of the chiral carbon furthest from the aldehyde group. The hydroxy is on the right and therefore all of the monosaccharides are of the D-configuration.

Likewise, there is a comparable group of L-aldoses, which are mirror images to those structures in Figure 10.1.

There is also a series of D- and L-ketoses, very few of which occur in any significant natural abundance. The exception is D-fructose, a 2-ketohexose that is the functional isomer of D-glucose. It will be discussed later.

You can readily see that aldohexoses in general have four chiral carbons and therefore 16 possible optical isomers. In nature, the D-series predominates almost exclusively, and among those only a few monosaccharides are found in significant amounts. D-Glucose is one of the most ubiquitous compounds in living organisms. It will be used to illustrate some other important considerations of carbohydrate structure.

PROBLEM

10.1. Draw the structures of L-ribose, L-glucose, and L-galactose. Compare to the D-series in Figure 10.1.

2. Cyclic Structures. Most monosaccharides exist primarily in a cyclic form created by the formation of a hemiacetal. Recall that in hemiacetal formation, an O—H bond of an alcohol adds to the carbon-oxygen double bond of an aldehyde or ketone (section 9.5.F). Since monosaccharides possess both alcohol and carbonyl functions, intramolecular hemiacetal formation can occur; a five- or six-membered ring is the result.

Let us apply this concept to D-glucose. The structure that appears in Figure 10.1 does not take into account the natural puckering that occurs in the chain

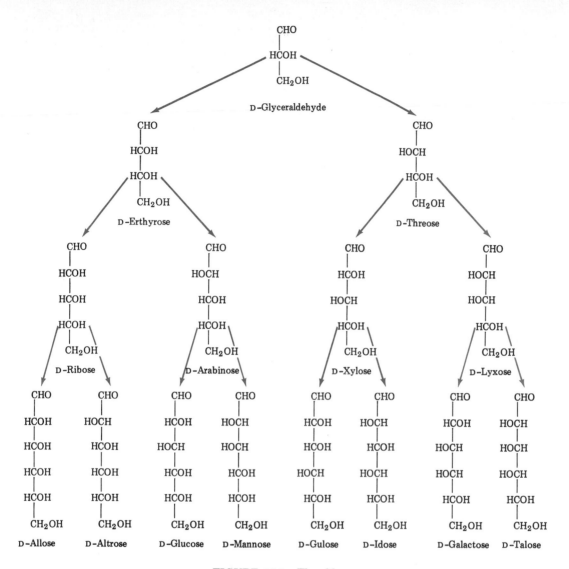

FIGURE 10.1. The aldoses.

because of the many tetrahedral carbons. When this is taken into consideration,
D-glucose could appear as in Figure 10.2a. Notice the proximity of the aldehyde
group and the alcohol group on carbon-5. Because of this closeness, a chemical
reaction between the C=O and the O—H group can occur, producing a hemiacetal
and a six-membered ring. Six-membered rings normally exist in the chair conformation (section 2.8.B) as shown in Figure 10.2C.

Two cyclic forms arise from the two possible orientations of the —OH on
carbon-1 (now a chiral carbon) during the hemiacetal formation. The designations "α" and "β" are used to describe the configuration at carbon-1 in the cyclic
structure. α-D-glucose has the —OH in the axial position and β-D-glucose has it in
the equatorial position. Since equatorial positions are usually less crowded than

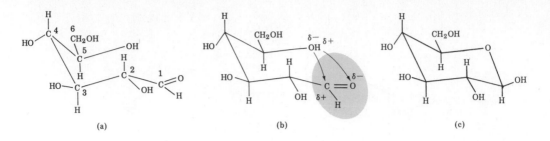

FIGURE 10.2. Puckered chain conformation of D-glucose. (a) Open-chain, free-aldehyde form. (b) Reaction of aldehyde and alcohol groups to form a hemiacetal. (c) Cyclic hemiacetal form.

axial (section 2.8.B), the β form is more stable and can be isolated from the α form. Note that all the hydroxy groups in β-D-glucose are equatorial (Figure 10.3). The α and β cyclic forms are optical isomers, more specifically, diastereomers. Because they differ in configuration at only one chiral carbon, they are called *anomers*.

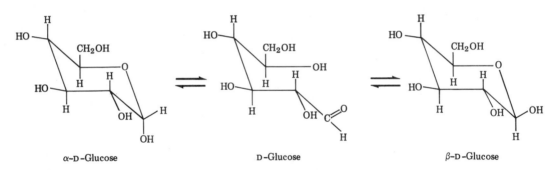

α-D-Glucose   D-Glucose   β-D-Glucose

FIGURE 10.3. The equilibrium mixture of D-glucose in solution. Less than 0.1% of the mixture is in the open-chain form.

PROBLEM

10.2. Using the puckered cyclic structure of D-glucose as a guide, draw the comparable forms for α- and β-D-galactose. Show conformation and stereochemistry.

3. Fischer Projections. The cyclic structure of D-glucose (and other monosaccharides) can be written in the straight-chain formula as shown.

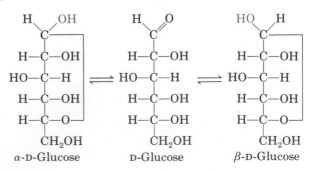

α-D-Glucose   D-Glucose   β-D-Glucose

These forms are often called Fischer projections in honor of Nobel Prize winner Emil Fischer, who in the late 1800s elucidated the structure of carbohydrates in addition to pioneering research in many other areas of organic chemistry.

4. Haworth Formulas. Another graphic representation of carbohydrate structure is the Haworth formula, a planar hexagon viewed as coming out of the page on its side.

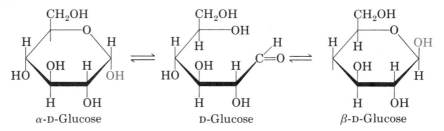

Comparing the Fischer, Haworth, and conformational structures, one can see that —OH groups located on the right in a Fischer projection are oriented downward in the conformational and Haworth structures. Since the —OH groups are more important in function than —H, the —H atoms are frequently omitted in writing the structure (see for example, α-D-mannose).

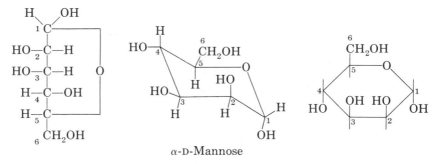

α-D-Mannose

PROBLEM

10.3. Draw the Fischer, Haworth, and conformational structures for the cyclic forms of **(a)** α-D-altrose; **(b)** β-D-glucose; and **(c)** β-D-idose.

---

5. Five- and Six-Membered Rings. Because the aldohexoses can form six-membered heterocyclic rings (*hetero*, "different") containing an oxygen atom, they are structurally similar to the compound pyran. As a result they are generally termed *pyranoses,* or more specifically, glucopyranose, galactopyranose, mannopyranose, and so on.

The three-dimensional conformations of monosaccharides also allow a second cyclic structure, that of a five-membered ring. Similar to furan, these forms are called *furanoses,* and most commonly are found with aldopentoses, such as ribose, and ketohexoses, such as fructose. Recall from section 2.8.A that five- and six-membered rings form the most easily. The naming of the carbohydrate is modified to indicate the ring size.

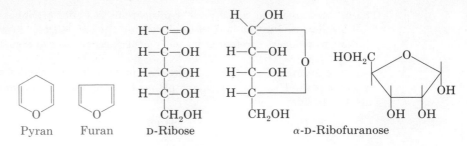

Pyran   Furan   D-Ribose   α-D-Ribofuranose

D-Fructose or fruit sugar (also found in honey and as the only saccharide in human semen) can be found in either a five- or six-membered ring.

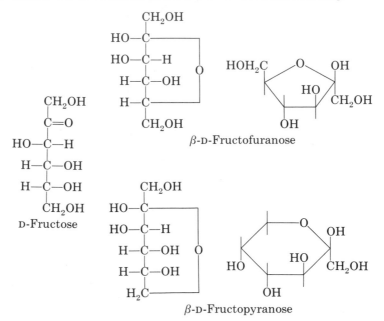

β-D-Fructofuranose

D-Fructose

β-D-Fructopyranose

Fructose also has the common name *levulose* because of its levorotatory optical activity. For a similar reason, glucose is often called *dextrose*.

PROBLEM

10.4. Draw the Haworth formulas for the α and β forms of D-mannose in the pyranose and furanose ring structures.

C. Mutarotation

Because most saccharides are asymmetric (chiral) molecules, their solutions exhibit optical activity; that is, they rotate the plane of plane-polarized light. α-D-Glucose, which can be isolated in pure form by crystallization from methanol, has rotation $[\alpha]_D^{20} = +112.2°$. It is often called dextrose due to this property. β-D-Glucose, crystallized from acetic acid, has rotation $[\alpha]_D^{20} = +18.7°$. If either of these pure forms or a mixture of the two is placed in water solution and observed with a polarimeter over a period, the optical rotation will change until a final value of $+52.7°$ is reached. This is the rotation of the equilibrium mixture of the α and β cyclic forms and the open-chain forms of D-glucose (Figure 10.3). The automatic alteration in structure and optical rotation that D-glucose and many other saccharides undergo in water solution is called *mutarotation*.

PROBLEM

10.5. Considering the mutarotation of α-D-glucose and β-D-glucose to 52.7° and assuming that the amount of open-chain form is negligible, state which form, α or β, is dominant in the equilibrium mixture.

REACTIONS OF MONOSACCHARIDES

10.2

As aldehydes, ketones, and alcohols, the saccharides will undergo the chemical reactions typical of these functional groups. The carbonyl moiety is subject to condensation, reduction, oxidation, addition, and so on. The alcohol groups may take part in ester formation and methylation, participate in hydrogen-bonding, and the like. We shall cover a few pertinent reactions, with particular interest in the structure identification these reactions afford.

A. Reducing and Nonreducing Sugars

1. Reducing Sugars. As aldehydes, the aldoses can readily undergo oxidation to acids, thereby acting as reducing agents (reducing sugars). Fehling's solution (section 9.4.B) of copper(II) (Cu^{2+}) ion complexed with tartrate in alkaline solution makes use of this fact. The initial complex is a deep blue color, and the reduced Cu^{+}, as Cu_2O, forms a red precipitate.

$$\overset{O}{\overset{\|}{R C H}} + Cu^{2+} \text{ (complex)} \xrightarrow{OH^-} \overset{O}{\overset{\|}{R C O^-}} + Cu_2O \downarrow$$

The aldehyde is oxidized to a carboxylic acid (in the salt form).

Benedict's solution (section 9.4.B) contains a copper(II) citrate complex rather than tartrate and can be purchased in tablet form for the clinical determination of glucose in body fluids.

Tollens' reagent (section 9.4.A) consists of a silver ion-ammonia complex in alkaline solution, which acts as an oxidizing agent. .

$$\overset{O}{\overset{\|}{R C H}} + Ag(NH_3)_2^+ \xrightarrow{OH^-} \overset{O}{\overset{\|}{R C}}-O^- + Ag \downarrow$$

The elemental silver plates out on the sides of the reaction container and hence the common name for this reaction, the *silver mirror test*.

Although ketones do not normally have the same susceptibility to oxidation as aldehydes, ketoses will give positive Fehling's, Benedict's and Tollens' tests. This is due to the OH groups located on carbons next to the ketone carbonyl group. The alkaline conditions of the reactions also allow the formation of enediols, which are susceptible to conversion to aldoses.

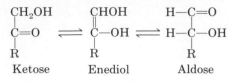

Ketose   Enediol   Aldose

Even though the monosaccharides are predominantly in the cyclic hemiacetal or hemiketal form, enough of the open-chain sugar exists in equilibrium to expose the free aldehyde or ketone group to reaction. As the open-chain aldehyde is oxidized, the equilibrium will shift to provide more open-chain form until all the sugar present has reacted (see Figure 10.3).

In summary, all monosaccharides in the open-chain or cyclic form can directly or indirectly reduce Fehling's, Benedict's, or Tollens' reagents and are termed *reducing sugars*.

2. Nonreducing Sugars. What then constitutes a nonreducing sugar? In some way, the hemiacetal or hemiketal group must be tied up so that it is not in equilibrium with an open-chain form of the sugar. This involvement that essentially "locks" the cyclic structure into an α or β position is called a *glycosidic bond*.

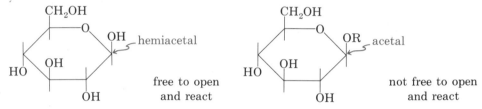

A glycoside linkage is formed by reaction of the hemiacetal or hemiketal hydroxy with another alcohol unit (section 9.5.F). An acetal or ketal results, which is not in the same type of equilibrium with the open chain as the hemiacetal.

B. Methylation: Glycoside Formation

Adding methanol to glucose in the presence of anhydrous HCl results in the formation of a methyl acetal derivative, or methylglycoside.

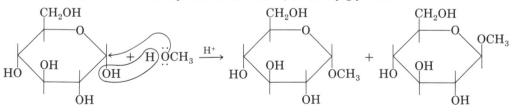

The general term for any such derivative is *glycoside*. The reaction illustrated above is with α-D-glucose, and the products are methyl α-D-glucoside and methyl-β-D-glucoside. If we take into consideration the pyranose ring structure, the more proper naming would be methyl α-D-glucopyranoside. Both anomers are formed because of the acidic condition of the reaction, which opens up the cyclic forms, and because of the carbocation mechanism of acetal formation (section 9.5).

Notice that glycoside formation ties up the hemiacetal —OH so that the α and β forms are no longer interconvertible by mutarotation. However, these glycosides can be hydrolyzed by acid solution or by enzymes to the original cyclic form which is in equilibrium with the open-chain form and the other anomer. The enzymes are specific for either the α or β forms. For example, the enzyme emulsion (from almonds) hydrolyzes β glycosides but not α glycosides. Amylase (in saliva and intestinal fluids), on the other hand, hydrolyzes only α glycoside linkages.

PROBLEMS

10.6. What are the structures of **(a)** methyl β-D-mannopyranoside, **(b)** methyl α-D-fructofuranoside, and **(c)** methyl β-D-galactopyranoside.

10.7. Are glycosides reducing or nonreducing sugars? Explain.

C. Esters

As alcohols, saccharides can condense with acids to form esters. The most important metabolic ester is the phosphate ester—for example, the product of the reaction of D-glucose with a special form of phosphoric acid called adenosine triphosphate (ATP), catalyzed by an enzyme (section 15.2.E). The difference in energy between ATP and its hydrolysis product, adenosine diphosphate (ADP), is relatively high. Adenosine triphosphate can, in effect, transfer this energy potential to glucose. Often the phosphate bonds in ATP are referred to as *high-energy* ($\sim$) *bonds*.

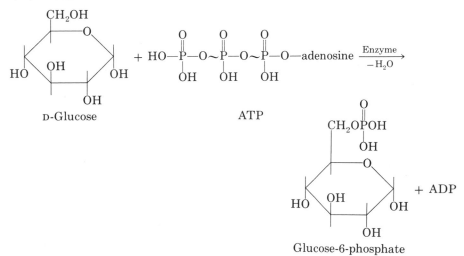

PROBLEM

10.8. Another important biochemical esterification involves the phosphorylation of galactose to form galactose-1-phosphate. Draw this structure with galactose in a pyranose ring.

DISACCHARIDES

10.3   The chemical reactivity of the hemiacetal and remaining —OH's in monosaccharides can give rise to various derivatives, as we have seen. This includes the interreaction of two monosaccharide units to form a disaccharide. Some of the common disaccharides are maltose, lactose, cellobiose, and sucrose. Two of the sugars occur naturally as disaccharides, while the other two are breakdown products of even larger carbohydrate structures, called *polysaccharides*.

A. Maltose and Cellobiose: The α and β of It

Two glucose units can react together to form a glycosidic bond in several ways depending upon which functional groups are involved and which anomer, α or β, predominates.

1. Maltose. Probably one of the most abundant glycosidic linkages found in nature occurs as a result of the reaction of the hemiacetal carbon of one glucose unit and the alcohol group of carbon-4 in a second glucose unit. Note that an acetal (glycoside) is formed at one glucose unit and the other still has a hemiacetal function. If the first unit is α-D-glucose, the bond is termed an α-1,4 linkage. The disaccharide as a whole has the common name *maltose*.

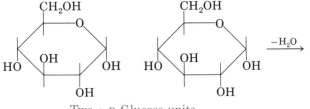

Two α-D-Glucose units

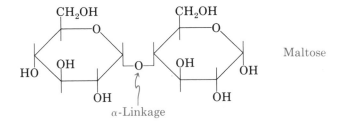

Maltose

α-Linkage

Maltose is a degradation product in the metabolism of starch by enzymes found in animal saliva and intestinal tracts and in yeast. The yeast cells will eventually produce ethyl alcohol from the sugar; animals burn the carbohydrate to produce CO_2, H_2O, and energy.

PROBLEM

10.9. Is maltose a reducing sugar? Why?

2. Cellobiose. If the β anomer, instead of the α anomer, of the first glucose unit had been a reactant, an entirely different and distinct disaccharide would have been formed, cellobiose. As with maltose, a glycoside linkage (acetal) is formed and one glucose unit still has a hemiacetal function, that is in equilibrium with the open-chain form.

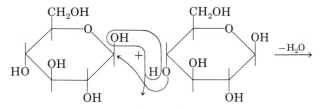

Two β-D-Glucose units

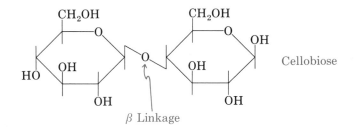

Cellobiose

β Linkage

Although it might seem that it should make no difference whether the glycosidic linkage is α or β, the properties, sources, and reactions of these disaccharides prove otherwise. *Cellobiose* is a small unit of the polysaccharide cellulose, the basic structural component of plants. Cellobiose and cellulose cannot be broken down by the intestinal enzyme maltase and are therefore indigestible to mammals. They can, however, be hydrolyzed to glucose by the enzyme emulsin and a class of enzymes commonly known as cellulases. These latter catalysts are produced by bacteria that inhabit the digestive tracts of herbivorous animals and such other species as the termite. Although both cellobiose and maltose exhibit mutarotation, their optical activities are intrinsically different; and so are other physical properties such as solubility and melting point. An interesting observation is that although maltose tastes sweet, cellobiose does not.

PROBLEM

10.10. Is cellobiose a reducing sugar? Explain your answer.

B. Sucrose: The Table Disaccharide

Everyday experience makes us most aware of sucrose. It is a disaccharide composed of a glucose unit and a fructose unit linked by a 1,2-glycosidic bond.

In sucrose the structure of fructose is that of a cyclic five-membered, or furanose, form. The anomeric carbon, that which forms the hemiketal, is located on carbon-2. Since the bond in sucrose between glucose and fructose is a 1,2 linkage, the glycoside must involve the anomeric (hemiacetal and hemiketal) carbons of both.

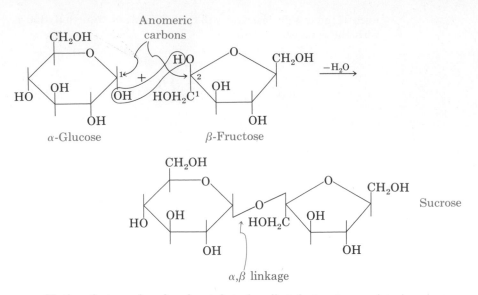

Notice that no free hemiacetal or hemiketal structure exists in sucrose. With the hemiacetal carbon of glucose and hemiketal carbon of fructose now in a glycosidic bond, sucrose lacks a group that could be in equilibrium with an open-chain form, as we saw with maltose. Therefore, sucrose cannot be oxidized by Fehling's or Benedict's solutions, and it is a nonreducing sugar.

PROBLEM

10.11. From the structure of sucrose, would you expect it to show mutarotation? Explain your answer.

Sucrose can be isolated from various sources including sugar cane (15%–20%), sugar beets (10%–17%), fruits, maple sap, seeds, and flowers. As the disaccharide, sucrose is dextrorotatory. On hydrolysis, either by acid or enzyme, the optical rotation changes to levorotatory ($+66.5°$ to $-20°$) owing to the production of equal amounts of glucose ($[\alpha]_D^{20} = +52°$) and fructose ($[\alpha]_D^{20} = -92°$). Since the sign of the optical activity goes from plus to minus, sucrose is said to invert, and the mixture of the two monosaccharides is called *invert sugar,* which is sweeter than sucrose itself. The bee secretes an enzyme, invertase, which accomplishes this hydrolysis during the production of honey.

Quite a bit of attention has been focused on the detrimental effects of eating large quantities of sucrose or "refined sugar." The furor centers about two issues: sucrose as a source of "empty calories" leading to obesity and its accompanying metabolic malfunctions, and sucrose as the main source of tooth decay.

First of all, sucrose, one of the purest chemicals commercially produced, is rapidly broken down by the body to glucose and fructose. Most of the fructose is then converted to glucose. Glucose is metabolized in a complex series of enzyme-catalyzed reactions to produce energy if needed. Otherwise it is converted to fat or polymerized to glycogen, both being the body's means for energy storage. Some glucose can also be excreted in the urine. So a large dose of sucrose will most likely end up where we need it least—as excess baggage. Glucose also requires insulin in order to be assimilated by the cells of the body. Insulin is a hormone present in intricate balance with other vital body regulators (see Panel 14 at the end of this chapter). A surge of glucose essentially throws everything off-balance metabolically. So while glucose is necessary in order to produce life-sustaining energy, it is preferable to have it enter the system at a more uniform rate, perhaps as a more complex and thereby less readily metabolized form of carbohydrate like that found in fruits and grains.

The second penalty for excess sucrose ingestion involves tooth decay. Bacteria known as *Streptococcus mutans* live in colonies called plaque on the surface of teeth and gums. They use sucrose both to produce an adhesive with which they stick to teeth and as a food. The end result of their digestion of sucrose is lactic acid. It is acidic and eats away at the surface of the teeth and gums. Brushing and flossing teeth frequently as well as avoiding sucrose-containing foods, especially those with a tendency to stick to the teeth, are recommended by dental experts.

C. Lactose: Mother's Disaccharide

Found exclusively in the milk of mammals, lactose (*lac*, Latin for "milk") or milk sugar makes up 4.5% of cow's milk and 6.7% of human milk. Lactose is composed of galactose and glucose linked by a β-1,4 glycosidic bond. This bond requires a specific β-cleaving enzyme just as cellobiose does. In fact, emulsin will act on both disaccharides. Lactase is an enzyme specific for lactose, which also breaks it into galactose and glucose.

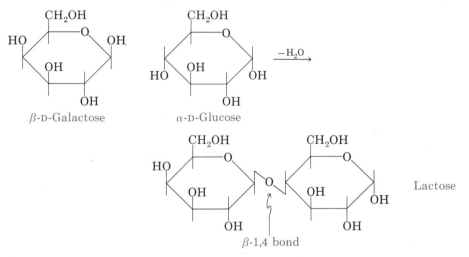

It is interesting to note that although lactase is secreted in the intestines of young mammals, as the infant is weaned the level of lactase being produced decreases markedly. The almost total absence of lactase activity is estimated to affect 70%–80% of the world's adult population. The result, lactose intolerance, is that lactose is not digested. Only those of northern European ancestry seem to be generally exempt from this deficiency. Lactose intolerance can bring about great discomfort and intestinal distress due to fermentation by bacteria in the intestine if certain dairy foods are ingested. Fermented milk products such as yogurt have little effect because the lactose has been already metabolized by resident bacteria. There has been much research into the possible elimination of lactose from some dairy products, especially milk, for adult consumption. It is also possible to purchase an enzyme additive (lactase) that can be mixed with milk in order to break down the lactose before the milk is drunk.

A few years ago a product called "sweet acidophilus" was introduced. Its basic premise is similar to that of other fermented milk derivatives. A bacterium, *Lactobacillus acidophilus,* which is fairly inactive at refrigerator temperatures, is added to regular milk. As it is warmed in the gastrointestinal tract it becomes active and begins to ferment the carbohydrates found in milk. Since the action does not take place until the milk is ingested, it tastes like normal milk— "sweet"—not sour like yogurt or sour cream, which have already been worked on by bacteria. Because most of the lactose is still present as it reaches the intestine, the benefits of "sweet acidophilus" for those with lactose intolerance are dubious.

PROBLEM

10.12. Will lactose exhibit mutarotation? Explain.

POLYSACCHARIDES

10.4   The disaccharides maltose and cellobiose are not found naturally as synthesis products of two monosaccharides, but are present as the breakdown products of extremely large carbohydrate molecules (macromolecules) called polysaccharides. While a hexose weighs in at 180 grams per mole, a respectable polysaccharide can tip the scales at anywhere from 25,000 to millions of grams per mole. Because polysaccharides are so large, they have many functions, depending on their size and composition.

A. Cellulose

The polysaccharide cellulose is a natural polymer well suited to its function as the structural skeleton of plants. Since cellulose constitutes approximately 50% of all plant life, it also is the most ubiquitous and abundant organic chemical on earth. Cellobiose has been mentioned as the disaccharide breakdown product of cellulose. This indicates that the polysaccharide is composed of glucose units joined between carbons 1 and 4 by β-glycosidic bonds. Recall that the β-linkage can only be digested by special enzymes like those found in the bacteria living in ruminants such as cows and sheep. Figure 10.4a illustrates the conformational structure of a partial cellulose polymer.

B. Starch

Just as cellobiose is a degradation product of cellulose, maltose is a hydrolysis product of the polysaccharide we know as starch. Starch is the main storage carbohydrate found in plants. As such, it also finds its way into the mammalian food chain, constituting the principal food-energy source for man. Like cellulose, the starch molecule has glucose as its monomer unit. Unlike cellulose, however, the molecule has α-glycosidic linkages and not all are between carbons 1 and 4. Some of it is also joined by α-1,6 bonds (Figure 10.4b). This allows branching of the molecule from the long polysaccharide chain.

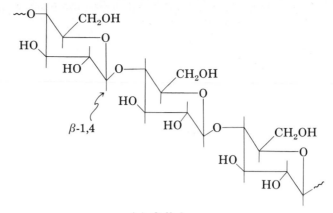

(a) Cellulose

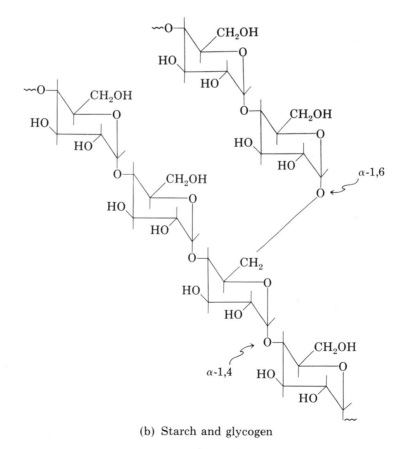

(b) Starch and glycogen

FIGURE 10.4. The conformational structures of the major polysaccharides.

When starch is placed in water, it will seem to partially dissolve. The portion that seems insoluble (about 27%), called *amylose,* is composed of the long-chained, linear α-1,4 glycosidic units; and the soluble material (73%) consists of the highly branched (α-1,6) chains, called *amylopectin.*

The overall structure of the starch polymer includes some long unbranched chains that can form a spiral. These areas as well as the rest of the huge molecule are stabilized by hydrogen-bonding among the alcohol groups not involved in acetal bonds. The same holds true for cellulose. The resulting open areas within the polysaccharide matrix can trap other molecules such as H_2O and I_2. The great water absorptiveness of cotton fibers is illustrative of this property. Cotton is more than 90% cellulose. And what child has not been awed at the characteristic purple I_2 test for starch in a potato? It is believed the I_2 becomes trapped within the matrix and will exhibit a hue anywhere from deep blue-black (long spiral, native starch) to light red (short spirals, degraded starch).

It is not practical to consider any chemical properties of starch, such as reducing power, since the molecule has few reducing ends compared to the size of the molecule itself. Chemical reactions would barely be noticeable using manageable quantities of the reactants.

C. Glycogen

Glycogen is the body's energy storage vault. A polymer of glucose, it is similar in structure to starch but more highly branched. This polysaccharide is especially preponderant in the liver and muscle tissue, sites where the mechanism for its metabolism exists. Enzymes that are present in these organs hydrolyze the glycogen to glucose and pass it into the cycles used for energy production.

D. Polysaccharide Variations

There exist a great number of other polysaccharides composed of the same repeating sugar unit, different units, or even modified units such as amine and sulfonic acid derivatives. These variations are indicative of differences in function and organism.

Polysaccharides are also used commercially as thickening agents, stabilizers, and water retainers. Sodium carboxymethylcellulose, gum acacia or arabic, guar gum, gum tragacanth, as well as agar agar can be found in everything from beer to cottage cheese, condiments to pickle relish. *Dextrins* are a partial chemical or enzymic breakdown product of starch. They vary in size and degree of branching and can be found in glues, printer's ink, and inert fillers in prescription medicines, to name a few uses. *Dextrans* are polysaccharides of glucose linked α-1,6 with α-1,3 branches. They are produced by certain bacteria and are used to expand the volume of blood plasma in cases of extreme loss of blood and also as the soft center of candies.

MODIFIED CARBOHYDRATES

10.5   Before human thought was perplexed with the idea of synthetic polymers, nature had devised a polymer of myriad uses, cellulose. One very important human application of this plant polymer was as cotton fiber. Cotton consists of over 90% pure cellulose. Its three-dimensional structure is stabilized by the formation of hydrogen bonds between the hydroxyl groups of adjacent polymer chains. Gaps that can trap water molecules exist between these bundles of chains, holding them there by additional hydrogen-bonding. Hence arises cotton's unique absorptivity. It is indeed fitting, then, that one of chemistry's first experiences with synthetic molecules had to do with the inadvertent chemical modification of cellulose.

A. Nitrated Celluloses

As with so many significant discoveries, serendipity smiled on Christian Schönbein in 1846 when he cleaned up a spill of nitric and sulfuric acids with his wife's cotton (90% cellulose) apron. After rinsing out the apron with water, he hung it to dry in front of a hot stove. To his utter surprise the cloth flashed up and disappeared with barely a trace. He had accidentally synthesized cellulose trinitrate—guncotton.

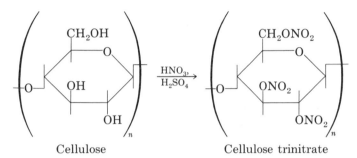

Cellulose   Cellulose trinitrate

If cellulose is not as thoroughly nitrated, another product is formed, called *pyroxylin*. Pyroxylin and camphor combined form celluloid, a plastic once used for film (celluloid screen), eyeglass frames, celluloid collars, dice, dominoes, and so on. Celluloid was made from collodion, nitrated cellulose dissolved in ether and alcohol. Because of the flammability of cellulose nitrate, this material gradually was replaced by petroleum-based polymers.

B. Viscose Rayon and Acetate Rayon: Artificial Silk

The nitration of cellulose converted a very insoluble material, cellulose, to a soluble one, nitrocellulose, which in turn could be forced through very small holes to form a silken thread. However, fabrics made of this material were short-lived. Quite literally, the fashion went up in a puff of smoke. Another, less flammable derivative had to be made or cellulose itself had to be rendered soluble so it could be regenerated into other shapes. Thus were born acetate rayon (or cellulose acetate) and viscose rayon (the xanthate process).

1. Viscose Rayon. In the viscose process the cellulose is solubilized by reaction with carbon disulfide to form cellulose xanthate, which is soluble as the sodium salt in basic solution.

$$\text{Cell}\!-\!\overset{-}{\text{O}}\overset{+}{\text{Na}} + \text{CS}_2 \longrightarrow \text{Cell}\!-\!\text{O}\!-\!\overset{\overset{\textstyle S}{\|}}{\text{C}}\!-\!\overset{-}{\text{S}}\overset{+}{\text{Na}}$$

Cellulose xanthate
(soluble)

The term *viscose* arises from the fact that cellulose xanthate dissolves in a NaOH bath to form a very viscous solution. The cellulose is regenerated in an acid bath after it has been forced through small openings, called spinnerets, to form threads.

$$\text{Cell—O—}\overset{\overset{\textstyle S}{\|}}{\text{C}}\text{—S}^-\text{Na}^+ \xrightarrow{\text{H}^+} \text{Cell—OH} + \text{CS}_2$$

The threads are then wound and further processed into practical items, such as clothing, tire cord, draperies, and carpeting. If the xanthate were extruded through a slit instead of a spinneret, a thin sheet of cellophane would result. Its uses are universally known.

2. Acetate Rayon. The physical process of making acetate rayon is similar in concept to the xanthate process. However, in this case insoluble cellulose is permanently derivatized and solubilized by causing it to react with acetic anhydride. The latter forms esters with the free —OH groups of the glucose units.

$$\text{Cell—OH} + \left(\text{CH}_3\overset{\overset{\textstyle O}{\|}}{\text{C}}\right)_2\text{O} + \text{CH}_3\overset{\overset{\textstyle O}{\|}}{\text{C}}\text{OH} \xrightarrow{\text{H}_2\text{SO}_4} \text{Cell—O}\overset{\overset{\textstyle O}{\|}}{\text{C}}\text{CH}_3$$

The cellulose acetate, in an acetone solution, is extruded through spinnerets to form threads. As the solution leaves the spinneret, hot air flash evaporates the acetone solvent.

C. Wood

Wood is primarily composed of cellulose and lignin, both of which are water insoluble. Lignin makes up about 30% of raw wood, and though its structure is not completely clear, it is very much in the forefront of research as a possible source of industrial chemicals. Paper is principally cellulose derived from wood chips freed of lignin.

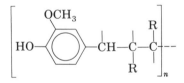

Lignin— a partial structure

Panel 14

METABOLIC OVERSIGHTS

There are numerous and varied pathways to the natural synthesis and degradation of saccharides, and more than an occasional foul-up can arise in any one or more of the individual steps. These deviations can be chemically induced; also they may occur because of a genetic malfunction, usually in the production of a task-specific enzyme or hormone.

[PANEL 14]   A. Diabetes

Glucose travels by way of the bloodstream to various organ cells, which can use it immediately for energy or shunt it elsewhere for storage. But before it can be utilized in either of these ways it must first be absorbed into the cells. It is one of the functions of the hormone insulin to facilitate the passage of glucose into the cell. The mechanism for this transfer is yet unknown. Should the insulin supply be negligible, a condition known as *juvenile diabetes mellitus* exists. If the insulin is present but for some reason is ineffective, *adult-onset diabetes* is the general diagnosis. Both conditions lead to the same clinical symptoms. The unabsorbed glucose exceeds its normal limits in the blood (65–95 mg/100 ml). The kidneys then attempt to excrete the excess glucose in the urine resulting in that fluid showing the sugar content above the renal threshold (180 mg/100 ml). The increased blood glucose content, termed *hyperglycemia,* is used as a diagnostic tool for the disease. The effects of unchecked diabetes are tissue starvation, especially of the brain, and the "dumping" of large quantities of glucose through the kidneys. Nervous system failures such as diabetic coma, blindness, and kidney disease are but a few of the fatal consequences.

Certain drugs, taken orally, are used to control some cases of adult-onset diabetes. These are believed to stimulate the production of more insulin. Their ingestion results in decreased blood-sugar levels.

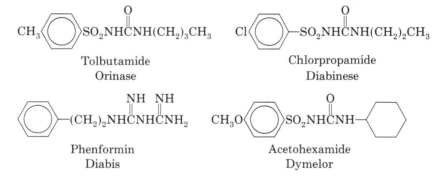

Generic name:	Tolbutamide	Chlorpropamide
Trade name:	Orinase	Diabinese

Generic name:	Phenformin	Acetohexamide
Trade name:	Diabis	Dymelor

Many adults, due to being overweight, will exhibit hyperglycemia. A glucose tolerance test will indicate whether they truly suffer from diabetes or are in a prediabetic state which is controllable by diet modification and exercise.

Research into the causes and control of diabetes is very active. The possible involvement of a virus as an inducer of juvenile-onset diabetes in genetically susceptible individuals is one such discovery. In an effort to reduce sucrose intake, and thereby glucose intake, diabetics and those on low-carbohydrate diets have turned to artificial sweeteners to maintain taste if not calories. Up to 1970, saccharin and particularly calcium cyclamate were the leading artificial sweeteners in beverages and other foods. Research into the carcinogenicity of cyclamates cast enough doubts about the safety of calcium cyclamate to result in a ban on its use by the FDA in 1970. Since March, 1977, saccharin, which is about 400 times sweeter than sucrose, has been under close scrutiny based on a Canadian study in which rats fed saccharin developed cancer. Products containing this compound are now labeled with a warning about the possible effects.

Compounds found in the skins of grapefruit and oranges are now in the forefront of research on nonnutritive sweeteners. Still under investigation are the derivatives of hesperetin. These compounds are about 1500 times sweeter than

[PANEL 14] sucrose and only absorbed to an extent of 1% in the intestine. Added to these are amino acid derivatives, the most promising of which is aspartame or L-aspartyl-L-phenylalanyl methyl ester, 200 times sweeter than sucrose. A major drawback lies in a relatively short shelf-life and heat lability.

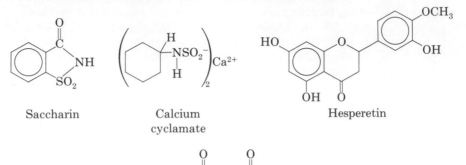

Saccharin   Calcium   Hesperetin
                   cyclamate

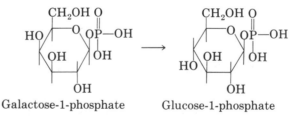

Aspartame

B. Galactosemia

Galactose is a hydrolysis product of lactose, most commonly entering an organism through the ingestion of milk or milk products. Some infants are born without the genetically inherited ability to synthesize an enzyme that will convert galactose-1-phosphate to glucose-1-phosphate and so allow it to be further metabolized.

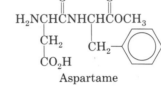

Galactose-1-phosphate   Glucose-1-phosphate

As the child matures to adolescence, the liver initiates a different set of reactions to effect this conversion. But the problem lies in eliminating galactose from the baby's diet during the early years of this galactosemia (-*emia*, "condition of having"). Symptoms of the enzyme deficiency are a buildup of galactose-1-phosphate in the tissues and galactose in urine. This can result in abnormal physical or mental development, or both. Genetic counseling can predict the presence of this condition even before birth.

C. Hypoglycemia

An excessive secretion of insulin will result in a *low* blood-sugar level and the overly efficient removal of glucose from the bloodstream. This again deprives certain cells of nourishment. This hypoglycemia, or hyperinsulinism, can be induced by diet, it may be symptomatic of inherent pancreatic malfunction, or it

[PANEL 14] may be caused by an excessive dose of insulin. The physical consequences can range from mild tiredness to the convulsions of "insulin shock."

D. Hormone-Induced Dysfunctions

A hormone is a chemical substance that is responsible for regulating a simple or a complex set of reactions in an organism. The structures of many hormones are known although the exact routes of their influence remain the subject of intensive research. The exact chemical interaction among these regulators within an organism is also often baffling even though the results of their interactions are documented. Insulin is the one example already mentioned for which effects are well known, but still the actual mechanism is highly debated. Glucagon is a polypeptide hormone that has an action antagonistic to that of insulin. It is produced by the α cells of the pancreas and raises blood sugar by breaking down existing glycogen. Somatostatin, another polypeptide hormone, acts as a regulator for both insulin and glucagon, suppressing their production as well as that of other hormones. Adrenalin, thyroxin, growth hormone, corticosterone—these and other regulators also interact in carbohydrate metabolism, either directly or indirectly.

Any trauma, physical or chemical, to the complex system of highly interrelated factors can in some way alter the cycles in question. Certain malfunctions, such as diabetes, galactosemia, lactase deficiency, are often rectifiable or controllable. Others are not so easily diagnosed or treated.

END-OF-CHAPTER PROBLEMS

10.13 Terms: Distinguish between the members of the following pairs of terms: **(a)** hexose, pentose; **(b)** aldose, ketose; **(c)** furanose, pyranose; **(d)** reducing sugar, nonreducing sugar; **(e)** monosaccharide, polysaccharide; **(f)** α-D-Glucose, β-D-glucose; **(g)** Haworth formula, Fischer projection; **(h)** maltase, emulsin; **(i)** amylose, amylopectin; **(j)** glycogen, cellulose; **(k)** insulin, glucagon; **(l)** viscose rayon, acetate rayon; **(m)** dextrin, dextran; **(n)** Fehling's and Tollens' tests.

10.14 Structure: How are the following pairs of saccharides different from each other structurally? Which are reducing and which are nonreducing? Explain. **(a)** cellobiose, maltose; **(b)** lactose, sucrose; **(c)** α-D-glucose, α-D-galactose; **(d)** α-D-glucose, α-D-fructose; **(e)** β-D-mannofuranose, β-D-mannopyranose; **(f)** maltose, lactose; **(g)** cellulose, starch.

10.15 Structure: Draw Haworth formulas for the following: **(a)** β-D-allopyranose; **(b)** β-D-fructopyranose; **(c)** β-D-ribofuranose; **(d)** methyl-α-D-galactopyranoside **(e)** α-D-glucofuranose; **(f)** β-D-xylofuranose; **(g)** α-D-deoxyribofuranose; (*deoxy*—"without oxygen"); **(h)** methyl-α-maltoside.

10.16 Terms: Briefly define the following: **(a)** mutarotation; **(b)** anomer; **(c)** hypoglycemia; **(d)** juvenile-onset diabetes; **(e)** invert sugar; **(f)** hemiacetal; **(g)** hormone; **(h)** glycoside; **(i)** hyperglycemia.

10.17 Structure: Draw the open chain forms of the following cyclic saccharides:

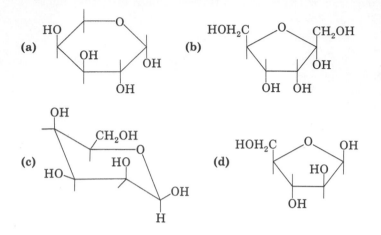

10.18 Optical Isomers: Draw the stereoisomers of 3-ketopentose. Which are the enantiomers, diastereomers, meso forms?

10.19 Reactions: Pure α-D-glucose or pure β-D-glucose in the presence of methanol (CH_3OH) and acid will give a mixture of α- and β-methyl glucosides. Why?

10.20 Structure: Starch and cellulose are both polymers of glucose. Why then can't mammals digest and use cellulose directly as they do starch?

10.21 Reactions: Draw the reactions and products of β-D-galactopyranose with **(a)** methanol (acidic solution); **(b)** α-D-mannose (β-1,4 bond); **(c)** copper (II) citrate (basic solution); **(d)** β-D-fructose (β-1,2 bond).

10.22 Reactions: Why do both glucose and fructose give positive Fehling's and Tollens' tests?

10.23 Reactions: After a few hours in dilute alkali, an originally pure D-glucose solution will show the presence of D-fructose and D-mannose. Why?

10.24 Reactions: Is a positive test for glucose in the urine a direct indication of diabetes? Explain your answer.

Chapter 11

Carboxylic Acids and Their Derivatives

STRUCTURE AND NOMENCLATURE OF CARBOXYLIC ACIDS

11.1   A. Structure

Carboxylic acids are structurally characterized by the carboxyl group, which can be represented in three ways:

$$-\overset{\overset{\displaystyle O}{\|}}{C}OH, \qquad -CO_2H, \qquad -COOH$$

Like inorganic acids, carboxylic acids often have an unpleasant or acrid odor and sour taste.

The simplest carboxylic acid, formic acid, is a dangerously caustic liquid with an irritating odor; it is a component of the sting of some ants. Acetic acid is responsible for the pungent taste and odor of vinegar (most vinegars are about 5% acetic acid) and finds extensive use in the industrial production of synthetic plastics, such as cellulose acetate (acetate rayon) and polyvinyl acetate. Butyric acid, from the Latin *butyrum* for "butter," contributes to the strong odor of rancid butter and other fats. Lactic acid is formed when milk sours, and as muscles tire. It is also a bacterial degradation product of sucrose by microorganisms in the plaque around teeth.

$$HCO_2H \qquad CH_3CO_2H \qquad CH_3CH_2CH_2CO_2H \qquad \underset{\underset{\displaystyle OH}{|}}{CH_3CHCO_2H}$$

| Formic acid | Acetic acid | Butyric acid | Lactic acid |

Caproic, caprylic, and capric acids (from the Latin *caper* for "goat") are present in the skin secretions of goats.

$$CH_3(CH_2)_4CO_2H \qquad CH_3(CH_2)_6CO_2H \qquad CH_3(CH_2)_8CO_2H$$

| Caproic acid | Caprylic acid | Capric acid |

The sour, biting taste of many citrus fruits is due to citric acid (6%–7% in lemon juice). Tartaric acid and its salts are found in grapes and tartar sauce.

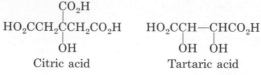

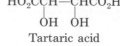

Citric acid   Tartaric acid

2,4-Dichlorophenoxyacetic acid (2,4-D) and 2,4,5-trichlorophenoxyacetic acid (2,4,5-T) are synthesized and used commercially as defoliants.

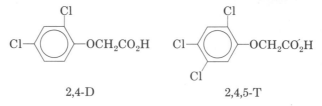

2,4-D   2,4,5-T

2,4-D is available to home gardeners as a broadleaf-weed killer. Broadleaf weeds have large surface areas and quickly absorb a lethal dose, whereas thin-bladed grasses remain unaffected. Compounds similar to 2,4-D and 2,4,5-T, such as *para*-chlorophenoxyacetic acid and β-naphthoxyacetic acid, are included in plant preparations used to aid tomato blossoms in setting fruit and to prevent the premature dropping of fruit.

p-Chlorophenoxyacetic acid   β-Naphthoxyacetic acid

A controversy surrounds the use of 2,4,5-T because of the suspicion that it becomes contaminated during production with 2,3,7,8-tetrachlorodibenzo-*p*-dioxin (2,3,7,8 TCDD), one of the most acutely toxic substances known. Massive defoliation projects have been blamed for causing illness, an excessive number of miscarriages, and increased incidence of birth abnormalities among residents and farm animals in affected areas.

2,3,7,8-Tetrachlorodibenzo-*p*-dioxin
("Dioxin")

B. IUPAC Nomenclature of Carboxylic Acids

1. Simple Carboxylic Acids. To name carboxylic acids, name the longest continuous carbon chain and replace the -*e* with the suffix for carboxylic acids, -*oic acid*. Number from the carboxylic acid group.

Ethanoic acid   Decanoic acid   1,6-Hexandioic acid   *p*-Chlorobenzoic acid

2. Polyfunctional Carboxylic Acids. Polyfunctional carboxylic acids contain other functional groups, such as aldehyde, ketone, or alcohol groups, in addition to the carboxylic acid group. In naming such compounds, the carboxylic acid takes the suffix ending, and all other functional groups are named by prefixes (aldehydes—*carboxaldo;* ketones—*keto;* alcohols—*hydroxy*).

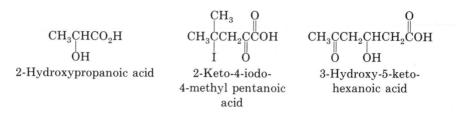

2-Hydroxypropanoic acid	2-Keto-4-iodo-4-methyl pentanoic acid	3-Hydroxy-5-keto-hexanoic acid

PROBLEM

11.1. Name by the IUPAC system of nomenclature:

(a) $CH_3(CH_2)_5CO_2H$ (b) $HO_2CCH_2CHCH_2CO_2H$ (c) $CH_3CCH_2CHCO_2H$
$\qquad\qquad\qquad\qquad\qquad\qquad\quad |$
$\qquad\qquad\qquad\qquad\qquad\qquad CH_3$
$\qquad\qquad\qquad\qquad\qquad\qquad\qquad\qquad\qquad\qquad\; O\quad OH$

3. Unsaturated Polyfunctional Carboxylic Acids. These compounds are named by the same procedure used for the corresponding aldehydes and ketones.

1. Name the longest continuous chain of carbons.
2. Designate carbon–carbon single bonds by -*an*, double bonds by -*en*, and triple bonds by -*yn*.
3. Name the most important functional group with a suffix (carboxylic acid > aldehyde > ketone > alcohol; see section 9.2.A.2).
4. Number the carbon chain, giving preference to the functional group named with a suffix. Identify with numbers all groups expressed in the name.
5. Name and number all other groups with prefixes.

Example: Name

$$\overset{10}{CH_3}\overset{9}{C}=\overset{8}{CH}\overset{7\ 6}{CCH}=\overset{5}{CH}\overset{4}{CH}\overset{3}{CH}=\overset{2}{CH}\overset{1}{COH}$$
$$\qquad\quad |\qquad\quad ||\qquad\qquad\quad |\qquad\qquad ||$$
$$\qquad\quad CH_3\quad O\qquad\qquad OH\qquad\quad O$$

1. Ten carbons is the longest chain, dec.
2. Three double bonds, decatrien.
3. The carboxylic acid is the most important functional group and is named with the suffix -*oic acid,* decatrienoic acid.
4. The carbon chain is numbered to give the lowest possible number to the functional group named by a suffix, the carboxylic acid. All parts of the suffix are identified by number, 2,5,8-decatrienoic acid. Since the carboxylic acid is on carbon-1, its position is understood without a number.
5. Other groups are named with prefixes. The complete name is

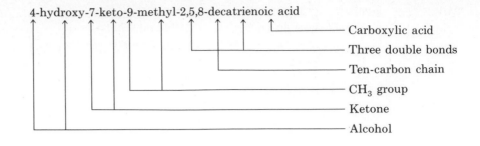

4-hydroxy-7-keto-9-methyl-2,5,8-decatrienoic acid

— Carboxylic acid
— Three double bonds
— Ten-carbon chain
— CH_3 group
— Ketone
— Alcohol

PROBLEM

11.2. Name by the IUPAC system of nomenclature:

(a) $CH_3CH{=}CHCO_2H$ (b) $CH_3C{\equiv}CCC{\equiv}CCHCO_2H$
                                                    $\underset{O}{\|}\quad\underset{Br}{|}$

C. Common Nomenclature of Carboxylic Acids

Carboxylic acids have long been referred to by common names that often describe familiar sources or properties of the compounds. Some of these names are summarized in Tables 11.1 and 11.2.

TABLE 11.1. PHYSICAL PROPERTIES OF CARBOXYLIC ACIDS

Structure	Common Name	Derivation of Name	Melting Point, °C	Boiling Point, °C	Water Solubility, g/100 g H_2O	Acidity Constant, K_a
HCO_2H	Formic acid	L. *formica*, "ant"	8	101	∞	1.77×10^{-4}
CH_3CO_2H	Acetic acid	L. *acetum*, "vinegar"	17	118	∞	1.76×10^{-5}
$CH_3CH_2CO_2H$	Propionic acid	Gr. *proto*, "first"; *pion*, "fat"	−22	141	∞	1.34×10^{-5}
$CH_3(CH_2)_2CO_2H$	Butyric acid	L. *butyrum*, "butter"	−8	164	∞	1.54×10^{-5}
$CH_3(CH_2)_3CO_2H$	Valeric acid	L. *valere*, "to be strong" (valerian root)	−35	187	4.97	1.51×10^{-5}
$CH_3(CH_2)_4CO_2H$	Caproic acid	L. *caper*, "goat"	−3	205	1.08	1.43×10^{-5}
$CH_3(CH_2)_5CO_2H$	Enanthic acid	Gr. *oinánth(ē)* the vine blossom	−8	223	0.24	1.42×10^{-5}
$CH_3(CH_2)_6CO_2H$	Caprylic acid	L. *caper*, "goat"	17	240	0.07	1.28×10^{-5}
$CH_3(CH_2)_7CO_2H$	Pelargonic acid	Pelargonium plant	13	253	0.03	1.09×10^{-5}

TABLE 11.2. PHYSICAL PROPERTIES OF DICARBOXYLIC ACIDS

Structure	Common Name	Derivation of Name	Melting Point, °C	Boiling Point, °C	Water Solubility, g/100 g H_2O	First Ionization, K_a
HO_2CCO_2H	Oxalic acid	Gr. *oxys*, "sharp" or "acid"; sorrels	190	...	14.3	5.9×10^{-2}
$HO_2CCH_2CO_2H$	Malonic acid	L. *malum*, "apple"	135	...	154	1.5×10^{-3}
$HO_2C(CH_2)_2CO_2H$	Succinic acid	L. *succinum*, "amber"	186	235	7.6	6.9×10^{-5}
$HO_2C(CH_2)_3CO_2H$	Glutaric acid	L. *gluten*, "glue"	98	303	64	4.6×10^{-5}
$HO_2C(CH_2)_4CO_2H$	Adipic acid	L. *adeps*, "fat"	152	338	1.44	3.7×10^{-5}

DERIVATIVES OF CARBOXYLIC ACIDS

11.2 A. Structure

Carboxylic acids and their derivatives can be expressed as variations of one formula, where Z is an electronegative group and R represents an alkyl or an aryl group or a hydrogen. The formula is

$$R-\overset{\overset{\displaystyle O}{\|}}{C}-Z$$

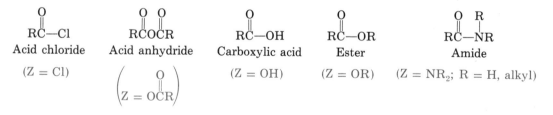

RC—Cl
Acid chloride
(Z = Cl)

RCOCR
Acid anhydride
$\left(Z = O\overset{\overset{\displaystyle O}{\|}}{C}R \right)$

RC—OH
Carboxylic acid
(Z = OH)

RC—OR
Ester
(Z = OR)

RC—NR
Amide
(Z = NR_2; R = H, alkyl)

Acid chlorides and acid anhydrides are extremely reactive compounds and are not found in nature. Acid anhydrides can be thought of as a combination of two carboxylic acid molecules minus the elements of water.

$$RC\boxed{OH \quad H}OCR \longrightarrow RC-O-CR + H_2O$$

Acids, esters, and amides are more stable and occur in large natural abundance. Esters usually have a pleasant odor, and in combination with other compounds they are responsible for the taste and fragrance of fruits and flowers.

$CH_3CH_2CH_2\overset{\overset{\displaystyle O}{\|}}{C}OCH_2CH_3$
Pineapple odor

$CH_3\overset{\overset{\displaystyle O}{\|}}{C}OCH_2CH_2\overset{\overset{\displaystyle CH_3}{|}}{C}HCH_3$
Banana odor

$CH_3\overset{\overset{\displaystyle O}{\|}}{C}O(CH_2)_7CH_3$
Orange odor

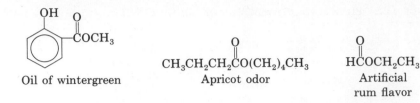

Oil of wintergreen

$CH_3CH_2CH_2\overset{O}{\overset{\|}{C}}O(CH_2)_4CH_3$
Apricot odor

$H\overset{O}{\overset{\|}{C}}OCH_2CH_3$
Artificial
rum flavor

The amide group is found in proteins and constitutes the linkage between amino acid units in these macromolecules.

B. Nomenclature of Carboxylic Acid Derivatives

Carboxylic acid derivatives are named by modifying the suffix ending on the name of the parent acid. To illustrate this principle, compare the following derivatives of propanoic acid and benzoic acid.

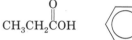

Propan*oic acid*   Benz*oic acid*

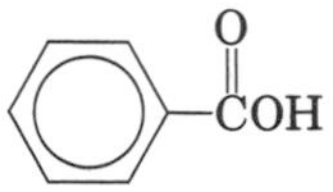

1. Acid Chlorides. Acid chlorides are named by changing the suffix -*ic acid* of the parent acid to -*yl chloride*.

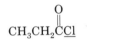

Propano*yl chloride*   Benzo*yl chloride*

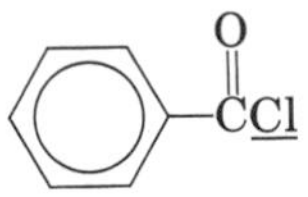

PROBLEM
_____

11.3. Name the following acid chlorides:

(a) $CH_3(CH_2)_3\overset{O}{\overset{\|}{C}}Cl$ **(b)** $CH_3CH{=}CH{-}CH{=}CH\overset{O}{\overset{\|}{C}}Cl$ **(c)** $O_2N{-}\underset{}{\bigcirc}{-}\overset{O}{\overset{\|}{C}}Cl$

_____

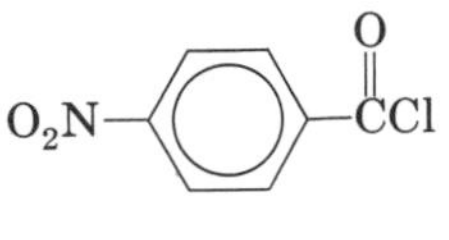

2. Acid Anhydrides. Acid anhydrides are named by changing the suffix *acid* of the parent acids to *anhydride*.

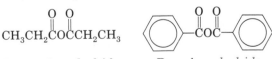

$CH_3CH_2\overset{O}{\overset{\|}{C}}O\overset{O}{\overset{\|}{C}}CH_2CH_3$

Propanoic *anhydride*   Benzoic *anhydride* 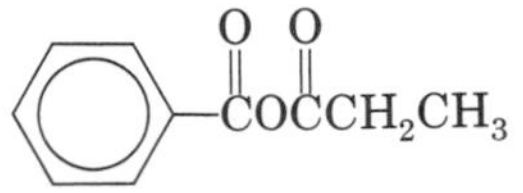   Benzoic propanoic *anhydride*

PROBLEM
_____

11.4. Name the following acid anhydrides:

(a) $CH_3\overset{O}{\overset{\|}{C}}O\overset{O}{\overset{\|}{C}}CH_3$ **(b)** $CH_3(CH_2)_3\overset{O}{\overset{\|}{C}}O\overset{O}{\overset{\|}{C}}(CH_2)_3CH_3$ **(c)** $CH_3\overset{O}{\overset{\|}{C}}O\overset{O}{\overset{\|}{C}}(CH_2)_3CH_3$

_____

3. Esters and Salts. The salts of many inorganic acids are named by changing the suffix *-ic acid* to *-ate* and prefixing the name with the name of the cation that replaced the acidic hydrogen.

HNO$_3$ Nit*ric acid*   H$_2$SO$_4$ Sulfu*ric acid*
NaNO$_3$ *Sodium* nitr*ate*   (NH$_4$)$_2$SO$_4$ *Ammonium* sulf*ate*

Similarly, organic esters and salts are named by changing the suffix *-ic acid* of the parent acid to *-ate* and prefixing the name with the name of the cation or organic group that replaced the hydrogen.

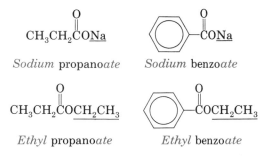

Sodium propano*ate*   *Sodium* benzo*ate*

Ethyl propano*ate*   *Ethyl* benzo*ate*

To name more complex esters and salts, mentally replace the cation or organic group with a hydrogen and name the parent acid. Then make the necessary changes to name the salt or ester. For example, let us name the following ester:

The parent acid is 3-methyl-2-buteno*ic acid,* and the ester is *isopropyl* 3-methyl-2-buteno*ate.*

11.5. Name the following salts and esters:

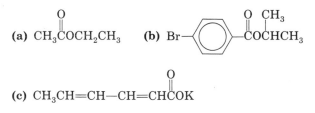

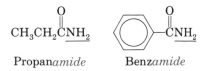

(c) CH$_3$CH=CH—CH=CHCOK

4. Amides. Amides are named by changing the suffix *-oic acid* to *-amide.*

CH$_3$CH$_2$CNH$_2$ —CNH$_2$

*Propan*amide*   *Benz*amide*

Substituted amides are named merely by locating the position of any substituents. For example, the following amide is a derivative of *para*-nitrobenzoic acid and is named as shown.

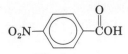

para-Nitrobenzoic acid

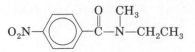

N-Ethyl-N-methyl-*para*-nitrobenzamide

PROBLEM

11.6. Name the following amides:

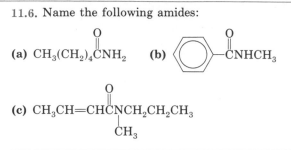

(a) $CH_3(CH_2)_4CNH_2$ (b) phenyl—$CNHCH_3$

(c) $CH_3CH\!=\!CHCNCH_2CH_2CH_3$ with CH_3 on N

PHYSICAL PROPERTIES

11.3　　　Carboxylic acids are capable of hydrogen-bonding, and they have higher boiling points and water solubilities than compounds of corresponding molecular weight that are incapable of hydrogen-bonding. Compare the compounds in Table 11.3. As is evident, carboxylic acids have higher boiling points than alcohols of the same molecular weight because of the greater polarity of the O—H bond (C=O is electron-withdrawing) and because carboxylic acids can hydrogen-bond in two places.

$$R—C \begin{matrix} O\cdots\cdots H—O \\ \diagdown\quad\quad\quad\diagup \\ O—H\cdots\cdots O \end{matrix} C—R$$

Hydrogen-bonding is not possible in esters and consequently they have significantly lower boiling points than acids. Tables 11.1 and 11.2 also summarize the physical properties of some carboxylic acids.

TABLE 11.3.　COMPARISON OF CARBOXYLIC ACIDS
AND OTHER COMPOUNDS IN PHYSICAL PROPERTIES

	MW	*BP, °C*	*Water solubility, g/100 ml*
$CH_3CH_2CH_2CH_2CH_3$	72	36	0.04
$CH_3CH_2CH_2CH_2OH$	74	117	7.4
CH_3CH_2COH (C=O)	74	141	∞
CH_3COCH_3 (C=O)	74	57	32

SOME PREPARATIONS OF CARBOXYLIC ACIDS

11.4

A. Oxidation of Alkylbenzenes (section 5.7)

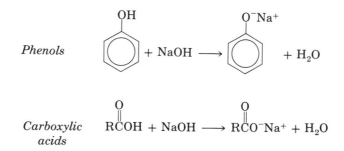

B. Oxidation of Primary Alcohols (section 8.6.E)

$$RCH_2OH \xrightarrow[Na_2Cr_2O_7]{KMnO_4 \text{ or}} R\overset{O}{\overset{\|}{C}}OH$$

C. Hydrolysis of Nitriles (section 9.5.B)

$$RC{\equiv}N \xrightarrow[H^+]{H_2O,} R\overset{O}{\overset{\|}{C}}OH + NH_4^+$$

D. Carbonation of Grignard Reagents (section 9.5.E.4)

$$RX \xrightarrow[\text{ether}]{Mg,} RMgX \xrightarrow{CO_2} R\overset{O}{\overset{\|}{C}}OMgX \xrightarrow[H^+]{H_2O,} R\overset{O}{\overset{\|}{C}}OH$$

ACIDITY OF CARBOXYLIC ACIDS

11.5

A. Salt Formation

1. Reaction of Acids with Base. Of the principal classes of organic compounds, only phenols and carboxylic acids are significantly acidic. This acidity can be detected by reaction with base, and in fact, these neutralization reactions are qualitative tests in organic analysis. Even though alcohols have a polar hydroxy group, they do not react with sodium hydroxide solution, whereas both phenols and carboxylic acids do.

Alcohols   ROH + NaOH $\longrightarrow$ no reaction

Phenols   (phenol) + NaOH $\longrightarrow$ (sodium phenoxide) + H_2O

Carboxylic acids $R\overset{O}{\overset{\|}{C}}OH$ + NaOH $\longrightarrow$ $R\overset{O}{\overset{\|}{C}}O^-Na^+$ + H_2O

Phenols, although definitely acidic compared with alcohols, are considerably weaker acids than carboxylic acids. Experimentally this is shown by the fact that carboxylic acids are neutralized by the weak base sodium bicarbonate but phenols are not.

Alcohols or phenols + $NaHCO_3 \longrightarrow$ no reaction

$$\underset{\text{Carboxylic acids}}{R\overset{\overset{\displaystyle O}{\|}}{C}OH} + NaHCO_3 \longrightarrow R\overset{\overset{\displaystyle O}{\|}}{C}O^-Na^+ + H_2O + \underset{(CO_2 \text{ bubbles})}{CO_2}$$

Even though carboxylic acids are very acidic compared with most organic substances, they are very weak acids in comparison with inorganic acids like HCl, HNO_3, and H_2SO_4.

2. Antibacterial Properties of Carboxylic Acid Salts. Salts of carboxylic acids, and sometimes the free acid, are added to a wide variety of processed foods as preservatives. They act to retard food spoilage by inhibiting or preventing bacterial and fungal growth. Some common food preservatives are shown below.

Sodium benzoate

$(CH_3CH_2CO_2^-)_2Ca^{2+}$

Calcium propionate

$CH_3CH{=}CH{-}CH{=}CHCO_2^-K^+$
Potassium sorbate

$$\underset{\underset{\text{(also a flavor enhancer)}}{\text{Monosodium glutamate}}}{HO_2C\underset{\underset{NH_2}{|}}{C}HCH_2CH_2CO_2^-Na^+}$$

Since carboxylate salts prevent bacterial growth in foods, it is conceivable that there are some that could prevent bacterial and fungal growth in other applications. Calcium and zinc undecylates, for example, are components of some foot and baby powders. Soaps are the sodium salts of long-chain fatty acids derived from fats and oils.

$\underset{\text{Zinc undecylate}}{(CH_3(CH_2)_9CO_2^-)_2Zn^{2+}}$

$\underset{\text{A soap (R = 12–18 carbons)}}{R{\sim\!\sim\!\sim}\overset{\overset{\displaystyle O}{\|}}{C}O^-Na^+}$

PROBLEM

11.7. Write balanced equations showing the preparations of the food preservatives **(a)** sodium benzoate, **(b)** calcium propionate, **(c)** potassium sorbate, and **(d)** monosodium glutamate from the free acids.

B. Reasons for Acidity

Both alcohols and carboxylic acids possess a hydroxy group with a polar O—H bond. Yet carboxylic acids are acidic and alcohols are not. Why? The answer is twofold and involves both the un-ionized acid and the carboxylate anion.

First, let us compare the polarity of the O—H bond in alcohols and carboxylic acids. An alcohol hydroxy group exhibits the polarity caused by the difference in electronegativity between the oxygen and hydrogen. Consider, though, that the hydroxy in a carboxylic acid is directly connected to a carbon-oxygen double bond, itself highly polarized. Because of the carbonyl carbon's positive character, this group is strongly electron-withdrawing and further polarizes the attached hydroxy group.

$$\overset{\delta-\delta+}{RO-H}$$

Alcohol, showing
normal polarity

$$\overset{\delta-}{\underset{\delta+}{R\overset{\displaystyle O}{\overset{\|}{C}}}}\overset{\delta-\delta+}{O-H}$$

Acid, showing
increased O–H polarity

As a result the O—H bond of an acid is significantly more polar and more easily ionizes.

Acidity is also influenced by the stability of the anion formed following ionization or reaction with a base. The negative charge that results is a destabilizing influence. The more the charge can be dispersed, the greater the stability of the anion. In the salt of an alcohol, the negative charge is localized on the oxygen, and this one atom must alone accommodate that charge.

$$R\overset{..}{\underset{..}{O}}:^-$$

Alkoxide ion, showing
concentrated negative charge

In marked contrast, the carboxylate anion has the negative charge dispersed by resonance over the entire carboxylate system, as shown in Figure 11.1. In reality, the resonance hybrid has no double bonds or single bonds. X-ray studies support this hypothesis. In formic acid, for example, the carbon-oxygen double bond is shorter than the carbon-oxygen single bond, whereas in sodium formate the two carbon-oxygen bond lengths are equivalent and intermediate between a single and a double bond.

$$\begin{array}{c} 1.23\,\text{Å} \quad O \\ H-C \\ 1.36\,\text{Å} \quad OH \end{array} \qquad \left.\begin{array}{c} 1.27\,\text{Å} \quad O \\ H-C \\ 1.27\,\text{Å} \quad O \end{array}\right\}^- Na^+$$

Formic acid   Sodium formate

This delocalization of π electrons and negative charge stabilizes the carboxylate ion relative to the alkoxide ion.

In summary, carboxylic acids are acidic due to (1) increased polarization of the O—H bond by the electron-withdrawing carbonyl group and (2) resonance stabilization of the carboxylate ion formed by ionization or reaction with base.

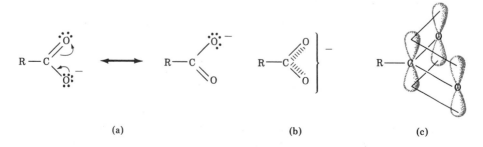

(a)   (b)   (c)

FIGURE 11.1. Resonance stabilization of carboxylate anion. (a) Resonance forms. (b) Resonance hybrid. (c) Bonding picture.

C. Acidity Constants

Acids that are completely ionized in aqueous solution, such as HCl and HNO_3, are termed *strong acids*. *Weak acids* are only partially ionized in aqueous solution and an equilibrium exists between the ionized and un-ionized forms. Carboxylic acids are generally weak acids and show only slight ionization.

$$\overset{O}{\overset{\|}{R C}} OH + H_2O \rightleftharpoons \overset{O}{\overset{\|}{R C}} O^- + H_3O^+$$

The extent of ionization is described by an equilibrium constant, K_a, which is known as the *acidity constant*. It is defined as the concentration of the products of ionization in moles per liter divided by the concentration of reactants.

$$K_a = \frac{[RCO_2^-][H_3O^+]}{[RCO_2H]}$$

Since water is the solvent and is present in large excess, its concentration remains essentially constant. Therefore, it is not included per se in the expression (as a constant, it does become part of K_a).

The acidity constant describes the relative strength of a weak acid. The stronger the acid, the greater the degree of ionization. Solutions of highly ionized acids have greater concentrations of RCO_2^- and H_3O^+ and lesser concentrations of un-ionized RCO_2H. Consequently, stronger acids will have numerically greater acidity constants. As you can see in Table 11.1, most carboxylic acids have an acidity constant of $K_a = 10^{-5}$, which corresponds to about 1% ionization in a 0.1 M solution. We have already seen that phenols are less acidic than carboxylic acids. This is further confirmed by the existence of acidity constants of 10^{-10} for phenols, which amounts to approximately 0.003% ionization for a 0.1 M solution. Carboxylic acids are approximately 100,000 times more acidic than phenols.

D. Structure and Relative Acidity

The presence of various substituents on a carboxylic acid molecule can measurably affect the acidity. Recall that carboxylic acids are acidic because of the electron-withdrawing effect of the carbonyl group and delocalization of the charge in the carboxylate anion. Any group that can enhance these effects, that is, increase the polarity of the O—H bond and disperse the negative charge of the ionized acid, will increase acidity. In general, electron-withdrawing groups increase acidity, whereas electron-releasing groups decrease acidity. For example, formic acid has neither electron-releasing nor electron-withdrawing substituents except for the acid group itself. In acetic and propionic acids, the alkyl groups are electron-releasing and decrease acidity relative to formic acid.

Electron-releasing groups

	HCO_2H	CH_3CO_2H	$CH_3CH_2CO_2H$
K_a	17.7×10^{-5}	1.76×10^{-5}	1.34×10^{-5}

These releasing groups (mainly alkyl groups) decrease O—H polarity and intensify the negative character of the carboxylate ion. In contrast, the two carboxyl groups on oxalic acid enhance the acidity of each other (compare oxalic and formic acids). Also note the effect of the —OH, —Br, —CN on acetic acid. All these electron-withdrawing groups increase O—H polarity and disperse the negative charge on the carboxylate ion.

Electron-withdrawing groups

	HCO$_2$H	HO$_2$CCO$_2$H	
K_a	1.77 × 10^{-4}	590 × 10^{-4}	

CH$_3$CO$_2$H	CH$_2$CO$_2$H $\mid$ OH	CH$_2$CO$_2$H $\mid$ Br	CH$_2$CO$_2$H $\mid$ CN
K_a 0.176 × 10^{-4}	1.48 × 10^{-4}	12.5 × 10^{-4}	36.5 × 10^{-4}

The strength of an electron-withdrawing group determines the magnitude of its effect on acidity. The electronegativity and thus the electron-attracting capability of halogens is of the order F > Cl > Br > I. This trend is exemplified in the haloacetic acids.

Strength of electron-withdrawing groups

	FCH$_2$CO$_2$H	ClCH$_2$CO$_2$H	BrCH$_2$CO$_2$H	ICH$_2$CO$_2$H	CH$_3$CO$_2$H
K_a	260 × 10^{-5}	136 × 10^{-5}	125 × 10^{-5}	67 × 10^{-5}	1.76 × 10^{-5}

As the number of electron-withdrawing substituents increases, so does acidity.

Number of electron-withdrawing groups

	CH$_3$CO$_2$H	ClCH$_2$CO$_2$H	Cl$_2$CHCO$_2$H	Cl$_3$CCO$_2$H
K_a	1.76 × 10^{-5}	136 × 10^{-5}	5,530 × 10^{-5}	23,200 × 10^{-5}

The proximity of the electron-withdrawing group is also important in considering acidity. The nearer the group to the carboxyl, the greater the effect.

Proximity of electron-withdrawing groups

CH$_3$CH$_2$CHCO$_2$H $\mid$ Cl	CH$_3$CHCH$_2$CO$_2$H $\mid$ Cl	CH$_2$CH$_2$CH$_2$CO$_2$H $\mid$ Cl	CH$_3$CH$_2$CH$_2$CO$_2$H
K_a 139 × 10^{-5}	8.9 × 10^{-5}	3.0 × 10^{-5}	1.5 × 10^{-5}

With some modification, these same principles apply to aromatic carboxylic acids. Electron-withdrawing groups enhance acidity. Because their effect is largely due to resonance, they have their greatest impact if positioned ortho or para to the acid group.

Aromatic carboxylic acids

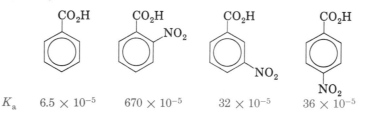

K_a	6.5 × 10^{-5}	670 × 10^{-5}	32 × 10^{-5}	36 × 10^{-5}

PROBLEM

11.8. Arrange in order of increasing acidity.
(a) K_a's (I) 5.9 × 10^{-2}, (II) 3 × 10^{-5}, (III) 6.9 × 10^{-5}, (IV) 1.5 × 10^{-3}, (V) 9.3 × 10^{-4}
(b) (I) F$_3$CCO$_2$H, (II) Br$_3$CCO$_2$H, (III) I$_3$CCO$_2$H, (IV) Cl$_3$CCO$_2$H
(c) (I) Cl$_2$CHCH$_2$CO$_2$H, (II) CH$_3$CCl$_2$CO$_2$H, (III) ClCH$_2$CHClCO$_2$H, (IV) ClCH$_2$CH$_2$CO$_2$H

CONDENSATION REACTIONS OF CARBOXYLIC ACIDS AND THEIR DERIVATIVES

11.6

A. General Reaction

Condensation reactions of carboxylic acids and their derivatives are essentially substitution reactions in which one acid derivative is converted into another. To master these reactions, one must be thoroughly familiar with the structures of acid derivatives. They are listed below in order of reactivity.

General Formula $\overset{\overset{\text{O}}{\|}}{\text{RC}}-\text{Z}$ Z = electronegative group

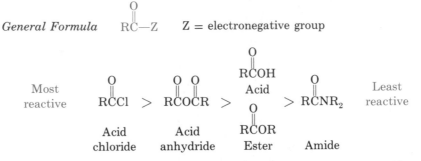

Note that the reactivity is directly proportional to the polarity of the C—Z bond.

The general condensation reaction involves cleavage of the C—Z bond. Using "lasso chemistry," we can express it in the following manner. The reagent H—A usually contains a very polar H—O or H—N bond.

$$\text{Z} = \text{Cl}, \overset{\overset{\text{O}}{\|}}{\text{OCR}}, \text{OH, OR, NR}_2$$

H—A = H—O or H—N bond

As you review examples of this reaction on the following pages, organize your thinking around the following facts:

1. The reactions involve simple substitution of a group A for a group Z.
2. Both the reactant and product are members of the group of five acid derivatives because the reaction involves conversion of one acid derivative to another.

B. Reactions of Acid Derivatives

1. Acid Chlorides. Acid chlorides are the most reactive of the acid derivatives and as such exhibit broad applications in synthesis. They can be transformed into all four of the other derivatives. Treatment with a carboxylic acid or preferably the salt of an acid produces acid anhydrides.

$$\overset{\overset{\text{O}}{\|}}{\text{RC}}-\text{Cl} + \quad \underset{\text{Acid or acid salt}}{\text{H}-\overset{\overset{\text{O}}{\|}}{\text{OCR}}} \quad \longrightarrow \quad \underset{\text{Anhydride}}{\overset{\overset{\text{O}\ \ \text{O}}{\|\ \ \|}}{\text{RCOCR}}} \quad + \text{HCl}$$

Acid chlorides are hydrolyzed in water to carboxylic acids.

$$\underset{\text{Water}}{\overset{\displaystyle O}{\overset{\|}{RC}}-Cl + H-OH \longrightarrow \underset{\text{Acid}}{\overset{\displaystyle O}{\overset{\|}{RC}OH}} + HCl}$$

Although acid chlorides have a sharp irritating smell, they are easily converted to pleasant-smelling esters when combined with alcohols.

$$\underset{\text{Alcohol}}{\overset{\displaystyle O}{\overset{\|}{RC}}-Cl + H-OR \longrightarrow \underset{\text{Ester}}{\overset{\displaystyle O}{\overset{\|}{RC}OR}} + HCl}$$

Amides result when acid chlorides and amines are allowed to react. The amine must have at least one hydrogen remaining on the nitrogen (NH_3, RNH_2, or R_2NH, but not R_3N).

$$\underset{\substack{\text{Amine}\\\text{(2 moles)}}}{\overset{\displaystyle O}{\overset{\|}{RC}}-Cl + H-\overset{\displaystyle R}{\underset{\displaystyle }{N}}-R \longrightarrow \underset{\text{Amide}}{\overset{\displaystyle O \quad R}{\overset{\| \quad |}{RC}-NR}} + HCl}$$

Normally two moles of amine are used in this reaction because amines are basic and the second mole reacts with the HCl formed to produce the salt $R_2NH_2{}^+Cl^-$.

Acid chlorides do not occur naturally because they are extremely reactive. They can be conveniently synthesized, however, by the reaction between carboxylic acids and thionyl chloride ($SOCl_2$).

$$\overset{\displaystyle O}{\overset{\|}{RC}OH} + SOCl_2 \longrightarrow \overset{\displaystyle O}{\overset{\|}{RC}Cl} + SO_2\uparrow + HCl\uparrow$$

Notice that this particular reaction involves the conversion of an acid derivative of low reactivity to a more reactive one. The reaction is irreversible, however, because the inorganic by-products, SO_2 and HCl, escape from the reaction as gases. Phosphorus pentachloride (PCl_5) is also used to make acid chlorides from acids.

PROBLEM

11.9. Write an equation showing the preparation of ethanoyl chloride ($CH_3\overset{\displaystyle O}{\overset{\|}{C}}Cl$) from ethanoic acid. Then write equations showing the reactions of ethanoyl chloride with the following: (a) H_2O; (b) CH_3CH_2OH; (c) $CH_3\overset{\displaystyle O}{\overset{\|}{C}}OH$(salt); (d) NH_3; (e) CH_3NH_2; (f) $CH_3CH_2NHCH_2CH_3$.

2. Acid Anhydrides. Acid anhydrides, like acid chlorides, react with water to form carboxylic acids, with alcohols to form esters, and with ammonia or amines to form amides. These reactions of acid anhydrides differ from similar ones of the acid chlorides only in the by-products—for the anhydrides the by-products are carboxylic acid molecules instead of HCl.

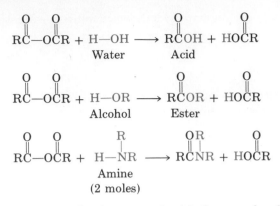

To make amides, two moles of amine are used, with the second mole forming a salt with the acid by-product.

PROBLEM

11.10. Write equations describing the reaction of ethanoic anhydride (CH_3COCCH_3, with O O double bonds) with **(a)** H_2O, **(b)** CH_3CH_2OH, **(c)** NH_3, and **(d)** CH_3NH_2.

3. Reactions of Carboxylic Acids and Esters. Both carboxylic acids and esters react with amines to produce amides. Water is a by-product in the acid-amine reaction, and an alcohol is produced in the ester-amine reaction. When an acid is treated with an amine, a salt ($RCO_2^-H_2\overset{+}{N}R_2$) forms initially; this decomposes to the amide when heat is applied.

$$RC(=O)-OH + H-NR \longrightarrow RC(=O)-NR + H_2O$$
Amine ; Amide

$$RC(=O)-OR + H-NR \longrightarrow RCNR + ROH$$
Amine ; Amide

Carboxylic acids are converted to esters when treated with an alcohol under acid catalysis.

$$RC(=O)-OH + H-OR \xrightarrow{H^+} RC(=O)OR + H_2O$$
Alcohol ; Ester

Conversely, an ester subjected to acid hydrolysis is transformed to an acid and alcohol.

$$RC(=O)-OR + H-OH \xrightarrow{H^+} RC(=O)OH + ROH$$

These last two reactions are obviously the reverse of one another and experimentally an equilibrium will be established between the reactants (acid and alcohol) and the products (ester and water).

$$\underset{\text{Acid}}{\text{RCOH}} + \underset{\text{Alcohol}}{\text{ROH}} \xrightleftharpoons{\text{H}^+} \underset{\text{Ester}}{\text{RCOR}} + \underset{\text{Water}}{\text{H}_2\text{O}}$$

If a high yield of ester is desirable, it is necessary to shift the equilibrium in that direction. This can be accomplished by applying a stress to the equilibrium system and allowing the system to relieve the stress (Le Chatelier's principle). Using excess alcohol or acid or removing one of the products will supply the needed stress. Any one (or all) of these will shift the equilibrium toward the ester. Because the ester is the only component not able to hydrogen-bond, it may have the lowest boiling point. This fact may sometimes facilitate its removal by distillation.

It is possible to convert one ester into another, a process called *transesterification*.

$$\text{RC}-\text{OR} + \text{H}-\text{OR}' \xrightleftharpoons{\text{H}^+} \text{RCOR}' + \text{HOR}$$

This is also an equilibrium reaction. So that the reaction will be driven toward the desired ester, either this ester must be removed as it is formed or a large excess of the reactant alcohol must be used.

PROBLEMS

11.11. Write equations indicating the reaction of ethanoic acid (CH_3CO_2H) with **(a)** NH_3; **(b)** CH_3NH_2; **(c)** CH_3CH_2OH/H^+.

11.12. Write equations indicating the reaction of ethyl ethanoate $CH_3\overset{O}{\overset{\|}{C}}OCH_2CH_3$ with **(a)** H_2O/H^+; **(b)** CH_3OH/H^+; **(c)** NH_3; **(d)** $CH_3CH_2NHCH_2CH_3$.

4. Amides. Amides are the least reactive of the acid derivatives and consequently do not engage in reactions as extensively as other acid derivatives. Amides can be hydrolyzed to acids, however, under either acidic or basic conditions.

$$\text{RC}-\text{NH}_2 + \text{H}-\text{OH} \longrightarrow \begin{cases} \xrightarrow{\text{H}^+} \text{RCOH} + \text{NH}_4^+ \\ \xrightarrow[\text{OH}^-]{} \text{RCO}^- + \text{NH}_3 \end{cases}$$

Acid hydrolysis gives an acid and an ammonium ion salt (since ammonia is basic), whereas in base, free ammonia and the acid salt result.

PROBLEM

11.13. (a) Write reaction equations showing the preparation of ethanamide from an acid chloride, acid anhydride, carboxylic acid, and an ester. **(b)** Write reaction equations for the acidic and basic hydrolysis of ethanamide.

C. Summary—Condensation Reactions of Acid Derivatives

Now that we have examined the condensations of acid derivatives, let us pause for a moment and summarize what we have seen.

1. First, all these reactions can be expressed by a general equation showing the conversion of one acid derivative into another.

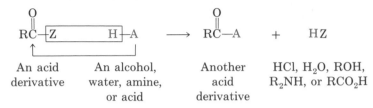

| An acid derivative | An alcohol, water, amine, or acid | Another acid derivative | HCl, H_2O, ROH, R_2NH, or RCO_2H |

2. H—A has an H—O, or H—N bond. Reaction (if one occurs) with an H—O bond (acid, water, or alcohol) produces acid anhydrides, acids, or esters; and with an H—N bond (amines) produces amides.

$$
\underset{RC-Z}{\overset{O}{\parallel}} + \quad
\begin{array}{l}
H-O-G \longrightarrow R\overset{O}{\overset{\parallel}{C}}OG \qquad \text{Acid (G = OH), acid anhydride} \left(G = \overset{O}{\overset{\parallel}{C}}R\right), \text{ or} \\
\qquad\qquad\qquad\qquad\qquad\qquad\qquad \text{ester (G = R)} \\[2ex]
\underset{R}{\overset{R}{\underset{|}{H-NR}}} \longrightarrow \underset{R}{\overset{O}{\underset{|}{RC-NR}}} \qquad \text{Amide}
\end{array}
$$

Assuming that a reaction occurs, we find that the only differences in the reactions are the facility with which the derivatives react and the by-products that are formed. The by-product of acid chlorides is HCl; of acid anhydrides, an acid; of acids, water; of esters, an alcohol; and of amides, ammonia or an amine.

D. Condensation Reactions: Reaction Mechanisms

To this point we have examined *what* happens in condensation reactions of acid derivatives, but we have yet to discuss *how* it happens. Now let us look at the reaction mechanisms. There are two fundamental types—acid-initiated and base-initiated.

1. Acid-Initiated Condensation Mechanisms. Condensation reactions catalyzed by acid are initiated by the binding of a proton to the acid derivative, forming a carbocation.

$$
\underset{RC-Z:}{\overset{:\overset{..}{O}}{\overset{\parallel}{}}} \xrightarrow{\text{H}^+} \underset{RC-Z:}{\overset{:\overset{..}{O}H}{\underset{+}{\overset{|}{}}}}
$$

The Lewis base H—A (with at least one lone pair of electrons) attacks the carbocation and the A becomes positive.

$$
\underset{+}{\overset{:\overset{..}{O}H}{\underset{RC-Z}{\overset{|}{}}}} + H\overset{..}{A} \longrightarrow \underset{+AH}{\overset{OH}{\underset{R-C-Z:}{\overset{|}{\underset{|}{}}}}}
$$

After a proton transfer from A to Z, HZ separates, leaving behind another carbocation.

$$RC\overset{\underset{+AH}{|}}{\underset{}{—}}Z: \longrightarrow R—\underset{\underset{A}{|}}{\overset{\overset{OH}{|}}{C}}—\overset{H}{\underset{\cdot\cdot}{Z}}{}^+ \longrightarrow R\overset{\overset{OH}{|}}{\underset{A}{C}}{}^+ + HZ$$

A proton initiated the reaction and loss of a proton concludes the sequence.

$$R—\underset{A}{\overset{\overset{:\overset{\cdot\cdot}{O}H}{|}}{C}}{}^+ \xrightarrow{-H^+} R\overset{\overset{:\overset{\cdot\cdot}{O}}{\|}}{C}—A$$

Now let us apply these concepts to the acid-catalyzed esterification of acetic acid to produce ethyl acetate, a solvent in fingernail polish remover. In this scheme, $Z = —OH$ and $HA = HOCH_2CH_3$.

Acid-Catalyzed Esterification

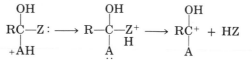

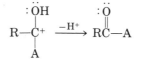

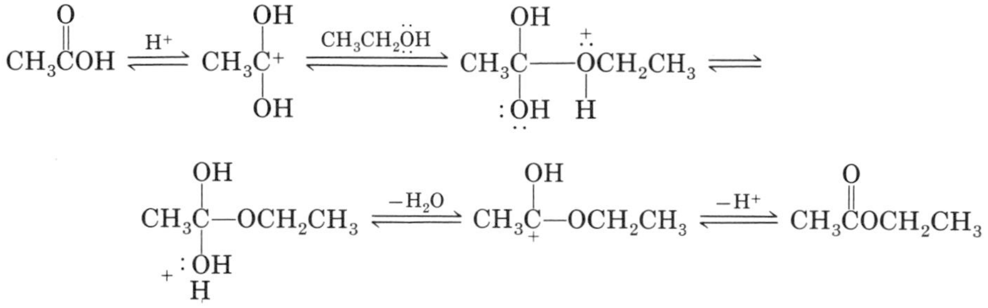

Note that this reaction is completely reversible at every step (see section 11.6.B.3).

PROBLEM

11.14. Write a step-by-step reaction mechanism for the acid-catalyzed hydrolysis of acetamide (CH_3CNH_2) with H_2O/H^+.
$$\overset{}{\underset{O}{\|}}$$

---

2. Base-Initiated Condensation Mechanisms. In base-initiated condensation reactions, the reactant HA is often in its salt form, Na^+A^-. The nucleophile, $A:^-$, attacks the positive carbonyl carbon (a natural attraction) to commence the reaction.

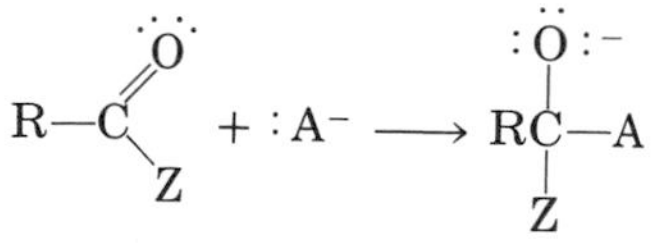

$$R—C\overset{\overset{\cdot\cdot}{O}\cdot}{\diagdown_Z} + :A^- \longrightarrow R\overset{\overset{:\overset{\cdot\cdot}{O}:^-}{|}}{\underset{Z}{C}}—A$$

Next $Z:^-$ leaves and the reaction is complete.

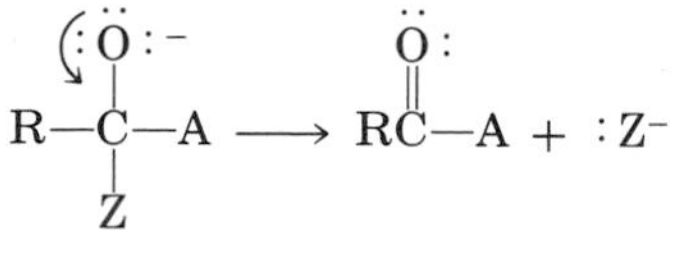

$$R—\underset{Z}{\overset{\overset{(:\overset{\cdot\cdot}{O}:^-}{|}}{C}}—A \longrightarrow R\overset{\overset{\overset{\cdot\cdot}{O}:}{\|}}{C}—A + :Z^-$$

Applying this sequence to base-initiated ester hydrolysis, we get the following mechanism:

Base-initiated ester hydrolysis

$$CH_3\overset{\displaystyle O}{\overset{\|}{C}}OCH_2CH_3 \xrightarrow{\;:\ddot{O}H^-\;} CH_3\overset{\displaystyle :\ddot{O}:^-}{\underset{\displaystyle OH}{\overset{|}{\underset{|}{C}}}}-OCH_2CH_3 \longrightarrow CH_3\overset{\displaystyle \ddot{O}:}{\overset{\|}{C}}OH + CH_3CH_2\ddot{O}:^- \longrightarrow$$

$$CH_3CH_2OH + CH_3\overset{\displaystyle O}{\overset{\|}{C}}\ddot{O}:^-$$

The negative hydroxide ion is attracted to the positive carbonyl carbon. Ethoxide ion then leaves and the acid produced is neutralized by the strongly basic ethoxide ion.

CONDENSATION POLYMERS

11.7   A variety of familiar polymers are prepared by esterification and amidification reactions such as those just covered in section 11.6. Recall that polymers are gigantic molecules that have a recurring unit or units. To build such molecules by condensation reactions requires that the two starting materials be at least difunctional (such as a diacid and dialcohol, or a diacid and diamine). In this way both ends of both starting molecules are reactive and the reaction proceeds continuously, producing long polymer chains called *polyesters* and *polyamides*.

A. Nylon

1. Nylon 66. Nylon 66 was developed by a research group at Du Pont headed by W. H. Carothers. The work, which began in 1928, culminated in a patent in 1937 and public sales three years later. Nylon stockings were introduced in May, 1940, and the initial production of four million pairs was sold within four days. Today, nylon is one of the most important synthetic fibers, finding use in among other things, clothing, sails, parachutes, fishing line, brushes, combs, gears, and bearings.

Nylon 66 derives its name from its starting materials, adipic acid and hexamethylene diamine, both of which have six carbons. The reaction is merely a condensation between an acid and an amine to form an amide. Since both ends of both molecules are reactive, however, a high-molecular-weight polyamide forms continuously.

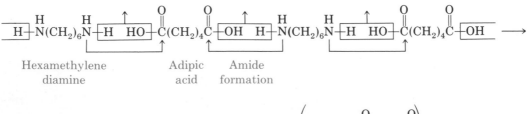

Hexamethylene diamine   Adipic acid   Amide formation

Nylon 66 (a polyamide)

Amide linkage

In the industrial production of Nylon 66, adipic acid and hexamethylene diamine are heated to 250–275 °C under pressure. The molten polymer is extruded and solidified in long cylinders, which are then broken into small chips. These are melted and shaped into the desired product. For nylon fibers, the molten polymer (from the chips) is extruded through tiny holes called spinnerets. As the nylon passes through the spinneret, it solidifies, is stretched, and is collected on rollers. The stretching allows the individual molecules in a strand to line up so that intermolecular hydrogen-bonding between the carbonyl groups and N—H bonds is maximized (Figure 11.2). This adds strength and durability to the final woven fiber.

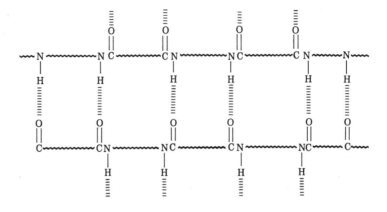

FIGURE 11.2. Hydrogen-bonding between nylon polymer molecules.

2. Nylon 6-10. Nylon 6-10, used for molded plastics, is made from hexamethylene diamine and 1,10-decandioic acid.

$$H_2N(CH_2)_6NH_2 \qquad HOC(CH_2)_8COH$$

Hexamethylene diamine   1,10-Decandioic acid

It is formed like Nylon 66. Again, the 6 and 10 correspond to the number of carbons in the starting materials.

PROBLEM

11.15. Write a structure for Nylon 6-10.

3. Nylon 6. One can think of Nylon 6 as being formed from 6-aminohexanoic acid. The amine end of one molecule reacts with the acid end of another. The remaining acid and amino ends react with still other molecules and the chain builds continuously.

$$H-N(CH_2)_5C-OH \quad H-N(CH_2)_5C-OH \quad H-N(CH_2)_5C-OH \quad H-N(CH_2)_5C-OH \longrightarrow$$

$$\left(N(CH_2)_5CNH(CH_2)_5C \right)_n$$

Nylon 6

Actually, Nylon 6 is formed by the ring opening and polymerization of caprolactam.

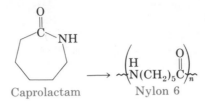

Caprolactam   Nylon 6

Nylon 6 and Nylon 66 are the most heavily used nylons for fiber manufacture.

B. Polyesters

1. Dacron-Mylar. Textile fibers known as Dacron and transparent plastic films marketed as Mylar are polyesters produced from the dimethyl ester of terephthalic acid and ethylene glycol. The reaction involves transesterification. Since both molecules are difunctional, a large polymer chain forms from continuous ester condensations.

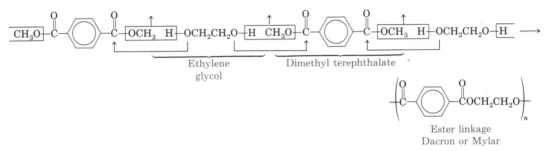

Ethylene glycol   Dimethyl terephthalate

Ester linkage
Dacron or Mylar

The linearity of the polymer, which allows it to pack into close-knit layers, is responsible for the strength and toughness of the plastic. Dacron and Mylar are used in clothing, sails, magnetic recording tape, packaging, and so forth.

2. Kodel Polyesters. Kodel polyester fabrics are spun from a polymer made from dimethyl terephthalate and 1,4-dihydroxymethylcyclohexane. The reaction is a condensation and a polyester is produced.

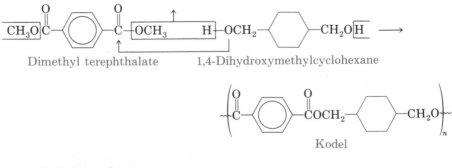

Dimethyl terephthalate   1,4-Dihydroxymethylcyclohexane

Kodel

C. Polyurethanes

Polyurethanes are large polymers consisting of both ester and amide units. The reaction involves the addition of an alcohol to an isocyanate to form a carbamate.

$$\underset{\text{An isocyanate}}{R-N=C=O} + R'OH \longrightarrow \underset{\text{A carbamate}}{R-\overset{H}{\underset{}{N}}-\overset{O}{\overset{\|}{C}}OR'}$$

The O—H bond adds to the N—C double bond. If a diisocyanate (such as toluene diisocyanate) is allowed to react with a diol or a polyhydroxy alcohol, a polyurethane is formed.

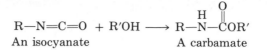

A polyurethane

Polyurethanes can be spun into elastic fibers such as Spandex. Or if they are formed in the presence of a low-boiling liquid that will readily vaporize as the reactants are heated, a semirigid polyurethane foam results. Polyurethane foams are used structurally in the construction and interior decoration of buildings and in other ways. Unfortunately, they have also been found to be extremely flammable.

REACTIONS OF ACID DERIVATIVES INVOLVING CARBANIONS

11.8

A. Malonic Ester Synthesis

The malonic ester synthesis is useful in preparing substituted acetic acids and their derivatives. Follow Figure 11.3 as this synthetic procedure is discussed for the preparation of 2-ethyl-5-methylhexanoic acid, a disubstituted acetic acid.

$$\underset{\text{Acetic acid}}{H-\overset{H}{\underset{H}{\overset{|}{C}}}-CO_2H} \qquad \underset{\text{A disubstituted acetic acid}}{(CH_3)_2CHCH_2CH_2-\overset{}{\underset{CH_2CH_3}{\overset{|}{C}}HCO_2H}}$$

Malonic ester is acidic and its α-hydrogens can be extracted by base (see step 1, Et = CH_2CH_3) because of the polarization of the carbon-hydrogen bonds by the adjacent carbonyl groups and because of the resonance stabilization of the resulting carbanion (see section 9.6.A for an explanation of the acidity of α-hydrogens).

$$\underset{\text{EtOC}-CH-COEt}{\overset{O}{\overset{\|}{}}} \qquad \text{Resonance hybrid of the diethyl malonate carbanion}$$

When this carbanion is treated with an alkyl halide (step 2), the halide is displaced by nucleophilic substitution (probably by an S_N2 mechanism, section 7.4.A and B). If the alkyl halide is CH_3CH_2X ($R_1 = CH_2CH_3$ in step 2), we will have inserted the ethyl group of the desired product. Steps 3 and 4 are a repeat of steps 1

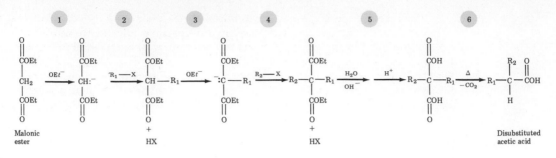

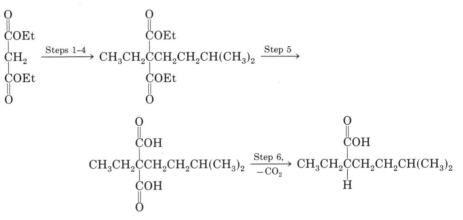

FIGURE 11.3. Malonic ester synthesis of disubstituted acetic acids. Numbers mark the successive steps as described in text. "Et" symbolizes CH_2CH_3.

and 2 and result in a disubstituted malonic ester. Use of $(CH_3)_2CHCH_2CH_2X$ as the alkyl halide ($R = (CH_3)_2CHCH_2CH_2$ in step 4) will provide the second alkyl group desired in the final product. In step 5, the diester is hydrolyzed to a dicarboxylic acid. Dicarboxylic acids in which the two acid groups are separated by one carbon atom decarboxylate (lose CO_2) when heated (step 6). The final product is the desired disubstituted acetic acid. If the disubstituted malonic acid or ester is isolated before decarboxylation, it can be used for the preparation of barbiturates (see Panel 15).

PROBLEM

11.16. Prepare the monosubstituted acetic acid hexanoic acid (caproic acid from goats) by the malonic ester synthesis. (Steps 3 and 4 would be eliminated in Figure 11.3.)

B. Claisen Condensation

The Claisen condensation is a carbanion-type reaction in which an ester is converted to a β-keto ester. Consider, for example, the condensation of ethyl acetate to ethyl 3-ketobutanoate:

$$2CH_3\overset{O}{\overset{\|}{C}}OCH_2CH_3 \xrightarrow[CH_3CH_2OH]{NaOCH_2CH_3,} CH_3\overset{O}{\overset{\|}{C}}CH_2\overset{O}{\overset{\|}{C}}OCH_2CH_3 + CH_3CH_2OH$$

The reaction mechanism involves initial abstraction of an α-hydrogen from an ester molecule by ethoxide ion. These hydrogens are acidic due to polarization of the carbon-hydrogen bond by the carbonyl group and resonance stabilization of the resulting carbanion (see section 9.6.A).

$$CH_2\overset{O}{\overset{\|}{C}}OCH_2CH_3 \ + \ :\overset{..}{\underset{..}{O}}CH_2CH_3 \ \longrightarrow \ \overset{-}{C}H_2\overset{O}{\overset{\|}{C}}OCH_2CH_3 \ + \ H:\overset{..}{\underset{..}{O}}CH_2CH_3$$

The carbanion attacks the positive carbonyl of an ester molecule and displaces an ethoxide ion.

$$CH_3\overset{:\overset{..}{O}}{\overset{\|}{C}}OCH_2CH_3 \ + \ \overset{O}{\overset{\|}{\underset{..}{C}}H_2COCH_2CH_3} \ \longrightarrow \ CH_3\overset{:\overset{..}{O}:^-}{\underset{\overset{|}{CH_2\overset{}{C}OCH_2CH_3}}{\overset{|}{C}}}OCH_2CH_3 \ \longrightarrow$$

$$CH_3\overset{O}{\overset{\|}{C}}CH_2\overset{O}{\overset{\|}{C}}OCH_2CH_3 \ + \ CH_3CH_2O^-$$

The product of this particular example is acetoacetic ester and is the starting material in the acetoacetic ester of ketones (see Problem 11.39).

Claisen-type condensations occur in some biological systems. For example, the biosynthesis of acetoacetyl coenzyme A, an intermediate in the biosynthesis of terpenes (see Panel 6 in Chapter 4), steroids (section 12.7.C), and fatty acids (Chapter 12), involves the Claisen-type condensation of two molecules of acetyl-CoA.

$$CH_3\overset{O}{\overset{\|}{C}}-SCoA \ + \ CH_3\overset{O}{\overset{\|}{C}}SCoA \ \xrightarrow{\text{Enzyme}} \ CH_3\overset{O}{\overset{\|}{C}}CH_2\overset{O}{\overset{\|}{C}}SCoA \ + \ HSCoA$$

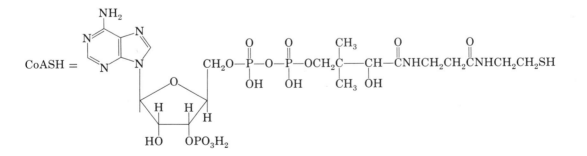

CoASH =

Panel 15

ACID DERIVATIVES AND DRUGS

Some of the most common medications used by today's society are derivatives of carboxylic acids.

[PANEL 15]   A. Pain Relievers

1. Salicylic Acid Derivatives.   Salicylic acid is a bifunctional molecule (acid and phenol) from which many familiar substances are derived.   Salicylic acid itself is used as a disinfectant in some first aid sprays and ointments.   When this substance is treated with methanol, it forms methyl salicylate, oil of wintergreen. Reaction with ammonia yields salicylamide, an over-the-counter drug found in some combination pain relievers.   And when salicylic acid is combined with acetic anhydride, acetyl salicylic acid, or aspirin, is formed.

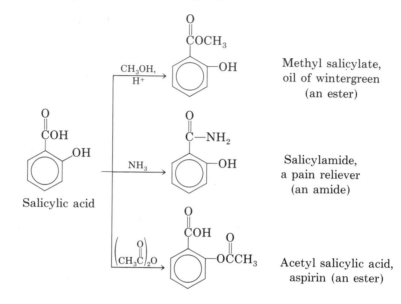

Methyl salicylate, oil of wintergreen (an ester)

Salicylamide, a pain reliever (an amide)

Acetyl salicylic acid, aspirin (an ester)

Salicylic acid

Aspirin is one of the most amazing medicines known to man.   As an antipyretic, it reduces fever but does not lower normal body temperature.   It has analgesic (pain-relieving) properties and is effective against pains accompanying headache, colds, flu, nervous tension, rheumatism, and arthritis.   The name *aspirin* comes from *Salix spirea,* a willow.   Jesuit missionaries in the Middle Ages used the bark of this tree for its medicinal properties.   In the 1600s, it was found that extracts of willow bark had fever-reducing properties, and in 1826 the active principle, salicylic acid, was isolated.   By 1852, salicylic acid had been independently synthesized, and by 1874 relatively large-scale production made it increasingly available as a medicine.   Although salicylic acid was an effective antipyretic, it caused severe stomach irritation in some people.   At the time, it was hypothesized that the neutralized acid would cause less gastric irritation.   So in 1875, sodium salicylate was introduced.

Sodium salicylate

[PANEL 15]This was not found to be any better than free salicylic acid. Salol, a phenol ester of salicylic acid, was introduced in 1886 and did show greatly decreased incidence of gastric distress. In the small intestine it hydrolyzed to sodium salicylate, which had previously been used as a pain reliever. The simultaneous liberation of phenol led to the danger of phenol poisoning, however.

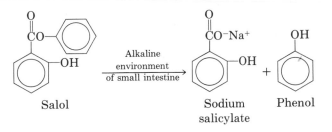

Salol   Sodium salicylate   Phenol

Toward the end of the nineteenth century, Felix Hofmann, who worked for the Bayer Company, investigated other derivatives of salicylic acid and tested acetyl salicylic acid on his father, who suffered from arthritis. This and other tests revealed its excellent medicinal properties and decreased frequency of gastric irritation. Acetyl salicylic acid, aspirin, was marketed in 1899 by the Bayer Company.

2. Combination Pain Relievers and Aspirin Substitutes. Combination pain relievers are preparations in which aspirin is combined with other pain relievers or stomach antacids or both. The most common alternative pain relievers are salicylamide, phenacetin, acetoaminophen, and caffeine.

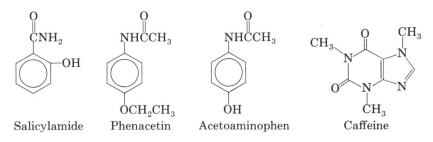

Salicylamide   Phenacetin   Acetoaminophen   Caffeine

Salicylamide is much less effective than aspirin and is too weak and unreliable to be useful alone as a pain reliever. Acetoaminophen and phenacetin are essentially equivalent to aspirin in their antipyretic and analgesic properties, but neither has a significant effect on inflamed joints caused by rheumatism and arthritis. Neither causes gastric irritation and they can be taken by those who experience unpleasant reactions to aspirin. Since phenacetin has been implicated in kidney damage (and has been removed from most combination pain relievers), acetoaminophen is probably the drug of choice. The rationale for the inclusion of caffeine in combination pain relievers is unclear; the amount present is too small to have stimulatory effects (as caffeine does in coffee, tea, cola, and sleep-prevention pills), and no enhancement effect on other pain relievers has been observed. APC tablets contain *a*spirin, *p*henacetin, and *c*affeine.

Antacids are added to pain relievers to raise gastric pH and thus minimize stomach upset (the extent of this effect is controversial) and to accelerate tablet

[PANEL 15] dissolution. Antacids found in pain relievers or over-the-counter antacid preparations include $NaHCO_3$ (baking soda, bicarbonate of soda), $CaCO_3$ (calcium carbonate), $Mg(OH)_2$ (milk of magnesia), $Al(OH)_3$ (aluminum hydroxide), $NaAl(OH)_2CO_3$ (dihydroxyaluminum sodium carbonate), and $Mg_2Si_3O_8$ (magnesium trisilicate).

B. Barbiturates

Barbiturates are made by condensing urea or thiourea with malonic ester and substituted malonic esters.

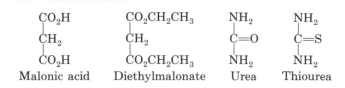

| Malonic acid | Diethylmalonate | Urea | Thiourea |

The reaction is a condensation between an ester and amine (urea is actually an amide) to form two new amide linkages.

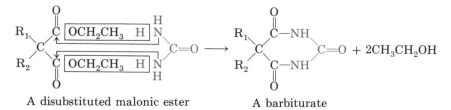

A disubstituted malonic ester → A barbiturate

Barbiturate Structures

$R_1 = CH_3CH_2-$
$R_2 = (CH_3)_2CHCH_2CH_2-$ } Amytal, amobarbital

$R_1 = -CH_2CH_3$
$R_2 = CH_3CHCH_2CH_2CH_3$ } Nembutal, pentobarbital

$R_1 = $ ⟨benzene ring⟩
$R_2 = -CH_2CH_3$ } Phenobarbital, luminal

$R_1 = CH_2=CHCH_2-$
$R_2 = CH_3CHCH_2CH_2CH_3$ } Seconal, secobarbital

Barbiturates have a depressing effect on the central nervous system and are useful as hypnotics and sedatives in both human and veterinary medicine. For example, in human medicine they are used (by prescription) as sleeping pills, to control blood pressure, to combat epileptic seizures, and to control colic in young babies. Since they also depress a wide range of biological functions, such as oxygen intake and heart activity, they must be used cautiously. When a barbiturate is applied as a general anesthetic, as sodium pentothal is, the effective dose is as much as 50%–75% the lethal dose.

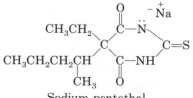

Sodium pentothal

[PANEL 15] Persons taking barbiturates as sleeping pills must be particularly careful to avoid drinking alcohol while under the influence of a barbiturate (or vice versa) because of their synergistic effect; that is, the combined effect of barbiturate and alcohol is greater than the expected sum of the two.

C. Local Anesthetics

Local anesthetics, especially benzocaine and lidocaine, are found in a tremendous number of over-the-counter medications purchased by the American consumer. These products include first aid sprays, sunburn sprays, itching creams, arthritis preparations, cough medicines, hemorrhoidal preparations, toothache medicinals, foot powders, and appetite control products. The most common local anesthetics are acid derivatives (esters or amides) and have similar structural features.

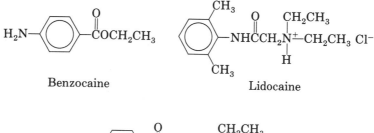

Benzocaine   Lidocaine

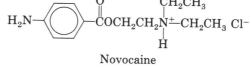

Novocaine

END-OF-CHAPTER PROBLEMS

11.17 **IUPAC Nomenclature:** Name by the IUPAC system of nomenclature

(a) $CH_3(CH_2)_6CO_2H$ **(b)** $CH_3CH{=}CHCO_2H$ **(c)** $CH_3CCH_2CH_2CHCO_2H$
                                                        $\underset{O}{\|}\quad\quad\underset{OH}{|}$

(d) Cl—⟨○⟩—CO_2H **(e)** $CH_3CCH_2CCH_2CO_2H$ **(f)** $HO_2C(CH_2)_7CO_2H$
    with CH_3 and CH_3 groups, and CH_3 on benzene ring

(g) ⟨○⟩—$\overset{O}{\overset{\|}{C}}Cl$ with O_2N **(h)** CH_3CH_2CHCCl with CH_3CH_2, and $\overset{O}{\|}$ **(i)** $CH_3C{\equiv}CCH_2\overset{O}{\overset{\|}{C}}Cl$

(j) $CH_3\overset{O}{\overset{\|}{C}}O\overset{O}{\overset{\|}{C}}CH_2CH_2CH_3$ **(k)** ⟨○⟩—$\overset{O}{\overset{\|}{C}}O\overset{O}{\overset{\|}{C}}CH_3$

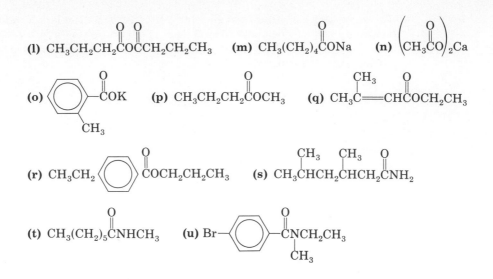

(l) $CH_3CH_2CH_2COCCH_2CH_2CH_3$ **(m)** $CH_3(CH_2)_4CONa$ **(n)** $(CH_3CO)_2Ca$

(o) aromatic ring with CH_3 and $\overset{O}{\overset{\|}{C}}OK$ **(p)** $CH_3CH_2CH_2\overset{O}{\overset{\|}{C}}OCH_3$ **(q)** $CH_3\overset{CH_3}{\overset{\|}{C}}\!\!=\!\!CH\overset{O}{\overset{\|}{C}}OCH_2CH_3$

(r) CH_3CH_2—ring—$\overset{O}{\overset{\|}{C}}OCH_2CH_2CH_3$ **(s)** $CH_3\overset{CH_3}{\overset{\|}{C}}HCH_2\overset{CH_3}{\overset{\|}{C}}HCH_2\overset{O}{\overset{\|}{C}}NH_2$

(t) $CH_3(CH_2)_5\overset{O}{\overset{\|}{C}}NHCH_3$ **(u)** Br—ring—$\overset{O}{\overset{\|}{C}}N\overset{}{\underset{CH_3}{}}CH_2CH_3$

11.18 IUPAC Nomenclature: Draw the following compounds: **(a)** 3-methyl-butanoic acid; **(b)** 5-bromo-3-hexynoic acid; **(c)** *meta*-chlorobenzoyl chloride; **(d)** octanoyl chloride; **(e)** pentanoic anhydride; **(f)** butanoic hexanoic anhydride; **(g)** sodium 2,4-heptadienoate; **(h)** ammonium acetate; **(i)** butyl *ortho*-nitroben-zoate; **(j)** dimethyl 1,4-butandioate; **(k)** pentanamide; **(l)** N-ethyl-N-methyl-*p*-isopropylbenzamide.

11.19 Preparations of Carboxylic Acids: Write reaction equations illustrating the preparation of benzoic acid by the following methods: **(a)** oxidation of alkyl-benzenes; **(b)** oxidation of primary alcohols; **(c)** hydrolysis of nitriles; **(d)** carbona-tion of Grignard reagents.

11.20 Physical Properties: Arrange the following in order of increasing boiling point.   Explain your answer.

(a) $CH_3\overset{O}{\overset{\|}{C}}OH$, $HOCH_2\overset{O}{\overset{\|}{C}}H$, $H\overset{O}{\overset{\|}{C}}OCH_3$

(b) $HO\overset{O}{\overset{\|}{C}}(CH_2)_3\overset{O}{\overset{\|}{C}}OH$, $CH_3O\overset{O}{\overset{\|}{C}}CH_2CH_2\overset{O}{\overset{\|}{C}}OH$, $CH_3O\overset{O}{\overset{\|}{C}}CH_2\overset{O}{\overset{\|}{C}}OCH_3$

11.21 Acidity: Arrange the following in order of increasing acidity:

(a) $CH_3\underset{Br}{\overset{}{C}}H\underset{Br}{\overset{}{C}}HCO_2H$, $CH_2\underset{Br}{\overset{}{C}}HCH_2\underset{Br}{\overset{}{C}}O_2H$, $CH_3CH_2\underset{Br}{\overset{}{C}}\!\!-\!\!CO_2H$, $CH_2\underset{Br}{\overset{}{C}}H_2\underset{Br}{\overset{}{C}}HCO_2H$

(b) $CH_3CH_2CH_2\overset{O}{\overset{\|}{C}}CO_2H$, $CH_3CH_2\overset{O}{\overset{\|}{C}}CH_2CO_2H$, $CH_3\overset{O}{\overset{\|}{C}}CH_2CH_2CO_2H$

(c) 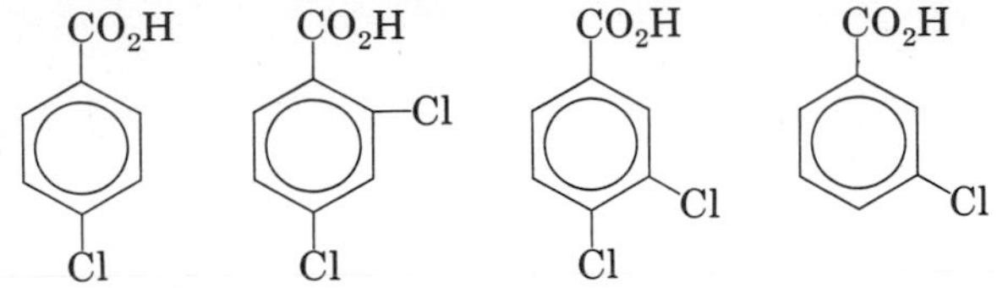

(d) 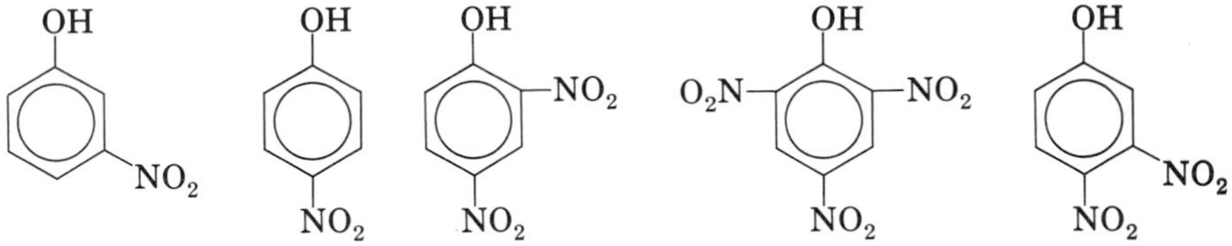

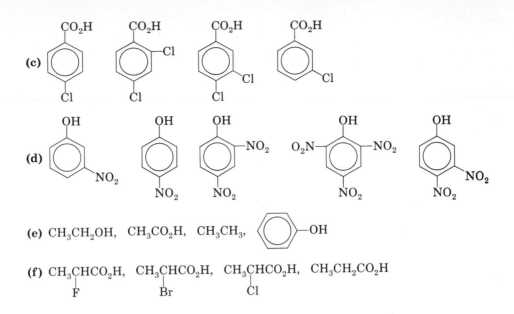

(e) CH_3CH_2OH, CH_3CO_2H, CH_3CH_3, 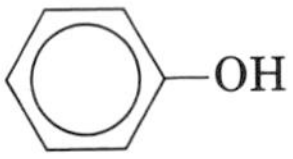

(f) $CH_3CH(F)CO_2H$, $CH_3CH(Br)CO_2H$, $CH_3CH(Cl)CO_2H$, $CH_3CH_2CO_2H$

11.22 Acidity: Explain the following facts:

(a) In Table 11.2, the order of acidity is oxalic > malonic > succinic > glutaric > adipic. Why? Why do the next three to four dicarboxylic acids all have about the same first ionization constants?

(b) The second ionization of oxalic acid (Table 11.2), 6.4×10^{-5}, is much smaller than the first, 5.9×10^{-2}.

(c) Nitric acid (HNO_3) is a strong acid; nitrous acid (HNO_2) is weak.

11.23 Reactions of Carboxylic Acids and Derivatives: Write products of the reactions between each of the following reactants:

(a) $CH_3(CH_2)_5CO_2H/NaOH$ **(b)** 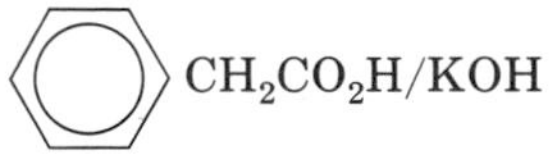

(c) $HO_2C(CH_2)_3CO_2H/Ca(OH)_2$ **(d)** $CH_3CH_2CO_2H/SOCl_2$

(e) 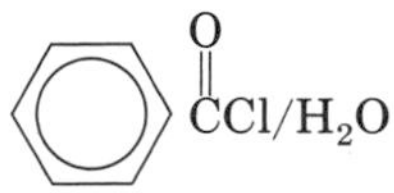

(f) $CH_3CH(CH_3)CH_2C(=O)Cl/CH_3CO_2H$

(g) $CH_3(CH_2)_4C(=O)Cl/CH_3OH$ **(h)** $CH_3C{\equiv}CC(=O)Cl/NH_3$

(i) 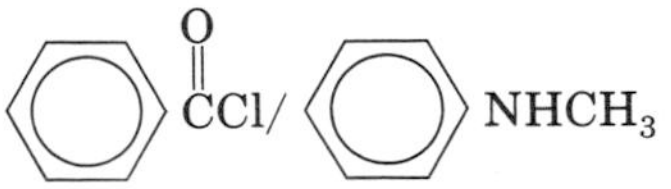

(j) $CH_3CH_2C(=O)OC(=O)CH_2CH_3/CH_3OH$

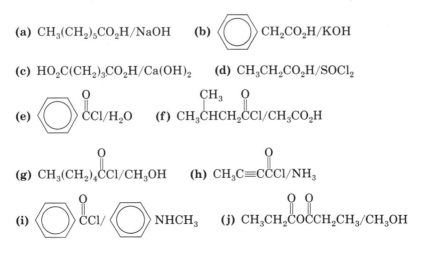

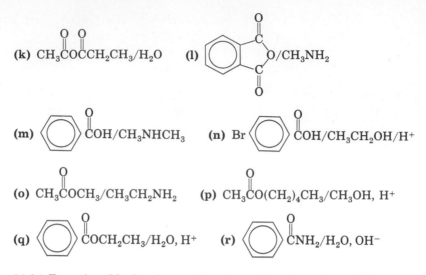

(k) $CH_3COCCH_2CH_3/H_2O$ **(l)**

(m) COH/CH_3NHCH_3 **(n)** Br—$COH/CH_3CH_2OH/H^+$

(o) $CH_3COCH_3/CH_3CH_2NH_2$ **(p)** $CH_3CO(CH_2)_4CH_3/CH_3OH, H^+$

(q) $COCH_2CH_3/H_2O, H^+$ **(r)** $CNH_2/H_2O, OH^-$

11.24 Reaction Mechanisms—Condensation Reactions: Write step-by-step reaction mechanisms for the following reactions:

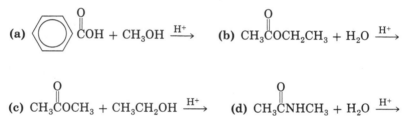

(a) $COH + CH_3OH \xrightarrow{H^+}$ **(b)** $CH_3COCH_2CH_3 + H_2O \xrightarrow{H^+}$

(c) $CH_3COCH_3 + CH_3CH_2OH \xrightarrow{H^+}$ **(d)** $CH_3CNHCH_3 + H_2O \xrightarrow{H^+}$

(e) $COCH_3 \xrightarrow[H_2O]{NaOH,}$ **(f)** $CH_3CCl + NH_3 \longrightarrow$

11.25 Reaction Mechanisms—Condensation Reactions: One of the important experiments used to elucidate the mechanism of acid-catalyzed esterification was to show whether the oxygen in the ester (the oxygen of —OR) came from the original acid or the alcohol. When ordinary benzoic acid is allowed to react with isotopically enriched methanol, $CH_3O^{18}H$, the methyl benzoate produced contains the labeled oxygen. Using words and reaction equations, show how this experiment answers the question.

11.26 Malonic Ester Synthesis: Prepare the following compounds from diethylmalonate using the malonic ester synthesis:

(a) $CH_3CH_2CH_2CO_2H$ **(b)** $CH_3CH_2CHCO_2H$
$\qquad\qquad\qquad\qquad\qquad\qquad\quad |$
$\qquad\qquad\qquad\qquad\qquad\qquad CH_3$

(c) $CH_3CH_2CH_2CNH_2$ **(d)** $CH_3CH_2CHCOCH_3$
$\qquad\qquad\qquad\qquad\qquad\qquad\qquad |$
$\qquad\qquad\qquad\qquad\qquad\qquad CH_3CH_2$

11.27 Preparation of Familiar Compounds: Complete the following reaction sequences showing all intermediate products:

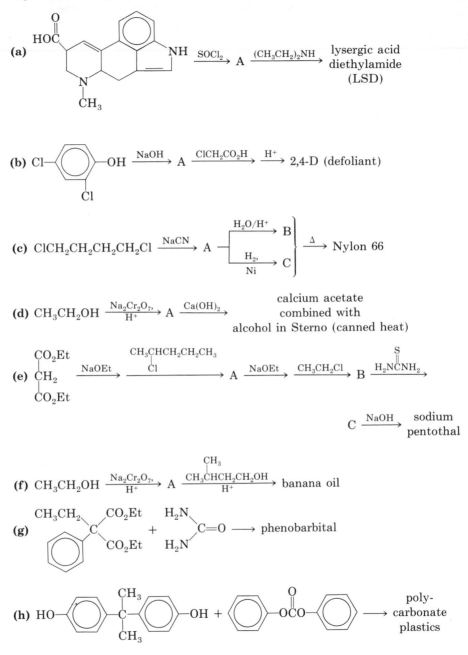

(a)

$\xrightarrow{\text{SOCl}_2}$ A $\xrightarrow{(\text{CH}_3\text{CH}_2)_2\text{NH}}$ lysergic acid diethylamide (LSD)

(b) Cl—⟨ ⟩—OH $\xrightarrow{\text{NaOH}}$ A $\xrightarrow{\text{ClCH}_2\text{CO}_2\text{H}}$ $\xrightarrow{\text{H}^+}$ 2,4-D (defoliant)

(c) ClCH₂CH₂CH₂CH₂Cl $\xrightarrow{\text{NaCN}}$ A ⎰ $\xrightarrow{\text{H}_2\text{O/H}^+}$ B ⎱ $\xrightarrow{\Delta}$ Nylon 66

with H₂, Ni → C

(d) CH₃CH₂OH $\xrightarrow[\text{H}^+]{\text{Na}_2\text{Cr}_2\text{O}_7,}$ A $\xrightarrow{\text{Ca(OH)}_2}$ calcium acetate combined with alcohol in Sterno (canned heat)

(e)

$\underset{\text{CO}_2\text{Et}}{\overset{\text{CO}_2\text{Et}}{\text{CH}_2}}$ $\xrightarrow{\text{NaOEt}}$ $\xrightarrow[\underset{\text{Cl}}{\text{CH}_3\text{CHCH}_2\text{CH}_2\text{CH}_3}]{}$ A $\xrightarrow{\text{NaOEt}}$ $\xrightarrow{\text{CH}_3\text{CH}_2\text{Cl}}$ B $\xrightarrow{\text{H}_2\text{NCNH}_2}$

C $\xrightarrow{\text{NaOH}}$ sodium pentothal

(f) CH₃CH₂OH $\xrightarrow[\text{H}^+]{\text{Na}_2\text{Cr}_2\text{O}_7,}$ A $\xrightarrow[\text{H}^+]{\overset{\text{CH}_3}{\text{CH}_3\text{CHCH}_2\text{CH}_2\text{OH}}}$ banana oil

(g)

$\underset{\text{CO}_2\text{Et}}{\overset{\text{CO}_2\text{Et}}{\text{C}}}$ CH₃CH₂ + $\underset{\text{H}_2\text{N}}{\overset{\text{H}_2\text{N}}{\text{C=O}}}$ ⟶ phenobarbital

(h) HO—⟨ ⟩—$\underset{\text{CH}_3}{\overset{\text{CH}_3}{\text{C}}}$—⟨ ⟩—OH + ⟨ ⟩—OCO—⟨ ⟩ ⟶ polycarbonate plastics

11.28 Preparation of Medicinal Compounds: Write a reaction equation showing the synthesis of the following materials from the indicated starting materials: **(a)** sodium salicylate from salicylic acid; **(b)** phenacetin from *p*-ethoxyaniline; **(c)** acetoaminophen from *p*-aminophenol; **(d)** benzocaine from *p*-aminobenzoic acid.

11.29 Reactions of Benzoic Acid: Show how each of the following can be made from benzoic acid: **(a)** benzoyl chloride; **(b)** isopropyl benzoate; **(c)** sodium benzoate; **(d)** N,N-dipropylbenzamide; **(e)** *m*-nitrobenzamide; **(f)** benzoic propanoic anhydride.

11.30 Polymers: From what monomers could the following polymers be prepared?

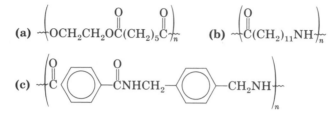

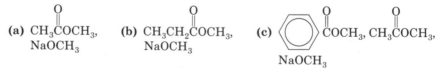

11.31 Organic Qualitative Analysis: Suggest a chemical method for separating the following mixture of compounds (all solids):

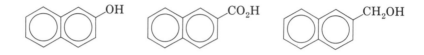

11.32 Claisen Condensation: Write a step-by-step reaction mechanism for the Claisen condensation of methyl ethanoate ($CH_3CO_2CH_3$) with sodium methoxide as the catalyst.

11.33 Claisen Condensation: Write reaction equations for the following Claisen condensations:

(a) $CH_3\overset{O}{\overset{\|}{C}}OCH_3$,   **(b)** $CH_3CH_2\overset{O}{\overset{\|}{C}}OCH_3$,   **(c)** ⬡$\overset{O}{\overset{\|}{C}}OCH_3$, $CH_3\overset{O}{\overset{\|}{C}}OCH_3$,
     $NaOCH_3$ $NaOCH_3$ $NaOCH_3$

11.34 Nomenclature of Familiar Acids and Esters: Name the following compounds by the IUPAC system: **(a)** esters in section 11.2.A; **(b)** acids in Table 11.1; **(c)** dicarboxylic acids in Table 11.2.

11.35 Lactones: A lactone is a cyclic ester formed by the internal esterification of an hydroxy acid. Write the structure of the lactone formed from 4-hydroxybutanoic acid.

11.36 Proteins: Proteins are large molecules composed of many amino acid units connected by amide bonds. Write the structure of a protein composed of the amino acids glycine, phenylalanine, and proline. See Table 14.1 for structures of amino acids.

11.37 Decomposition of Aspirin: On prolonged standing, aspirin tablets sometimes take on the odor of vinegar. Explain. Illustrate your explanation with a chemical equation.

11.38 Reading Labels:
(a) Read the labels of a variety of food products in a local grocery and record the presence of salts of carboxylic acids as food preservatives (see section 11.5.A.2).
(b) Examine the labels of several pain-relieving preparations in a drug store and record the name of the pain relievers present (see Panel 15).
(c) Examine the labels on a variety of products used as local anesthetics and record the anesthetic present as the active ingredient (see Panel 15).

11.39 Acetoacetic Ester Synthesis: A synthetic method similar to the malonic ester synthesis is called the acetoacetic ester synthesis. The starting material is ethyl acetoacetate and the final products are mono- and disubstituted acetones. The reaction sequence is analogous to that shown in Figure 11.3 for the malonic ester synthesis.
(a) Fill in the following sequence, illustrating the acetoacetic ester synthesis.

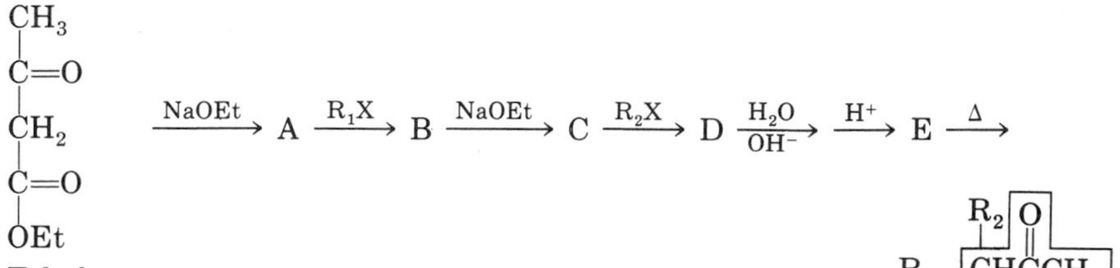

$$
\begin{array}{c}
CH_3 \\
| \\
C=O \\
| \\
CH_2 \\
| \\
C=O \\
| \\
OEt
\end{array}
\xrightarrow{\text{NaOEt}} A \xrightarrow{R_1X} B \xrightarrow{\text{NaOEt}} C \xrightarrow{R_2X} D \xrightarrow[\text{OH}^-]{H_2O} \xrightarrow{H^+} E \xrightarrow{\Delta}
$$

Ethyl
acetoacetate

$$
R_1 - \boxed{CH} \overset{R_2}{\underset{}{\text{C}}} \overset{O}{\overset{\|}{C}} CH_3
$$

A disubstituted
acetone

(b) Prepare by the acetoacetic ester synthesis

(1) $CH_3CH_2CH_2\overset{O}{\overset{\|}{C}}CH_3$ and   (2) $CH_3CH_2\overset{CH_3}{\underset{|}{C}H}-\overset{O}{\overset{\|}{C}}CH_3$

11.40 Barbiturate Synthesis: In section 11.4.A, diethylmalonate was alkylated using chloroethane and 1-chloro-3-methylbutane (steps 1–4 of Figure 11.4, malonic ester synthesis). Treatment of this alkylated malonic ester with urea in the presence of a base produces the barbiturate amytal, a sedative. Write equations showing the preparation of amytal from diethylmalonate as described above.

Chapter 12

Lipids

LIPIDS

12.1   Although living organisms are composed primarily of water, they are built of and run by carbohydrates, lipids, and proteins. Lipids have several important biological functions including acting as chemical storehouses of energy, as the structural component of membranes, and as a protective coating on some organisms. Lipids have an imprecise definition, and the classification embraces a variety of substances. They are water-insoluble biological organic molecules that are soluble in nonpolar solvents such as ether, chloroform, carbon tetrachloride, and benzene (they can thus be extracted from an organism by these solvents). Lipids are further classified as saponifiable (subject to basic hydrolysis) or nonsaponifiable. Fats, oils, and waxes are saponifiable lipids, whereas terpenes and steroids are examples of nonsaponifiable lipids.

WAXES

12.2   Structurally, waxes are defined as esters of long-chain carboxylic acids and long-chain alcohols.

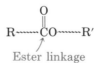

Natural waxes differ from paraffin wax in that they are high-molecular-weight esters produced directly by living organisms, whereas paraffin wax is a mixture of high-molecular-weight hydrocarbons separated during the fractionation of petroleum. Following are some representative natural waxes and the structures of their principal components.

Spermaceti

$$C_{15}H_{31}\overset{\displaystyle O}{\overset{\|}{C}}OC_{16}H_{33}$$

Spermaceti is a soft wax which is obtained from the head of the sperm whale; it has a melting range of 42–50 °C. It consists largely of cetyl palmitate (above). Because of its softness, it is used as a base emollient for ointment medications and cosmetics. Also, like paraffin wax, it is used in the production of candles.

Beeswax

Beeswax is taken from the honeycomb and is a mixture of esters of alcohols and acids having carbon numbers up to 36 and some high-molecular-weight hydrocarbons.

$$CH_3(CH_2)_{14}\overset{\displaystyle O}{\overset{\|}{C}}O(CH_2)_{29}CH_3 \qquad CH_3(CH_2)_{24}\overset{\displaystyle O}{\overset{\|}{C}}O(CH_2)_{25}CH_3$$

Beeswax has a melting range of 62–65 °C and is used in shoe polishes, candles, wax paper, and the manufacture of artificial flowers.

Carnauba Wax

Carnauba wax is a very hard wax capable of producing a high polish; it has a melting range of 82–86 °C. It is obtained from the leaves of the Brazilian palm tree and is used in automobile and floor waxes, as a coating on carbon paper and mimeograph stencils, and in deodorant sticks. When carnauba wax is hydrolyzed, some hydroxy acids are produced, indicating the presence of large polyesters in the wax; these could contribute to its hardness and durability.

STRUCTURE OF FATS AND OILS

12.3 Fats and oils of either animal or vegetable origin are triesters of glycerol and are consequently called glycerides, triglycerides, or more properly triacylglycerols. Esters are derivatives formed between carboxylic acids and alcohols. Since glycerol is a trihydroxy alcohol, a triester is possible. The following reaction is a general expression for the formation of a triacylglycerol. The natural synthesis is catalyzed by enzymes and proceeds through intermediate stages.

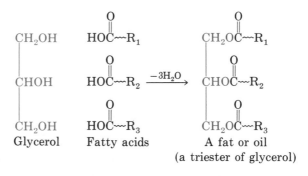

The acids making up the triester are known as fatty acids and have very long chains, most in the range of 12–18 carbons. Since the biosynthesis of fatty acids involves acetate units, naturally occurring fatty acids usually have an even number of carbons, although those with an odd number do exist. A glyceride can have all three fatty acid units identical (simple glycerides) or as many as two or three different ones (mixed glycerides).

Simple Glycerides

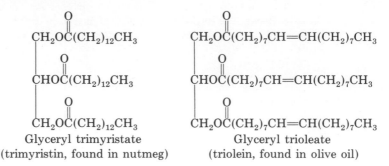

Glyceryl trimyristate
(trimyristin, found in nutmeg)

Glyceryl trioleate
(triolein, found in olive oil)

Mixed Glycerides

Glyceryl dipalmitostearate

Glyceryl oleopalmitostearate

The mixed glycerides shown are possible glycerides in tallow, lard, and butter.

Fats and oils are alike in that they are both triesters of glycerol with long straight-chained, fatty acids having an even number of carbons. They differ in two aspects: (1) Fats are solid and oils are liquid at normal room temperature. (2) Although both show some unsaturation in the long alkyl chains, oils show a greater degree of unsaturation than fats do. The greater the degree of unsaturation, the lower is the melting point of the fat or oil. The properties of fats and oils, then, depend on the structure of the long-chain alkyl group of the fatty acid, which can be either saturated or unsaturated. Structures of the more commonly occurring fatty acids are presented in Table 12.1, and the fatty acid compositions of some familiar fats and oils are described in Table 12.2. For the most part, fats and oils are mixtures of triacylglycerols. Note in Table 12.2 that the vegetable oils have a greater proportion of unsaturated fatty acids than the animal fats.

TABLE 12.1. COMMON FATTY ACIDS

$CH_3(CH_2)_{10}CO_2H$	Lauric acid
$CH_3(CH_2)_{12}CO_2H$	Myristic acid
$CH_3(CH_2)_{14}CO_2H$	Palmitic acid
$CH_3(CH_2)_{16}CO_2H$	Stearic acid
$CH_3(CH_2)_7CH{=}CH(CH_2)_7CO_2H$	Oleic acid (cis)
$CH_3(CH_2)_4CH{=}CHCH_2CH{=}CH(CH_2)_7CO_2H$	Linoleic acid (cis, cis)
$CH_3CH_2CH{=}CHCH_2CH{=}CHCH_2CH{=}CH(CH_2)_7CO_2H$	Linolenic acid (cis, cis, cis)
$CH_3(CH_2)_3CH{=}CH{-}CH{=}CH{-}CH{=}CH(CH_2)_7CO_2H$	Eleostearic acid

TABLE 12.2. FATS AND OILS

| | | | | Percent Fatty Acid Composition[a] | | | | | | | |
| | | | | Saturated Acids | | | | Unsaturated Acids | | | |
Fat or Oil	Iodine Number	Saponification Number	Melting Point, °C	C_{12} Lauric Acid	C_{14} Myristic Acid	C_{16} Palmitic Acid	C_{18} Stearic Acid	C_{18} Oleic Acid	C_{18} Linoleic Acid	C_{18} Linolenic Acid	C_{18} Eleostearic Acid
Animal Fats											
Beef tallow	31–47	190–200	40–46		3–6	24–32	20–25	37–43	2–3		
Lard	46–66	193–200	36–42		1	25–30	12–16	41–51	3–8		
Butter	36	227	32	1–4	8–13	25–32	8–13	22–29	0.2–1.5	3	
Marine Animals											
Cod liver oil	145–180	180–190			2–6	7–14	0–1	25–31	27–32		
Whale oil	120	195		0.2	9.3	15.6	2.8		35.8		
Vegetable Oils											
Coconut oil	10	255–258	25	44–51	13–18	7–10	1–4	5–8	0–1		
Corn oil	109–133	187–196	−20		0.1–1.7	8–12	2.5–4.5	19–49	34–62	1–3	
Cottonseed oil	105–114	190–198	−1		0–3	17–23	1–3	23–44	34–55		
Olive oil	79–90	187–196	−6			9.4	2.0	83.5	4.0		
Palm oil	54	199	35		1–6	32–47	1–6	40–52	2–11		
Peanut oil	84–102	188–195	−5			8.3	3.1	56	26		
Soybean oil	127–138	185–195	−16		0.3	7–11	2–5	22–34	50–60		
Linseed oil	179	190	−24		0.2	5–9	4–7	9–29	8–29	45–67	
Tung oil	168	193	−3					4–13		8–15	74–91

[a] These percentages do not include short-chain fatty acids nor fatty acids present in minute amounts.

PROBLEM

12.1. Draw a triacylglycerol with a lauric acid, myristic acid, and palmitic acid unit.

REACTIONS OF FATS AND OILS

12.4   A. Addition Reactions

Most fats and oils are composed of unsaturated fatty acids as well as the saturated variety. Because of the presence of carbon-carbon double bonds, fats and oils undergo addition reactions characteristic of alkenes. We consider here the addition of halogens and hydrogen, where X_2 = hydrogen, halogen.

$$-\overset{|}{C}=\overset{|}{C}- \ + \ X_2 \ \xrightarrow{\text{Addition}} \ -\overset{|}{\underset{X}{C}}-\overset{|}{\underset{X}{C}}-$$

1. Iodine Number. The iodine number is a measure of the extent of unsaturation in fats and oils. It is expressed as the number of grams of iodine that will add to 100 grams of the fat or oil being tested. The greater the number of double bonds in a lipid, the greater the amount of iodine that adds to the 100 grams. Thus, high iodine numbers indicate a high degree of unsaturation and low iodine numbers indicate low unsaturation. In Table 12.2 note that the animal fats have low iodine numbers relative to the more highly unsaturated marine animal and vegetable oils.

PROBLEM

12.2. Arrange the following in order of increasing iodine number: **(a)** trimyristin, triolein, glyceryl oleopalmitostearate; **(b)** stearic, oleic, linoleic, and linolenic acids.

2. Hydrogenation. In the presence of a metal catalyst, such as nickel, hydrogen adds to the double bonds of fats and oils producing more highly saturated glycerides. Consider, for example, the hydrogenation of the following oil, a possible component of soybean oil.

This general reaction is used in the production of shortenings and margarine. Cooking shortenings and margarine differ from lard and butter in that they are derived from vegetable oils, whereas lard and butter are natural animal fats. In the production of shortening or margarine, liquid vegetable oils are partially hydrogenated until the desired consistency is achieved. Enough unsaturation is left to create a low-melting, soft product. Complete hydrogenation (as in the sample equation) would produce a hard, brittle fat.

In the manufacture of margarine, these partially hydrogenated vegetable oils (often soybean, corn, and safflower oils) are mixed with water, salt, and nonfat dry milk. Other liquid glycerides are added to achieve the desired consistency and homogeneity. Vitamins, especially vitamin A, are added along with artificial

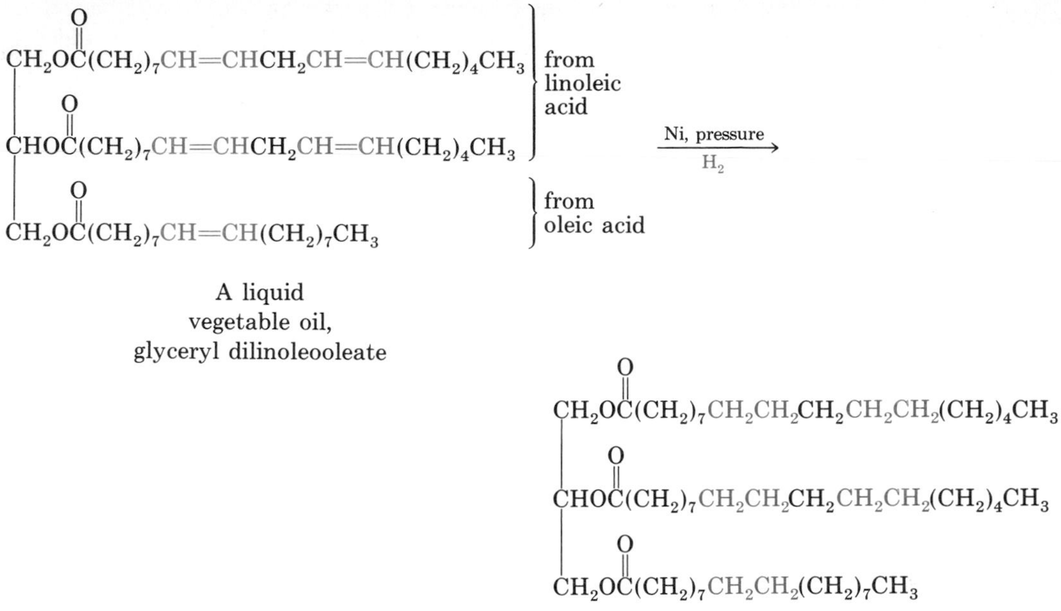

$$CH_2O\overset{O}{\overset{\|}{C}}(CH_2)_7CH=CHCH_2CH=CH(CH_2)_4CH_3$$

$$CHO\overset{O}{\overset{\|}{C}}(CH_2)_7CH=CHCH_2CH=CH(CH_2)_4CH_3$$ from linoleic acid

$$CH_2O\overset{O}{\overset{\|}{C}}(CH_2)_7CH=CH(CH_2)_7CH_3$$ from oleic acid

$$\xrightarrow[\text{H}_2]{\text{Ni, pressure}}$$

A liquid
vegetable oil,
glyceryl dilinoleooleate

$$CH_2O\overset{O}{\overset{\|}{C}}(CH_2)_7CH_2CH_2CH_2CH_2CH_2(CH_2)_4CH_3$$

$$CHO\overset{O}{\overset{\|}{C}}(CH_2)_7CH_2CH_2CH_2CH_2CH_2(CH_2)_4CH_3$$

$$CH_2O\overset{O}{\overset{\|}{C}}(CH_2)_7CH_2CH_2(CH_2)_7CH_3$$

A solid fat

flavoring and coloring. Diacetyl and methyl acetyl carbinol, which are responsible for the characteristic taste of butter, are common flavorings.

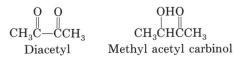

$$\underset{\text{Diacetyl}}{CH_3\overset{O}{\overset{\|}{C}}-\overset{O}{\overset{\|}{C}}CH_3} \qquad \underset{\text{Methyl acetyl carbinol}}{CH_3\overset{OH}{\overset{|}{C}}H\overset{O}{\overset{\|}{C}}CH_3}$$

Finally, preservatives such as potassium sorbate (and other salts of carboxylic acids, section 11.5.A) and antioxidants such as butylated hydroxy toluene (and other phenols, section 8.5.C) are added.

B. Oxidation Reactions

1. Rancidification. Fats and oils, when exposed to air, tend to oxidize or hydrolyze to produce volatile carboxylic acids. These have a sour, unpleasant taste and aroma. The process, called rancidification, makes lard, shortenings, butter, margarine, cooking oils, and milk unpalatable and unusable.

Oxidative rancidification involves the oxidation of carbon-carbon double bonds to produce carboxylic acids.

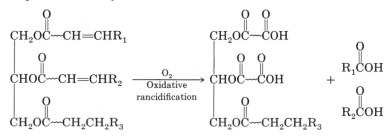

In hydrolytic rancidification, one or more of the ester units of triacylglycerol are hydrolyzed back to the original acid. Antioxidants (section 8.5.C) are added to many edible fat and oil products to retard rancidification.

We are often warned by safety officials that rags containing unsaturated oils are subject to spontaneous combustion. When such rags are stored in a warm, poorly ventilated area, heat builds up as the exothermic oxidation (rancidification) occurs, eventually causing ignition.

2. Drying Oils. When highly unsaturated oils are exposed to air, they undergo an alternative form of oxidation called *drying,* which causes them to harden. This process involves the attack of oxygen at allylic positions in the oil to form intermolecular peroxide linkages. As oil molecules are drawn into close proximity, the double bonds can also polymerize, forming a gigantic, interlinking hard mass.

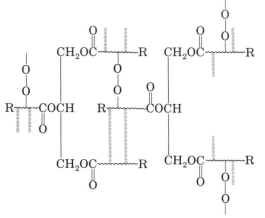

This principle governs the drying action of an oil-based paint. Commercial oil-based paints consist of a pigment dispersed in a drying oil, such as linseed oil. Turpentine is added as a thinner and cobalt and lead salts as oxidation catalysts. When the paint is applied, the volatile turpentine evaporates and the oil begins to polymerize, eventually forming a hard, protective surface.

Linoleum is made from a thick suspension of cork and rosin in linseed oil. The suspension is pressed and allowed to "dry" (the linseed oil oxidizes). A similar process is used to make oilcloth. Tough, durable surface coatings result.

C. Saponification

Fats and oils are acid derivatives, triesters of glycerol. When any acid derivative reacts with water, the products are an acid and an alcohol. *Saponification* is the basic hydrolysis of esters, resulting in the production of glycerol and the salts of the constituent fatty acids (since basic conditions are employed).

$$
\begin{array}{l}
CH_2OC\!\!\sim\!\!R \\
\qquad\overset{\parallel}{O} \\
CHOC\!\!\sim\!\!R + 3NaOH \xrightarrow{H_2O} \\
\qquad\overset{\parallel}{O} \\
CH_2OC\!\!\sim\!\!R
\end{array}
\qquad
\begin{array}{l}
CH_2OH \\
CHOH \quad + \quad 3R\!\!\sim\!\!CO^-Na^+ \\
CH_2OH
\end{array}
$$

Fat or oil   Glycerol   Fatty acid salt

1. Saponification Number. The *saponification number* is defined as the number of milligrams of potassium hydroxide required to saponify one gram of a fat or an oil. On a molecular basis, one molecule of fat or oil requires three units of KOH for complete saponification because there are three ester linkages in a fat or an oil molecule. Because each molecule of a high-molecular-weight fat and a low-molecular-weight fat requires three KOH units for saponification, the weight of KOH needed per gram of fat will be lower for the heavier fat than what is needed to saponify the low-molecular-weight glyceride. High-molecular-weight fats and oils have lower saponification numbers than oils and fats of lower molecular weight. Table 12.2 lists saponification numbers for some common fats and oils.

2. Production of Soap. The term *saponification* means "soap making." The salts of long-chain fatty acids produced by the saponification of fats and oils are soaps.

$$
\begin{array}{l}
\text{CH}_2\text{OC}{\sim}\text{R} \\
\quad\;\overset{\displaystyle O}{\|} \\[2pt]
\text{CHOC}{\sim}\text{R} \xrightarrow[\text{H}_2\text{O}]{\text{NaOH,}} \;\begin{array}{l}\text{CH}_2\text{OH}\\ \text{CHOH}\\ \text{CH}_2\text{OH}\end{array}\; + \; 3\text{R}{\sim}\overset{\displaystyle O}{\underset{\text{Soap}}{\|}}\text{CO}^-\text{Na}^+ \\[2pt]
\text{CH}_2\text{OC}{\sim}\text{R}
\end{array}
$$

Soap has its origins in antiquity. It was prepared for over two thousand years by mixing fire ashes, which are quite basic, with tallow and water. Today, soap is made by two processes.

In the boiling, or kettle, process, up to 50 tons of rendered fat are melted in steel tanks three stories high and then injected with steam and sodium hydroxide solution. Following saponification, brine is added to salt out the soap; this forms an upper curdy layer. The soap is then separated, purified, and cut into bars or chips. Glycerol, for use in the plastics and explosives industries, is recovered from the lower layer, the aqueous salt solution.

The modern continuous soap-making process involves high-temperature water hydrolysis of fats and oils to fatty acids and glycerol. The fatty acids are vacuum-distilled, mixed in specific ratios, and neutralized with alkali to form the soap.

Tallow and coconut oil are frequently the initial glycerides used in the soap industry. Tallow is rendered by heating to produce a liquid. The unmelted protein material is filtered away; the melt is termed *lard*. Tallow or lard produces a good-cleansing but slow-lathering soap. Soaps from coconut oil form better lathers, so that some coconut oil is often included in the lipid material to be saponified. Coloring, perfumes, disinfectants, and deodorants can also be added to body soaps. Heavy-duty hand soaps may contain scouring powders, sand, or volcanic pumice, for an abrasive effect. Glycerol confers transparency to bar soaps, while air beaten into the soap will allow it to float. Shaving cream is made using the potassium salts of fatty acids colloidally dispersed into a foam.

SOAPS AND DETERGENTS

12.5

A. Structure of Soaps

Dirt adheres to our bodies and clothes by a thin film of fat, oil, or grease. So that this dirt can be removed, the oily materials must first be dissolved. The most abundant liquid on earth, and the only one economically feasible for day-to-day washing, is water. But water is a polar liquid—and fats and oils, because of their long hydrocarbon chains, are nonpolar, that is, water-insoluble.

Soaps are structurally capable of solving this dilemma. Recall that soaps are salts of long-chain fatty acids. The long alkyl group has 12–18 carbons, is completely nonpolar, and consequently is soluble in fats and oils but insoluble in water. The other end of the molecule, a carboxylic acid salt, is very polar, in fact ionic, and is water-soluble. A soap then has two diverse solubility properties—it has a hydrophilic end (water-loving), soluble in water, and a hydrophobic end (water-fearing), soluble in fats and oils.

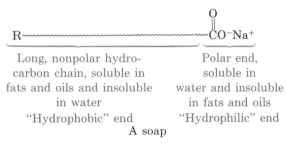

Long, nonpolar hydro-
carbon chain, soluble in
fats and oils and insoluble
in water
"Hydrophobic" end

Polar end,
soluble in
water and insoluble
in fats and oils
"Hydrophilic" end

A soap

Soap, by simultaneously dissolving in oils and water, removes oil from dirty clothes and emulsifies the droplets in water.

B. Mechanism of Soap Action

Let us take a closer look at what is happening in a soap solution on a molecular basis (Figure 12.1). As a soap dissolves in water, the molecules orient themselves on the water's surface with the ionic end submerged and the nonpolar hydrocarbon chain bobbing above the surface like a buoy on the ocean. In this manner, the soap molecule satisfies its opposing solubility characteristics—the water-soluble, hydrophilic end is in the water and the nonpolar, hydrophobic end is not in contact with the water. This molecular orientation lowers the surface tension of water. The liquid surface is no longer strongly associated, hydrogen-bonded water molecules, but nonpolar, nonassociated, and hydrocarbon in nature, somewhat like gasoline. This gives the water a better wetting capacity, allowing it to spread out and penetrate fabrics rather than bead up on the surface. Soaps or detergents are often added to herbicide and pesticide sprays to aid in the emulsification of the active ingredient in the water carrier and to promote better spreading of the solution over the leaves of the treated plants.

What happens to the soap molecules for which there is no room on the water's surface? They will have to orient themselves in such a way beneath the surface that the hydrophobic portions of the molecules have minimal contact with water. The soap molecules achieve this by grouping in three-dimensional clusters, with the nonpolar hydrocarbon chains filling the interior of the cluster and the water-soluble ionic ends composing the outer surface. These molecular conglomerations are called *micelles*. The solubility characteristics of the soap molecules are satisfied in that all the hydrocarbon chains are grouped together away from water (a hydrophobic core) and the ionic portions are in contact with water (Figure 12.1).

Now if we submerge some soiled clothing in the water, the nonpolar oil films are loosened and they dissolve in the nonpolar hydrocarbon center of the mi-

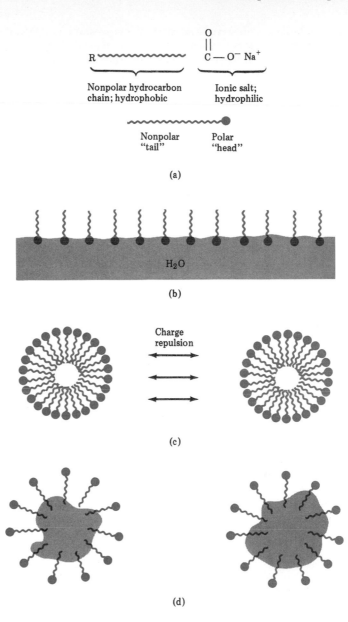

FIGURE 12.1. Mechanism of soap action. (a) Soap molecule. (b) Water surface. (c) Micelle formation. (d) Oil drop dispersal.

celles. The micelles remain colloidally dispersed in the water, with no tendency to coagulate, since there is an ionic repulsion between their charged outer surfaces. The oily films are thus washed away as finely dispersed oil droplets.

C. Detergents

Soaps, the sodium and potassium salts of long-chain fatty acids, have one serious disadvantage, they are insoluble in hard water. Hard water is water containing dissolved salts of calcium, magnesium, and iron picked up as water trickles over and filters through soil, rocks, and sand. Soaps react with these ions to form insoluble scums (the familiar bathtub ring).

$$2R\text{—}\overset{\overset{\displaystyle O}{\|}}{C}O^-Na^+ + Ca^{2+} \longrightarrow (R\text{—}\overset{\overset{\displaystyle O}{\|}}{C}O^-)_2Ca^{2+} + 2Na^+$$

Water-soluble   Water-insoluble

Detergents, first introduced in 1933, are considerably more effective in hard water.
Detergents have the same two structural characteristics that soap does:

1. They possess a long, nonpolar, hydrophobic, hydrocarbon chain, which is soluble in fats, oils, and greases.
2. They possess a polar, hydrophilic end soluble in water.

Furthermore, in the way they work detergents are analogous to soaps (as described in section 12.5.B and Figure 12.1).
Synthetic detergents, syndets, fall into three main categories, determined by the structure of the water-soluble portion of the molecule. Anionic detergents have an ionic water-soluble end, in which the portion attached to the hydrocarbon chain is negative. Alkyl sulfates and alkyl benzene sulfonates (ABS) are the two most common anionic detergents.

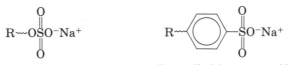

A sodium alkyl sulfate   A sodium alkyl benzene sulfonate

The water-soluble end of a cationic detergent is a positive quaternary ammonium salt.

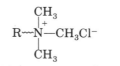

A quaternary ammonium salt

These detergents have significant germicidal properties, and similar compounds such as the ones shown are used in shampoos, mouthwashes, germicidal soaps, and disinfectant skin sprays.

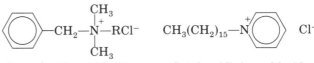

Benzalkonium chlorides   Cetylpyridinium chloride

R = 8–14 carbons

In nonionic detergents, the water-soluble end is polar and can hydrogen-bond with water, but it is not ionic.

$$R\text{—}\overset{\overset{\displaystyle O}{\|}}{C}OCH_2\overset{\overset{\displaystyle CH_2OH}{|}}{\underset{\underset{\displaystyle CH_2OH}{|}}{C}}\text{—}CH_2OH \qquad R\text{—}\overset{\overset{\displaystyle O}{\|}}{C}(OCH_2CH_2)_nOH$$

Most detergents today are biodegradable. This means that they can be quickly metabolized by microorganisms in a sewage disposal plant and not be released into the environment. For a detergent to be biodegradable, the long alkyl chain must be unbranched. Detergents used in the 1950s and early 1960s had branched chains, were not readily biodegradable, and foamed when the water discharged from sewage plants was agitated.

$$CH_3(CH_2)_{15}CH-\langle\bigcirc\rangle-SO_3^-Na^+ \qquad CH_3CHCH_2CHCH_2CHCH_2CH-\langle\bigcirc\rangle-SO_3^-Na^+$$
$$\quad\quad\quad CH_3 \qquad\qquad\qquad\qquad CH_3\;\;CH_3\;\;CH_3\;\;CH_3$$

A biodegradable detergent   A nonbiodegradable detergent

BIOLIPIDS

12.6

As we have seen, the term *lipid* is a generalization that fits more than one chemical structure. The property common to all lipids, though, is their insolubility in water. In a living organism this property allows lipids to serve in many capacities. They are found in cell membranes, insulation, fuel storage, organ cushions, body regulators, and so on. Let us consider the structures of some complex biolipids.

A. Phosphoglycerides

Recall the discussion of fats and oils in section 12.3 where these substances were defined as triesters of glycerol. Phosphoglycerides, a predominant component of most cell membranes, are variations of the triester structure already presented. A phosphoglyceride contains only two long-chained fatty acids and a phosphate group esterified with glycerol. The phosphate group is itself further esterified by another group, X.

Where X =	*Name*
—H	Phosphatidic acid
—$CH_2CH_2\overset{+}{N}H_3$	Phosphatidyl ethanolamine (cephalin)
—$CH_2CH_2\overset{CH_3}{\underset{CH_3}{N^+}}$—$CH_3$	Phosphatidyl choline (lecithin)
—$CH_2\overset{+}{C}HNH_3$, CO^-, O	Phosphatidyl serine
	Phosphatidyl inositol

Frequently, the R_2 group will be saturated and the R_1 group unsaturated. Although pure phosphoglycerides are waxy, white solids, they rapidly darken when exposed to air because of the peroxidation of the fatty acid side chains.

Notice that the phosphate group, even though it is esterified through two of its oxygens, still possesses an ionizable hydrogen. Phosphoglycerides will therefore have a nominal polarity and a high degree of nonpolarity.

PROBLEM

12.3. Draw out the products of the complete hydrolysis of phosphatidic acid.

B. Sphingolipids

Sphingolipids derive the name from one of their component units, sphingosine. Sphingosine forms an amide bond with a long-chain fatty acid. The result is called ceramide. There are three structural units constituting a sphingolipid: sphingosine, a fatty acid, and a third variable group analogous to the —X group in phosphoglycerides. This third group forms a covalent bond with the free alcohol group of the ceramide.

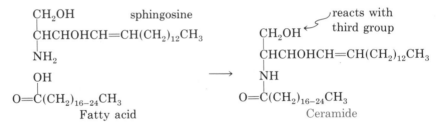

If the third group is phosphatidyl choline or ethanolamine (as occurs in phosphoglycerides), the whole molecule is called a sphingomyelin. Sphingomyelin is the main lipid component of the myelin sheath surrounding nerve cells. The presence of one to three carbohydrate units as the third group attached to ceramide produces a cerebroside which imparts cell specificity to the outer portion of a cell membrane. Carbohydrate derivatives such as sialic acid can also be found as the third group, and the resulting molecule will be known as a ganglioside, found mainly in the grey matter of the brain.

With sphingosine and the fatty acid contributing long, nonpolar R groups, a curious but familiar representation arises. Recall that in our description of soap molecules the two distinct features were a polar "head" and nonpolar "tail". So too with sphingolipids and for that matter phosphoglycerides—a polar portion and two polar tails are evident. This structure affects their function, that of a cell membrane component (section 12.7).

C. Steroids

The diversity of the term *lipid* is evident in the class of steroids. Although these compounds are relatively water-insoluble, they do not owe this property to long, straight carbon chains. Rather they are members of a family with a multicyclic ring structure. All steroids possess this fused four-ring system with variations in the degree of unsaturation and differences in substituents.

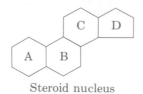

Steroid nucleus

Specific steroids will be discussed later.

D. Prostaglandins

The prostaglandins are cyclic derivatives of 20-carbon fatty acids. Although first found in the prostate gland, they are ubiquitous in animal organisms, are present in very small amounts (about 10^{-9} grams), and are extremely potent. All prostaglandins seem to stimulate smooth muscle contraction and lower blood pressure, some more than others. Among the varied hormone-like functions of specific prostaglandins are the control of gastric acid secretion, inflammation, vascular permeability, blood aggregation, food intake, nasal congestion, asthma, and body temperature. They can be made synthetically on a large commercial scale and have been used to induce labor and stimulate menstruation as well as to induce abortion. It is now believed that the action of aspirin involves the inhibition of an enzyme that is involved in the biosynthesis of a prostaglandin responsible for fever.

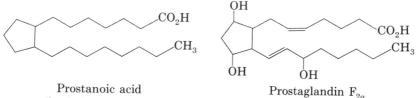

Prostanoic acid
(basis for structure)

Prostaglandin $F_{2\alpha}$

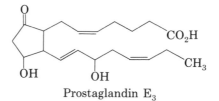

Prostaglandin E_3

THE FUNCTIONS OF BIOLIPIDS

12.7   Lipids perform myriad functions within living organisms. We shall consider only a few. Keep in mind the general structural considerations of biolipids and the many variations they allow.

A. Storage and Energy

The body contains areas of fat deposit called adipose tissue, and lipids, in general, are stored in this tissue. The abdominal area as well as other parts of the body have adipose tissue. Its function is to provide insulation to maintain the proper internal temperature for the body's chemical reactions. It also cushions our vital organs, so sensitive to trauma. While performing these important tasks, lipid deposits are a storage vault for energy. Overall, lipids store more calories of energy per gram than carbohydrates or proteins do, and this energy reserve can be called on when other sources are exhausted.

B. Membranes

The cells of living organisms are self-contained biochemical factories; some cells are highly specialized and others perform hundreds, perhaps thousands, of different chemical reactions. Oxygen and carbon dioxide, as gases, should fairly readily diffuse through any part of the body they come in contact with. Ions, such as Na^+ and K^+, are ubiquitous in the water solution that constitutes over 90% of the body. Yet their concentrations are very different inside the cell when compared to outside the cell. Carbohydrates, proteins, lipids, vitamins, and so on are stored in some cells and metabolized in others. What constitutes the material that can both confine these cellular components and yet allow certain substances to pass back and forth to the cell's environment? The answer is the cell membrane.

Although the way in which cell membranes work is still a topic of earnest research, their fundamental structural units are known. Most are composed of approximately 60% protein and 40% lipid. Large variations from these figures can exist in certain cells. For example, the myelin of nerve cells can contain up to 75% lipid material. Carbohydrates can also be complexed with the lipids and proteins. The physical arrangement of the lipid material depends on its structure. Most of it is phospholipid, with some sphingolipids. We are aware that these two types of compounds can be represented as having a polar head and a pair of nonpolar tails.

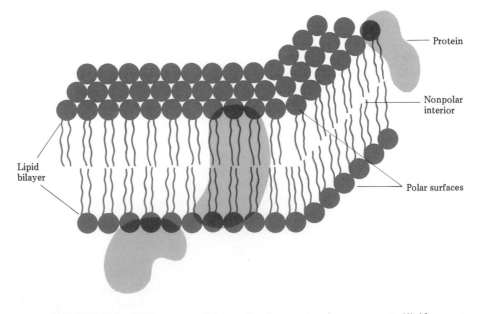

FIGURE 12.2. Cell membrane diagram showing structural arrangement of lipids.

The biolipids, like soap, form micelles in the water solution of an organism. These gigantic micelles are living cells—they can be blood cells, kidney cells, liver cells, etc. The membrane of the cell consists of two layers of lipid material known as a bilayer. The lipids in the bilayer arrange their polar heads toward the water surface, or outside and inside surfaces, of the membrane, while their tails form a hydrophobic, nonpolar interior (Figure 12.2). Protein molecules are believed to be embedded in the lipid membrane where they aid in the passage of substances through the membrane.

Lipids, such as the cerebrosides, are embedded in the cell membrane, acting to impart blood-group specificity or tissue recognition. Cancer cells are known to have characteristic cerebrosides different from these structures in normal cells. This may be part of the reason for the metabolic difficulties that cancer imparts.

The gangliosides, complex acidic sphingolipids, are situated in the membranes of brain (nerve) tissue and aid in transmitting nerve signals. Tays-Sachs disease, which seems to arise from a recessive gene in persons of Eastern European Jewish ancestry, is the result of an accumulation of ganglioside G_{M_2} in the brain. An enzyme necessary for its degradation is inherently defective. The symptoms of this fatal disease are severe brain damage, paralysis, and blindness, with death occurring between the ages of two and four years.

C. Steroids

Naturally occurring steroids perform diverse functions—and may be regulatory hormones, emulsifying agents, or poisons. But even with this variety they all still possess the four-ring steroid nucleus as a common feature. Those of animal origin also have the same metabolic precursor, cholesterol.

1. Cholesterol

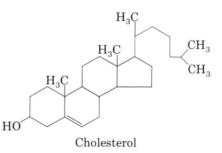

Cholesterol

Found only in animal organisms, cholesterol is concentrated mainly in the brain and spinal cord. It arises from the biological polymerization of two-carbon acetate units. From it are derived all the other steroids that can be biosynthesized. In an individual who has proper weight and diet, cholesterol is present because it is taken into the body through foods. Its biosynthesis is actually inhibited by this dietary intake. Fasting will also inhibit its formation. Should the organism begin a diet high in fat content, however, especially if fats of higher saturation are consumed, cholesterol synthesis and retention in the blood can accelerate. High concentrations of cholesterol in the blood can lead to its deposit, along with other lipid material, on the inner walls of blood vessels, a condition called *atherosclerosis*. When such deposits eventually impair the blood's circulation, usually in the heart or brain, a heart attack or stroke will occur. Cholesterol can also precipitate in the gallbladder as gallstones, which prove to be highly uncomfortable. Some persons suffer from a genetic predisposition to accumulate a large cholesterol content in the blood with all its accompanying symptoms. This is termed *familial hypercholesterolemia*. Should the elevated amounts of cholesterol occur in skin or bone tissue, a tumorous condition called *xanthomatosis* exists. A lifelong diet as devoid of cholesterol and other lipids as possible is so far the only way to resolve these disorders.

2. Steroid Hormones. Hormones are the reaction regulators of the body. They trigger, moderate, and stop biological activity in numerous interconnected ways. The steroid hormones, though composed of a large ring system, are some of the simplest hormones structurally. Several protein regulators such as insulin are far larger and more complex. Table 12.3 summarizes the steroid sex hormones and some pertinent synthetic derivatives.

The adrenal cortex, the outer portion of the adrenal glands which are located near the kidneys, produces a variety of hormones. A deficiency in these secretions of the adrenals is known as *Addison's disease*. It results in a severe condition with such symptoms as anemia, gastrointestinal disturbances, and low resistance to infection among other potentially fatal imbalances. One of these hormones, cortisone, is effective in alleviating the symptoms of rheumatoid arthritis and various skin irritations. Cortisol is even more active than cortisone. Many synthetic derivatives of cortisone have been developed that have enhanced therapeutic results and decreased side effects.

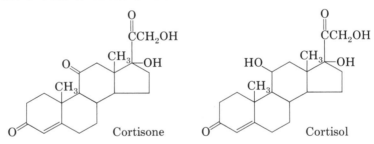

Cortisone   Cortisol

Steroids serve a variety of diverse functions. For example, glycocholic and taurocholic acids along with their salts are produced in the liver and secreted through the bile duct into the intestines. There they act as emulsifying agents for dietary lipids, facilitating the absorption of those lipids through the intestinal walls.

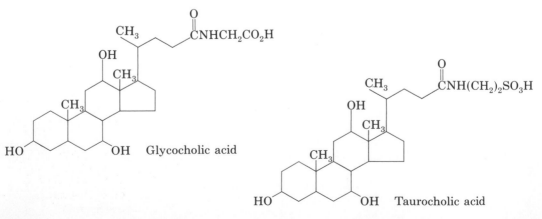

Glycocholic acid   Taurocholic acid

TABLE 12.3. STEROID SEX HORMONES

Type	Example	Structure	Function
Male			
Androgens	Testosterone		The most important of the androgens, this steroid is necessary for development of male sex organs as well as secondary characteristics such as facial hair, voice quality, and body musculature.
Female			
Estrogens	Estradiol		Secreted by the ovaries and placenta, these compounds direct female sexual maturation, inhibit male sex hormonal responses, and direct sex characteristics such as distribution of body fat.
	Progesterone		This steroid is responsible for the preparation of the uterus for the implantation of a fertilized egg and maintenance of the uterus during pregnancy.
Synthetic			
	Diethylstilbestrol (DES)		This steroid-like compound has estrogenic potency and was once used to stimulate female fertility. The daughters of women so treated have been found to have a high incidence of vaginal cancer. Once used in cattle feed to stimulate weight gain, DES is under government investigation.

TABLE 12.3. STEROID SEX HORMONES   (*Continued*)

Type	Example	Structure	Function
Contraceptive	Mestranol	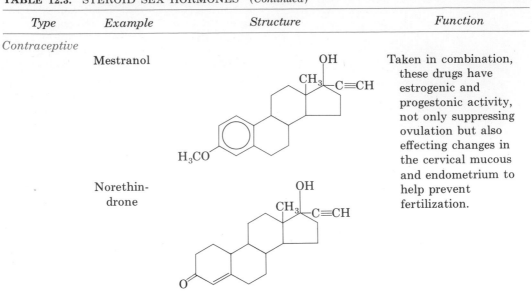	Taken in combination, these drugs have estrogenic and progestonic activity, not only suppressing ovulation but also effecting changes in the cervical mucous and endometrium to help prevent fertilization.
	Norethindrone		

Steroids also act as toxins (digitoxigenin, bufotalin), antibiotics (helvolic acid), and insect-molting hormones (ecdysone)—to name but a few.

D. Fat Soluble Vitamins

As the term *vitamin* suggests, vitamins are substances essential to life (*vita* is Latin for "life"). They cannot be produced by the normal metabolism of the body. Like hormones, vitamins can take many chemical forms, including those that are water-soluble (the B-complex vitamins and vitamin C) and others that are fat-soluble (A, D, E, and K). Although water-soluble vitamins must be supplied frequently, the fat-soluble ones are stored within the body until needed. This can lead to an overdose.

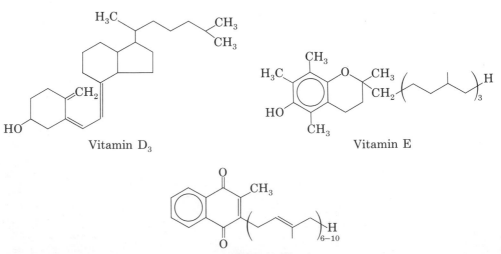

Vitamin D$_3$ Vitamin E

Vitamin K

Vitamin A and its role in the visual cycle and development have been discussed previously (Panel 1, Chapter 2; Panel 9, Chapter 6).

There are several forms of vitamin D of which D_3 or cholecalciferol is one. It is formed from a precursor in the skin by the action of sunlight. Milk is supplemented with D_3 and D_2, activated ergosterol, obtained from yeast. Vitamin D facilitates the absorption of calcium and phosphorous from the small intestine and their incorporation into bone. A deficiency of vitamin D leads to a condition known as rickets and is evidenced by bone malformations such as bowlegs and extreme tooth decay. Overdoses result in hypercalcification and kidney problems.

Vitamin E, tocopherol, is rarely deficient in diets since it is found in most foods in sufficient quantities. Not much is known about its role in the human body except that it helps to maintain cell membranes by acting as an antioxidant.

The K vitamins are produced by the bacteria inhabiting the intestinal tract. They aid in the complex mechanism of blood clotting; the rare deficiency results in a tendency to hemorrhage. Aspirin and related compounds are antagonistic to the K vitamins.

END-OF-CHAPTER PROBLEMS

12.4 Structures of Fats and Oils: Draw structures for the following fats and oils:
(a) a glyceride with three lauric acid units, trilaurin
(b) a glyceride with a myristic acid, a palmitic acid, and a stearic acid unit
(c) a glyceride with two myristic acid units and one oleic acid
(d) a glyceride likely to be found in corn oil
(e) a glyceride likely to be found in soybean oil
(f) a glyceride likely to be found in tung oil

12.5 Reactions of Fats and Oils: Write chemical equations using the following glyceride to describe the reactions indicated:

$$CH_2O\overset{\overset{O}{\|}}{C}(CH_2)_7CH{=}CH(CH_2)_7CH_3$$

$$CHO\overset{\overset{O}{\|}}{C}(CH_2)_7CH{=}CHCH_2CH{=}CHCH_2CH{=}CHCH_2CH_3$$

$$CH_2O\overset{\overset{O}{\|}}{C}(CH_2)_{14}CH_3$$

(a) saponification with NaOH; **(b)** hydrogenation; **(c)** Br_2/CCl_4.

12.6 Reactions of Soaps: Write chemical equations showing the reaction of a soap such as sodium stearate with the following: **(a)** hard water containing Mg^{2+}; **(b)** hard water containing Fe^{3+}; **(c)** an acid solution (HCl).

12.7 Structures of Soaps and Detergents: Which of the following would be an effective soap or detergent in water? For those that are effective and for those not effective, explain why.
(a) $CH_3(CH_2)_{14}CO_2^-Na^+$
(b) $(CH_3(CH_2)_{16}CO_2^-)_2Ca^{2+}$
(c) $CH_3CH_2CO_2^-Na^+$
(d) $CH_3(CH_2)_{14}CH_2N(CH_3)_3^+Cl^-$
(e) $CH_3(CH_2)_{16}CH_3$
(f) $CH_3(CH_2)_{14}CO_2H$
(g) $CH_3(CH_2)_{14}CH_2OSO_3^-Na^+$

12.8 Consumer Chemistry: In a grocery store or drugstore, examine the labels on the following products:

(a) Margarine. Make a list of the vegetable oils used to produce various brands of margarine.

(b) Shortenings and oils. Make a list of the vegetable oils used to produce various brands.

(c) Oils. Make a list of the various types of oils (from different plant sources) available for sale in your local supermarket.

(d) Detergents. Determine if possible the type of detergent, and additives; if the selection is phosphate-based, record the percentage of phosphorus.

(e) Disinfectants. Find some products containing benzalkonium chlorides or cetylpyridinium chloride as antiseptics.

(f) Biolipids. Check various products as to biolipid content and consider the purpose of the compounds noted in that product.   Your pharmacist should be able to help you with steroids.

12.9 Properties of Fats and Oils: What is the relationship between the melting point of a triacylglycerol and iodine number?

12.10 Structure: How are detergents and phospholipids and sphingolipids alike in structure and function?   How do they differ?

12.11 Structures of Biolipids: Point out one characteristic structural feature unique to **(a)** androgens, **(b)** estrogens, **(c)** progesterone, **(d)** contraceptives.

12.12 Structures of Biolipids: Cholesterol has how many chiral carbons?   How many optical isomers are possible?

Chapter 13

Amines

STRUCTURE OF AMINES

13.1   Amines are nitrogen-containing compounds that can be described as derivatives of the inorganic compound ammonia. If one hydrogen of ammonia is replaced by an organic group, a primary amine results; if two hydrogens are replaced, the amine is secondary; and if all three hydrogens are replaced, the amine is tertiary.

$$
\begin{array}{cccc}
\text{H} & \text{H} & \text{H} & \text{R} \\
| & | & | & | \\
\text{H—N—H} & \text{R—N—H} & \text{R—N—R} & \text{R—N—R} \\
\text{Ammonia} & \text{Primary, 1°} & \text{Secondary, 2°} & \text{Tertiary, 3°}
\end{array}
$$

Should one or more of the organic groups be aromatic, the compound is an aromatic amine.

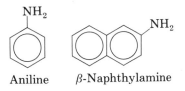

Aniline   β-Naphthylamine

Volatile amines generally have an unpleasant, fishy odor, as illustrated by the common names for the following examples.

$H_2NCH_2CH_2CH_2CH_2NH_2$ Putrescine ⎫

$H_2NCH_2CH_2CH_2CH_2CH_2NH_2$ Cadaverine ⎭   Bacterial decomposition products of protein

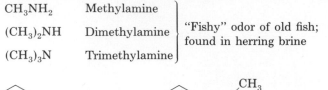

$$CH_3NH_2 \qquad \text{Methylamine}$$
$$(CH_3)_2NH \qquad \text{Dimethylamine}$$
$$(CH_3)_3N \qquad \text{Trimethylamine}$$

"Fishy" odor of old fish; found in herring brine

Indole

Skatole

Formed during putrefaction of protein; contribute to odor of feces

Organic nitrogen compounds are found in all living organisms in the form of proteins, genetic material (DNA, RNA), hormones, vitamins, and so on. For this reason, they are also found in urea (H_2NCNH_2, in urine) and in waste and decom-

$$\underset{\displaystyle \overset{\|}{O}}{}$$

position products, as just described.

NOMENCLATURE OF AMINES

13.2   A. IUPAC Nomenclature

1. Simple Amines. Aliphatic amines are named by adding the suffix *-amine* to the name of the longest continuous carbon chain. The base name of the simplest aromatic amine is *aniline*.

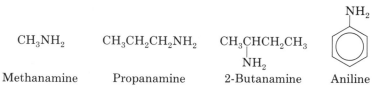

$$CH_3NH_2 \qquad CH_3CH_2CH_2NH_2 \qquad \underset{\displaystyle \overset{|}{NH_2}}{CH_3CHCH_2CH_3}$$

Methanamine   Propanamine   2-Butanamine   Aniline

2. Substituted Amines. To name a substituted amine, both the name and location of the substituent must be identified. Substituents on a carbon chain are located by a number, whereas those on a nitrogen are identified by a capital N. These two principles are illustrated in the following examples:

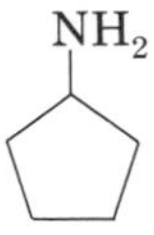

Cyclopentanamine 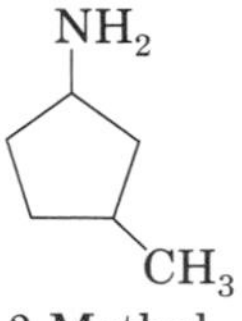   3-Methyl-cyclopentanamine 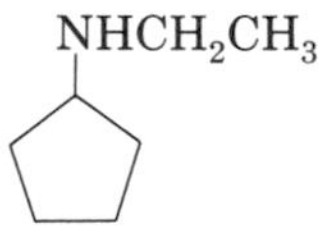   N-Ethyl-cyclopentanamine 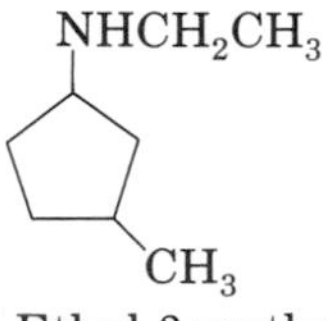   N-Ethyl-3-methyl-cyclopentanamine

$$CH_3CH_2CH_2CH_2CH_2\underset{\displaystyle \overset{|}{\overset{\displaystyle CH_3}{}}}{N}CH_3$$

N,N-Dimethylpentanamine   3-Bromo-4-chloro-N-methyl-N-propylaniline

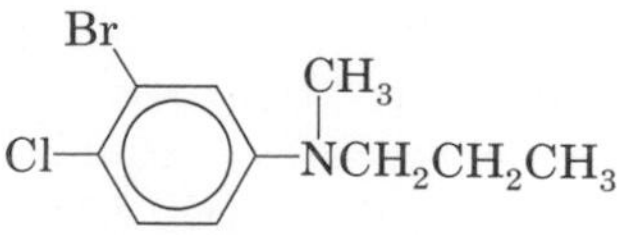

PROBLEM

13.1. Name the following by the IUPAC system:

(a) $CH_3(CH_2)_7CH_2NH_2$ **(b)** $CH_3CH(CH_2)_3CH_3$ **(c)** $CH_3CH_2CHCH_2CH_3$

$\qquad\qquad\qquad\qquad\qquad\qquad$ NH$_2$ $\qquad\qquad\qquad\qquad$ N(CH$_3$)$_2$

(d) $CH_3CH_2NCH_2(CH_2)_6CH_3$ **(e)** 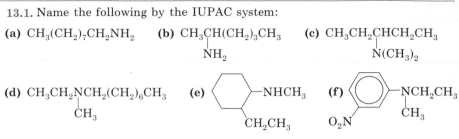 **(f)**

$\qquad\qquad$ CH$_3$

---

3. Polyfunctional Amines. Of the major functional groups discussed in this text, amines have the lowest priority in terms of taking a suffix designation (priority order: carboxylic acids > aldehydes > ketones > alcohols > amines). Thus, when present in a molecule with other functional groups, such as in amino acids, the amine is described by the prefix *amino-*.

$$CH_2CHCO_2H \qquad\qquad CH_2CH_2CH_2\overset{\overset{\displaystyle O}{\|}}{C}CH_3 \qquad\qquad H_2N(CH_2)_6NH_2$$

$\quad$ OH NH$_2$ $\qquad\qquad\qquad$ NHCH$_3$

2-Amino-3-hydroxypropanoic acid $\quad$ 5-N-Methylamino-2-pentanone $\quad$ 1,6-Hexandiamine

PROBLEM

13.2. Name the following amino acids (lysine, glutamic acid, phenylalanine, sacrosine) by the IUPAC system of nomenclature:

(a) $H_2N(CH_2)_4CHCO_2H$ **(b)** $HO_2CCH_2CH_2CHCO_2H$ **(c)** CH_2CHCO_2H

$\qquad\qquad\qquad$ NH$_2$ $\qquad\qquad\qquad\qquad\qquad$ NH$_2$ $\qquad\qquad\qquad\qquad\qquad$ NH$_2$

(d) CH_2CO_2H

$\quad$ NHCH$_3$

B. Common Nomenclature

Simple amines have common names derived by attaching the suffix *-amine* to the name of the alkyl group or groups.

$\qquad\qquad\qquad\qquad$ NH$_2$

$CH_3NH_2 \qquad CH_3CHCH_3 \qquad CH_3CH_2CH_2NHCH_3 \qquad (CH_3CH_2)_3N$

Methylamine $\quad$ Isopropylamine $\quad$ Methylpropylamine $\quad$ Triethylamine

PHYSICAL PROPERTIES OF AMINES

13.3 $\quad$ Like most other classes of compounds, amines have melting points and boiling points that generally increase with molecular weight (Table 13.1). However, the magnitude of the boiling points and the fact that lower-molecular-weight amines are water-soluble can be attributed to their ability to participate in hydrogen-bonding (Figure 13.1). Notice that methylamine, due to its ability to hydro-

TABLE 13.1. PHYSICAL PROPERTIES OF AMINES

Structure	Molecular Weight	Melting Point, °C	Boiling Point, °C	Basicity Constant, K_b
CH_3NH_2	31	-94	-6	3.7×10^{-4}
$CH_3CH_2NH_2$	45	-81	17	6.4×10^{-4}
$CH_3CH_2CH_2NH_2$	59	-83	48	5.1×10^{-4}
$(CH_3CH_2)_2NH$	73	-48	56	9.6×10^{-4}
$(CH_3)_3N$	59	-117	3	6.5×10^{-5}
⬡—NH_2	93	-6	184	4.3×10^{-10}

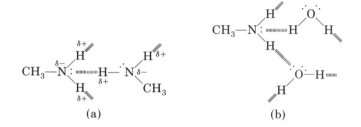

(a)   (b)

FIGURE 13.1. Hydrogen-bonding in methanamine. (a) Pure liquid. (b) Water solution.

gen-bond, is considerably higher-boiling than ethane. But it is not as high-boiling as methanol, since the N—H bond is not so polar as the O—H bond.

	CH_3CH_3	CH_3NH_2	CH_3OH
MW	30	31	32
BP	-89 °C	-6 °C	65 °C

Figure 13.1a shows hydrogen-bonding between methylamine molecules. As the ability to hydrogen-bond decreases, so does the boiling point, as illustrated by the following compounds of similar molecular weight:

	$\overset{H}{H}NCH_2CH_2N\overset{H}{H}$	$CH_3CH_2CH_2N\overset{H}{H}$	$CH_3\overset{H}{N}CH_2CH_3$	$(CH_3)_3N$
MW	60	59	59	59
BP	117 °C	48 °C	37 °C	3 °C

PROBLEM

13.3. Arrange in order of increasing boiling point. Explain.
(I) $CH_3CH_2NH_2$; **(II)** CH_3NHCH_3; **(III)** $CH_3CH_2CH_3$; **(IV)** CH_3CH_2OH.

BASICITY OF AMINES

13.4

A. Salt Formation

Basicity and the ability to react with acids are the most characteristic properties of amines. The presence of a nonbonding electron pair on the nitrogen makes the amines Lewis bases, and, like ammonia, they can share this pair of electrons with hydrogen ions from strong mineral acids.

$$NH_3 + HNO_3 \longrightarrow NH_4^+NO_3^- \quad \text{Ammonium nitrate}$$
$$CH_3NH_2 + HCl \longrightarrow CH_3NH_3^+Cl^- \quad \text{Methyl ammonium chloride}$$

PROBLEM

13.4. Write equations illustrating the reaction of the following amines with acids:

(a) $(CH_3CH_2)_3N$, HCl; (b) $CH_3CH_2NHCH_3$, HNO_3; (c) 2 ⟨◯⟩—NH_2, H_2SO_4

B. Ionization and Basicity Constants

Although amines are more basic than water, they are still weak bases and show only slight ionization in aqueous solution. This basicity is described by a basicity constant K_b, just as acidity is described by an acidity constant K_a.

$$R_3N + H_2O \rightleftharpoons R_3NH^+ + OH^- \qquad K_b = \frac{[R_3NH^+][OH^-]}{[R_3N]}$$

Because water is in great excess, it is not included in the equilibrium expression. Its concentration is essentially constant and thereby part of the constant K_b. Because the concentrations of ionized products appear in the numerator and unionized base in the denominator of the equilibrium expression, large K_b's indicate high basicity; low constants are characteristic of weaker bases.

PROBLEM

13.5. Write a reaction equation illustrating the ionization of methanamine in water solution. Then write the equilibrium expression K_b.

C. Relative Basicities

The basic nature of an amine depends on the availability of its lone pair of electrons for reaction with a Lewis acid.

1. Aliphatic Amines. Aliphatic amines are generally more basic than ammonia, since the electron-releasing alkyl groups increase the electron density around the nitrogen, thereby increasing the availability of the lone pair of electrons. In addition, the ammonium ion formed on reaction has its positive charge partially dispersed and stabilized by the electron-releasing capacity of the alkyl substituents.

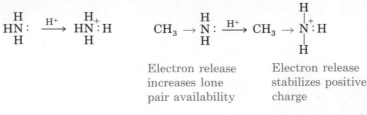

Electron release
increases lone
pair availability

Electron release
stabilizes positive
charge

Dimethylamine, a 2° amine, has two electron-releasing alkyl groups and is more basic than methylamine, a 1° amine, which has only one. Trimethylamine and other 3° amines are less basic than 1° and 2° amines, however. Even though the electron density is further increased in a tertiary amine, the steric crowding created by three alkyl groups makes approach and bonding by a Lewis acid relatively difficult. The electrons are there but the path is blocked.

	NH_3	CH_3NH_2	CH_3NHCH_3	$CH_3-\overset{\overset{\displaystyle CH_3}{\displaystyle |}}{N}-CH_3$
	Ammonia	1° Amine	2° Amine	3° Amine
K_b	1.8×10^{-5}	37×10^{-5}	54×10^{-5}	6.5×10^{-5}

2. Aromatic Amines. Aromatic amines are considerably less basic than aliphatic amines due to resonance of nitrogen's lone pair with the benzene ring. This occurs through p orbital overlap (Figure 13.2). The resonance not only decreases the availability of nitrogen's lone pair but also adds stability to aromatic amines, making them less reactive.

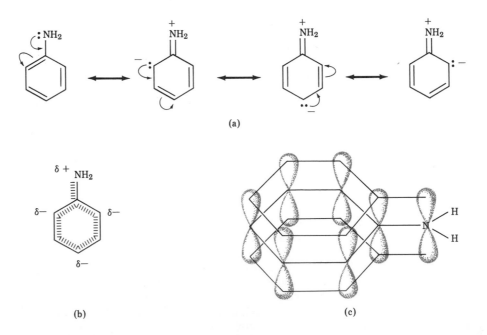

FIGURE 13.2. Resonance in aniline and aromatic amines. (a) Resonance forms. (b) Resonance hybrid. (c) Pi-bonding picture.

R—NH$_2$	—NH$_2$
Aliphatic amines	Aniline, an aromatic amine
$K_b = 10^{-4}$ or 10^{-5}	$K_b = 4.3 \times 10^{-10}$

As much as the electron density around the nitrogen is decreased by resonance, it is increased in the benzene ring. While benzene requires pure bromine, heat, and an iron catalyst for monobromination, aniline tribrominates spontaneously at room temperature with a dilute aqueous bromine solution.

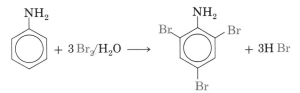

Since resonance makes the ring more negative (Figure 13.2), it is readily attacked by the positive electrophile, Br$^+$.

Electron-withdrawing groups on the benzene ring, especially if ortho or para, further decrease the basicity of aromatic amines.

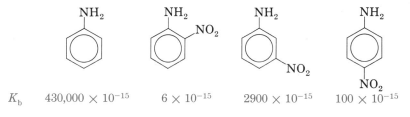

| K_b | $430{,}000 \times 10^{-15}$ | 6×10^{-15} | 2900×10^{-15} | 100×10^{-15} |

PROBLEM

13.6. Arrange the following amines in order of increasing basicity:

(a) (I) CH$_3$CH$_2$CH$_2$CH$_2$NH$_2$; (II) (CH$_3$CH$_2$)$_2$NH; (III) CH$_3$CH$_2$N(CH$_3$)$_2$;

(IV) —NH$_2$

(b) (I) —NH$_2$; (II) —NHCH$_3$; (III) —CH$_2$NH$_2$;

(IV) O$_2$N— —NH$_2$

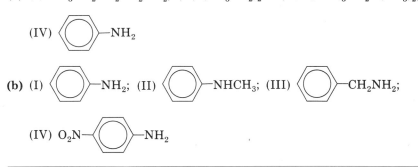

PREPARATIONS AND REACTIONS OF AMINES

13.5   **A. Reduction**

Almost any nitrogen-containing compound that is not already an amine can be reduced to an amine under appropriate conditions.

1. Reduction of Nitro Compounds. Aliphatic or aromatic nitro compounds can be reduced using hydrogen and a metal catalyst, or iron or tin in an acid solution.

$$R{-}NO_2 \xrightarrow[\substack{Fe/HCl \\ \text{followed by base}}]{H_2/Ni \text{ or}} R{-}NH_2$$

2. Reduction of Nitriles. Addition is the characteristic reaction of most multiple bonds. Addition of 2 moles of hydrogen to the carbon-nitrogen triple bond of nitriles produces primary amines.

$$RC{\equiv}N + 2H_2 \xrightarrow{Ni} RCH_2NH_2$$

3. Reduction of Amides. Amide reduction with lithium aluminum hydride can be used to prepare 1°, 2°, and 3° amines.

$$\overset{\displaystyle O}{\overset{\|}{R C} N R_2} \xrightarrow{LiAlH_4} RCH_2NR_2 \qquad (R = H \text{ or alkyl})$$

PROBLEM

13.7. Show how ethanamine could be prepared by the three reactions just described.

B. Alkylation of Amines by Nucleophilic Substitution

Most of the reactions of amines are due to the presence of a nonbonding pair of electrons on the nitrogen. We have already seen that amines are Lewis bases and react with acids because of this lone pair (section 13.4). Amines are also effective nucleophiles and will react with alkyl halides.

Consider, for example, the reaction of methylamine with chloromethane. The lone pair of electrons on nitrogen is attracted to the positive carbon of the polar carbon-chlorine bond. A new bond forms between the nitrogen and carbon using this pair of electrons, and the chloride is displaced, probably by an S_N2 mechanism (section 7.4.B).

$$\underset{\substack{| \\ H}}{\overset{\substack{H \\ |}}{CH_3N{:}}} \quad \overset{\substack{\delta^+ \quad \delta^-}}{CH_3{-}Cl} \longrightarrow \underset{\substack{| \\ H}}{\overset{\substack{H \\ |}}{CH_3N{:}}}{\cdots\cdots}CH_3{\cdots\cdots}Cl \longrightarrow \underset{\substack{| \\ H}}{\overset{\substack{H^+ \\ |}}{CH_3NCH_3}}\ \overset{-}{Cl}$$

Transition state

Dimethylammonium chloride results, an amine salt. But a hydrogen ion on this salt can be transferred to the basic methylamine, which, especially early in the reaction, is in high concentrations.

$$(CH_3)_2NH_2{}^+Cl^- + CH_3NH_2 \longrightarrow (CH_3)_2NH + CH_3NH_3{}^+Cl^-$$

This frees dimethylamine to react with methyl chloride, as does methylamine.

$$(CH_3)_2NH + CH_3Cl \longrightarrow (CH_3)_3NH^+Cl^-$$

The salt formed can also be neutralized by methylamine. The resulting trimethylamine can react with yet another molecule of methyl chloride to form what is called a quaternary ammonium salt, a positive nitrogen with four bonded alkyl groups. Because there are no more replaceable hydrogens on the nitrogen, the reaction stops here.

$$(CH_3)_3NH^+\ Cl^- + CH_3NH_2 \longrightarrow (CH_3)_3N + CH_3NH_3{}^+\ Cl^-$$

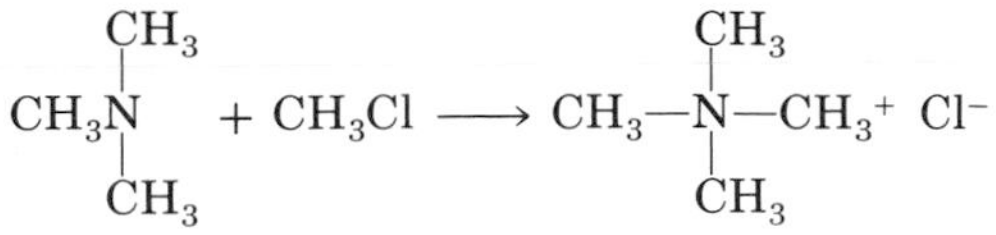

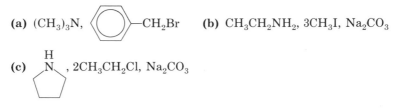

Quaternary ammonium salts are easily prepared by this method from ammonia or an amine (a weak base such as $CO_3{}^{2-}$ is needed to free the amine from the salt in each step). For example, ammonia can be completely alkylated by this reaction.

$$NH_3 \xrightarrow[\text{2) } CO_3{}^{2-}]{\text{1) RX}} RNH_2 \xrightarrow[\text{2) } CO_3{}^{2-}]{\text{1) RX}} R_2NH \xrightarrow[\text{2) } CO_3{}^{2-}]{\text{1) RX}} R_3N \xrightarrow{\text{RX}} R_4N^+\ X^-$$

However, any of the intermediate amines are difficult to isolate in reasonable yields because they tend to become further alkylated very easily.

PROBLEM

13.8. Write the products of the following alkylation reactions:

(a) $(CH_3)_3N$, [benzene ring]—CH_2Br **(b)** $CH_3CH_2NH_2$, $3CH_3I$, Na_2CO_3

(c) [pyrrolidine ring with N—H] , $2CH_3CH_2Cl$, Na_2CO_3

C. Reactions of Amines to Form Amides

1. Carboxylic Acid Amides. Treatment of a carboxylic acid or carboxylic acid derivative with ammonia or a primary or secondary amine can produce an amide (section 11.6). The positive carbon of the polar carbon-oxygen double bond is attacked by the nucleophilic nitrogen of the amine (the lone pair of electrons is used for the new bond). The electronegative leaving group (Z) and a hydrogen from the amine form the accompanying product.

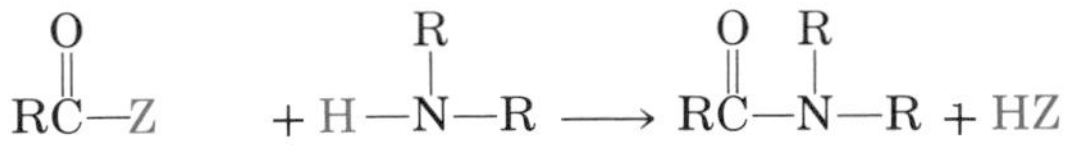

$$\underset{\text{Acid derivative}}{\overset{\overset{\displaystyle O}{\|}}{RC-Z}} + \overset{\overset{\displaystyle R}{|}}{H-N-R} \longrightarrow \underset{\text{Amide}}{\overset{\overset{\displaystyle O}{\|}\ \overset{\displaystyle R}{|}}{RC-N-R}} + HZ$$

$$Z = Cl,\ O\overset{\overset{\displaystyle O}{\|}}{C}R,\ OH,\ OR \qquad R = H\ \text{or alkyl}$$

PROBLEM

13.9. Write the products of the following reactions:

(a) 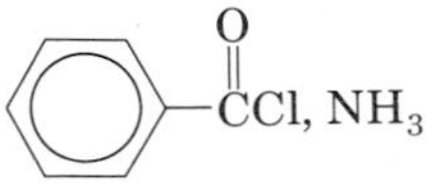 [benzene ring]—$\overset{\overset{\displaystyle O}{\|}}{C}Cl$, NH_3 **(b)** $CH_3CH_2\overset{\overset{\displaystyle O}{\|}}{C}OH$, CH_3NH_2 **(c)** $CH_3\overset{\overset{\displaystyle O}{\|}}{C}O\overset{\overset{\displaystyle O}{\|}}{C}CH_3$, $(CH_3)_2NH$

2. Sulfonamides. Just as there are amides of carboxylic acids, there are also amides of sulfonic acids (RSO_3H, section 5.4.B). Sulfonamides are usually prepared from the corresponding sulfonyl chloride. The reaction is the basis for the preparation of sulfa drugs and for a qualitative test for amines known as the Hinsberg test.

The Hinsberg test can be used to distinguish between 1°, 2°, and 3° amines using benzenesulfonyl chloride as the test reagent. Primary and secondary amines react with benzenesulfonyl chloride to form benzenesulfonamide precipitates. Since tertiary amines do not possess a replaceable hydrogen, they do not react. The sulfonamide from a primary amine still has a replaceable hydrogen on the nitrogen. The strong electron-withdrawing property of the sulfonyl group makes this hydrogen sufficiently acidic to be abstracted by base. The sulfonamides from primary amines dissolve in base. Sulfonamides from secondary amines do not have a remaining nitrogen-hydrogen bond and do not dissolve in base.

Primary amines

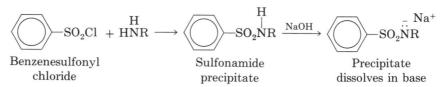

Benzenesulfonyl   Sulfonamide   Precipitate
chloride   precipitate   dissolves in base

Secondary amines

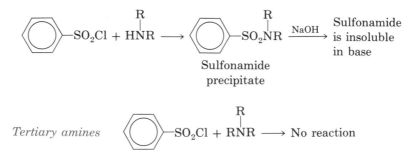

Sulfonamide
precipitate

Tertiary amines

To summarize the Hinsberg test: Amines are treated with benzenesulfonyl chloride. (1) If no precipitate forms the amine is tertiary. (2) If a sulfonamide precipitate forms that is soluble in base, the amine is primary. (3) If a sulfonamide precipitate forms that is not soluble in base, the amine is secondary.

PROBLEM

13.10. Write reaction equations as shown above to illustrate the Hinsberg test in distinguishing between propylamine, ethylmethylamine, and trimethylamine.

3. Sulfa Drugs. Research on sulfa drugs began in 1935 when a physician, Gerhard Domagk, gave his young daughter an oral dose of a sulfonamide dye in a desperate attempt to save her from death from a streptococcal infection. The discovery that sulfonamides retard bacterial growth led to the synthesis and testing of over 5000 sulfonamides, particularly those related to sulfanilamide, during the ensuing dozen years.

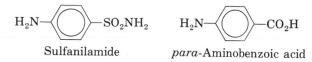

Sulfanilamide   *para*-Aminobenzoic acid

Sulfa drugs do not kill bacteria but only inhibit their growth. This limits the infection to a small colony, which can be destroyed by natural body mechanisms. To reproduce, some bacteria require a common body chemical, *p*-aminobenzoic acid (PABA, used as a sunscreen in some antisunburn creams). Sulfa drugs chemically resemble *p*-aminobenzoic acid, and the bacteria mistakenly absorb the sulfa drug instead of the needed material. Sulfa drugs are effective only on bacteria requiring *p*-aminobenzoic acid for growth. The drugs are used to treat a variety of bacterial infections, including respiratory, gastrointestinal, and urinary tract infections, gonorrhea, and some eye and skin infections. Some common examples of sulfa drugs follow:

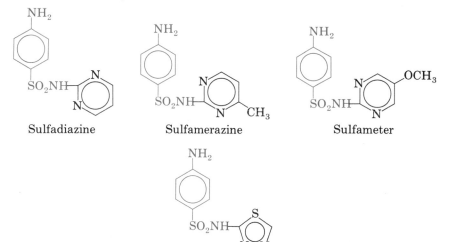

Sulfadiazine   Sulfamerazine   Sulfameter

Sulfamethizole

Although sulfa drugs are still used in veterinary medicine, their use with humans has declined with the advent of antibiotics.

D. Reaction of Aliphatic Amines with Nitrous Acid

Nitrous acid is a weak, unstable acid that must be generated as needed. It is usually prepared by combining aqueous sodium nitrite and a strong acid in the reaction container.

$$NaNO_2 + HCl \longrightarrow HONO + NaCl$$

Nitrous
acid

As nitrous acid is generated, it quickly reacts with amines. The reaction, however, is different with primary, secondary, and tertiary amines and can be used to distinguish among them.

1. Primary Amines. Primary amines form a highly unstable diazonium salt on reaction with nitrous acid. This compound quickly loses nitrogen as a gas, and a carbocation is formed. The carbocation can either eliminate a proton, thereby producing an alkene, or it can combine with a nucleophile for neutralization. Consider, for example, the reaction of ethylamine.

$$CH_3CH_2NH_2 \xrightarrow[\text{HCl}]{\overset{\text{(HONO)}}{NaNO_2,}} CH_3CH_2\overset{+}{N}\equiv N \ Cl^-$$

Aliphatic diazonium salt

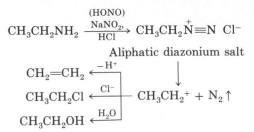

The evolution of nitrogen gas is a visual indication of the presence of a 1° amine.

2. Secondary Amines. Secondary amines react with nitrous acid to produce N-nitrosoamines. These appear as yellow oils.

$$R_2NH \xrightarrow[\text{HCl}]{\overset{\text{(HONO)}}{NaNO_2,}} R_2N-N=O \ + H_2O$$

N-nitrosoamine
(yellow oil)

Nitrosoamines are carcinogenic. The use of sodium nitrite as a food preservative for many processed meat products has caused concern. Since meat contains protein (macromolecules composed of amino acids), it is feared that free amino groups could be nitrosated as the meat is cooked, and carcinogenic materials thereby produced. The possibility also exists that nitrous acid is formed from sodium nitrite as it enters the acidic environment of the stomach. Again nitrous acid could produce nitrosoamines with food proteins or even body proteins. Nitrous acid can react detrimentally with nucleic acids, which act as blueprints for the biosynthesis of proteins and genetic material. The public became especially aware of nitrosoamines in 1979 when they were discovered in low concentrations in many brands of beer.

3. Tertiary Amines. Tertiary amines dissolve in nitrous acid solutions by producing salts, some of which are complex. The simplest salt is illustrated by the following reaction:

$$R_3N \xrightarrow[\text{NaNO}_2/\text{HCl}]{\text{(HONO)}} R_3\overset{+}{N}:H + ONO^-$$

AROMATIC DIAZONIUM SALTS

13.6   A. Preparation

Like primary aliphatic amines, primary aromatic amines react with nitrous acid to form diazonium salts. Aromatic diazonium salts are considerably more stable, however, and can be kept in a cold solution without decomposition. The diazonium salts are prepared by combining a cold solution of sodium nitrite with a cold acidic solution of the primary aromatic amine.

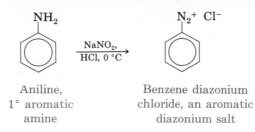

Aniline,	Benzene diazonium
1° aromatic	chloride, an aromatic
amine	diazonium salt

Although these diazonium salts are stable in cold solutions, they are dangerously explosive if isolated. Consequently, they are used for immediate reaction as the diazonium salt solution.

Diazonium salts are important for their synthetic utility. They undergo two general types of reactions: replacement reactions, in which nitrogen is evolved; and coupling reactions, with the retention of nitrogen. The following resonance forms are useful when considering these reaction types.

Resonance forms

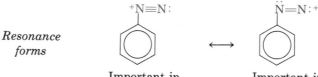

| Important in | Important in |
| replacement reactions | coupling reactions |

B. Replacement Reactions

If a solution of benzene diazonium chloride at 0 °C is allowed to warm to room temperature, nitrogen gas is evolved and phenol and chlorobenzene are formed in the solution.

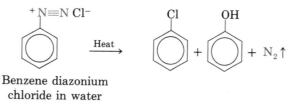

Benzene diazonium
chloride in water

The reaction is thought to involve the elimination of the very stable nitrogen molecule, with a phenyl carbocation left behind that can be neutralized by either chloride ion or water.

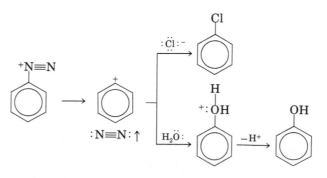

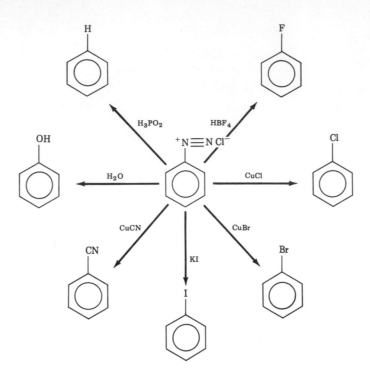

FIGURE 13.3.   Replacement reactions of diazonium salts.

This reaction is known as a replacement reaction and is very characteristic of diazonium salts. Figure 13.3 summarizes some of the most useful of these reactions.

One should be aware that the groups in Figure 13.3 can ultimately replace a nitro group. Consider the preparation, for example, of *meta*-bromoiodobenzene, which cannot be prepared by a simple electrophilic aromatic substitution.

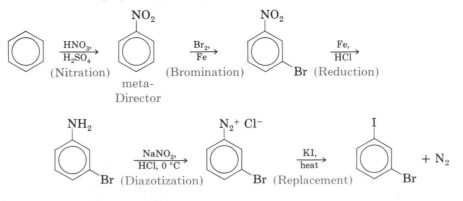

PROBLEM

13.11. Show how the following compounds can be prepared: **(a)** phenol from aniline; **(b)** *para*-methylfluorobenzene from *para*-methylaniline; **(c)** benzonitrile (cyanobenzene) from nitrobenzene; **(d)** *meta*-dichlorobenzene from nitrobenzene.

C. Coupling Reactions

Diazonium salts couple with highly activated aromatic rings to form azo compounds. The general reaction can be summarized as follows. Note that the reaction involves aromatic substitution and nitrogen is retained in the product.

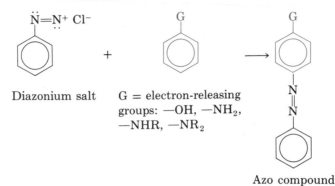

Diazonium salt   G = electron-releasing groups: —OH, —NH$_2$, —NHR, —NR$_2$

Azo compound

The reaction mechanism is electrophilic aromatic substitution. The positive nitrogen of the diazonium salt, the electrophile, is attracted to the π cloud of the activated ring. It usually bonds to the para position (which is less crowded), forming a carbocation. Loss of a hydrogen ion re-forms the aromatic ring.

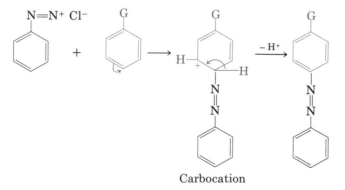

Carbocation

Azo compounds are highly colored and they constitute an important part of the dye industry. The diazonium coupling reaction is the basis for the ingrain dyeing method (section 13.7.D.5). Consider, as an example, the synthesis of the dye and acid-base indicator methyl orange.

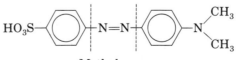

Methyl orange

One might consider making this by coupling a diazonium salt to benzene sulfonic acid or to N,N-dimethylaniline (see dashed lines of structure). The sulfonic acid group is deactivating toward electrophilic substitution, and attempting to couple *para*-dimethylaminobenzene diazonium chloride to benzene sulfonic acid would be unproductive. The dimethylamino group is strongly activating, however, and the following sequence can produce methyl orange in excellent yield:

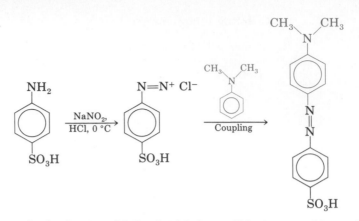

To apply the dye to a fabric, the fabric would be immersed in a solution of N,N-dimethylaniline and then in a solution of the diazonium salt. The two reactants meet deep in the fabric and the dye is synthesized.

PROBLEM

13.12. Synthesize the following compound by a coupling reaction, starting with available compounds:

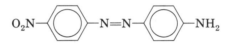

DYES AND DYEING

13.7   With eyesight as a primary sense, humans have always been fascinated by color. Throughout history, ways have been devised to extract dyes from their natural sources and then cause them to adhere to textile materials. This has been a formidable task, as most natural dyes have little direct affinity for fabrics. The extraction of natural dyes became a flourishing industry until the latter part of the nineteenth century, when the synthetic dye industry had its beginnings. Today, several thousand synthetic dyes are known, over a thousand of which are actively manufactured.

A. Color

The electromagnetic spectrum consists of energy from wavelengths as short as 10^{-12} cm (gamma rays) to radiowaves that can be kilometers in length (Figure 16.2). Only a tiny portion of this spectrum, energy with wavelengths 4000–8000 Å (angstrom, 10^{-10} m), is visible to the human eye. Light possessing all wavelengths in the visible range appears white; and the complete absence of these wavelengths is darkness, black.

When a beam of white light strikes a colored surface, certain wavelengths are absorbed and others are reflected. Those reflected compose the color observed. For example, if an object absorbs wavelengths in the blue and green region of the spectrum, the object will appear red because this color constitutes the remaining wavelengths. Conversely, if red light is absorbed, the object will

TABLE 13.2. RELATIONSHIP OF COLOR ABSORBED AND COLOR OBSERVED

Light Absorbed		*Complementary Color Observed*
Wavelength, Å	Color	
4000–4350	Violet	Yellow-green
4350–4800	Blue	Yellow
4800–4900	Green-blue	Orange
4900–5000	Blue-green	Red
5000–5600	Green	Purple
5600–5800	Yellow-green	Violet
5800–5950	Yellow	Blue
5950–6050	Orange	Green-blue
6050–7500	Red	Blue-green
None		White
All wavelengths		Black

appear blue-green. Table 13.2 correlates wavelengths absorbed with complementary colors observed.

B. Structural Characteristics of Dyes

To be an effective dye, an organic chemical must have two properties: it must be able to exhibit color and to adhere to fabrics.

1. Color and Organic Structure. To show color, a compound must absorb visible light. This absorption involves the promotion of a bonding electron to a higher energy level. Visible light has energy to promote only loosely held π electrons to higher energy levels. It has been observed that organic compounds that can absorb visible light and show color have (1) a chromophore group conjugated with (2) an extensive network of alternating single and double bonds or an aromatic system. Some common chromophore (color-bearing) groups follow:

Chromophore Groups

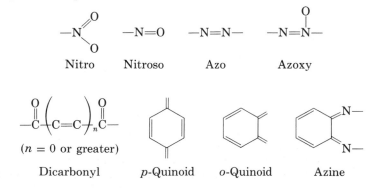

If the conjugated network of double and single bonds is extensive enough, this alone may be sufficient for color. Beta-carotene, for example, is the principal orange pigment in carrots.

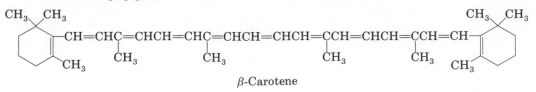

β-Carotene

2. Auxochrome Groups. Exhibiting color is not a sufficient property for a molecule to be a dye. The molecule must also be able to adhere to fiber either directly or indirectly. Auxochrome groups are acidic or basic functions that influence the color of the molecule and cause it to bind to the fabric.

Auxochrome Groups

Basic $-\overset{..}{N}H_2, -\overset{..}{N}HR, -\overset{..}{N}R_2$

Acidic $-CO_2H, -SO_3H, -OH$

Dyes can bind to fibers via ionic attractions, hydrogen-bonding, van der Waals' forces, or actual covalent bonds, or some combination. Ionic attractions can occur between an acidic or basic group on the fabric and an acidic or basic group on the dye. Wool and silk are proteinaceous materials composed of amino acids linked by amide linkages. The salt of an acid or base group on the fiber has an electrostatic interaction to a salt of opposite charge on the dye.

Ionic Attractions

$$\text{fiber}-\overset{\overset{O}{\|}}{C}O^{-}\cdots\cdots H_3N^{+}-\text{dye}$$

$$\text{fiber}-\overset{+}{N}H_3\cdots\cdots \bar{O}_3S-\text{dye}$$

Opportunities for hydrogen-bonding are abundant between auxochrome groups and fibers with polar bonds, especially wool and silk.

Hydrogen-Bonding

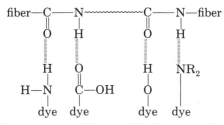

Van der Waals' forces are weak intermolecular attractions between nonpolar molecules. They are especially effective if the dye molecule is linear and if it has a large surface area to interact with the surface to be dyed. In the case of "reactive dyes," an actual covalent bond linkage causes the dye to adhere to fabric.

C. Classification of Dyes by Structure

Dyes are classified both by their chemical constitution and by method of application. Following are some of the chief classes of dyes by chemical struc-

ture. Identify the chromophore and auxochrome groups and networks of extensive conjugation as you go along. Many dyes are named according to the Color
Index (CI) reference tables.

1. Nitro and Nitroso Dyes. The NO_2 and NO groups are chromophores in
this class of nitro and nitroso dyes.

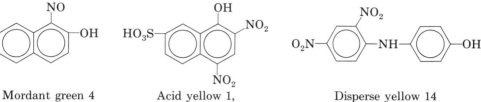

Mordant green 4   Acid yellow 1,   Disperse yellow 14
                        Naphthol yellow S

2. Azo Dyes. Azo dyes have one (*monoazo-*) or more (*diazo-*, *triazo-*,
polyazo-) azo groups, —N=N—, as the primary chromophore.

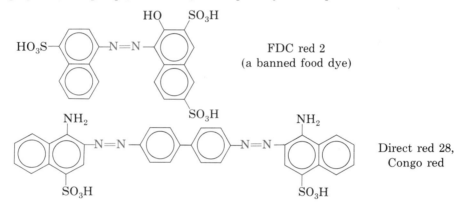

FDC red 2
(a banned food dye)

Direct red 28,
Congo red

3. Triarylmethane Dyes. In triarylmethane dyes, a central carbon is
bonded to three aromatic rings, one of which is in the quinoid form (the chromophore).

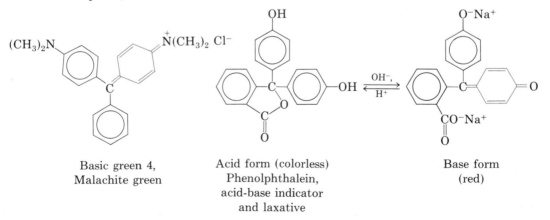

Basic green 4,   Acid form (colorless)   Base form
Malachite green   Phenolphthalein,   (red)
                        acid-base indicator
                        and laxative

Note the absence of extensive conjugation and the quinoid chromophore in the
acid form of phenolphthalein as opposed to its presence in the base form.

4. Azine and Related Dyes. Most of these dyes possess some modification of the ortho quinoid-azine chromophore.

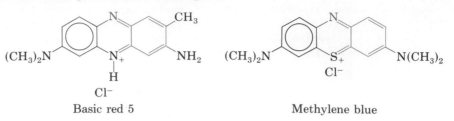

Basic red 5   Methylene blue

5. Anthraquinone Dyes. The para quinoid chromophore is present in these anthracene-type dyes.

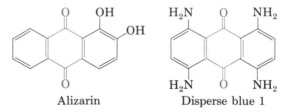

Alizarin   Disperse blue 1

6. Indigo Dyes. Notice the dicarbonyl chromophore.

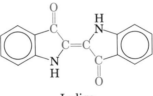

Indigo

D. Application of Dyes to Fabrics

A second method of classifying dyes involves method of application.

1. Direct Dyes. To apply a direct dye to a fabric, one immerses the material in a water solution of the dye (which obviously must be water-soluble). The dye must contain acidic or basic auxochrome groups to interact with polar groups of the fabric. Wool and silk are readily dyed by this method because of the presence in their chemical structures of acid and base salts, which are electrostatically attracted to the auxochromes in the dye (section 13.7.B.2).

2. Disperse Dyes. Disperse dyes are almost completely insoluble in water but "soluble" in the fiber. The dye is finely milled and colloidally dispersed in water with the help of dispersing agents. Sometimes heat and pressure are used to aid in the absorption of the dye by the fabric. Disperse dyes are used with modern synthetic fabrics such as nylon, Orlon, cellulose acetates, and polyesters.

3. Mordant Dyes. In the mordant dyeing method, a dye can be applied to a fabric for which it has no natural affinity or the process can strengthen the attachment of a dye that does adhere. A mordant (Latin *mordere,* "to bite") is capable of binding both to the dye and to the fiber, thus acting as a "glue" between the two. In some applications, a fiber such as cotton is treated with a mordant, that is, the oxides or salts of aluminum, copper, chromium, and cobalt, and then the dye is applied to the mordanted fiber. Whether or not the oxides or salts are used, the mordant forms a coordination complex between the fiber and dye, as when alizarin binds to cotton.

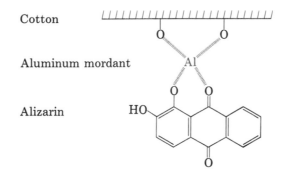

Cotton

Aluminum mordant

Alizarin

4. Reactive Dyes. A reactive dye actually forms covalent bonds between the dye and the fiber. In general, either the dye or the fiber is covalently bound to a trichlorotriazine ring and the other is bound in the actual dye application process. Both reactions involve nucleophilic substitution. Reactive blue 4 is an example of a reactive dye used on cotton.

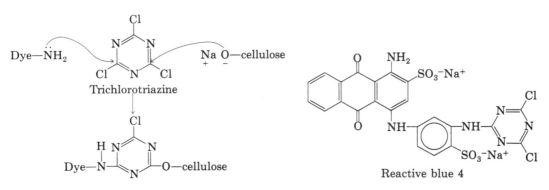

Trichlorotriazine

Reactive blue 4

5. Ingrain Dyes. Ingrain dyes are dyes synthesized directly in the fabric. Often the reactants are water-soluble, but the actual dye has less water-solubility and in addition binds to the fabric via auxochrome groups.

Azo dyes can be applied by the ingrain dyeing method with a diazonium coupling reaction (section 13.6.C).

6. Vat Dyes. Vat dyes are insoluble in water but can be reduced to a water-soluble form, which may be colorless or have a different shade. After this kind of dye is applied, oxidation (sometimes merely in the air) re-forms the original insoluble dye, which is now bound fast to the fabric. Indigoid, thioindigoid, and anthraquinoid dyes are often applied by this method. Indigo itself is a good example. The colorless reduced form is called indigo white, whereas on oxidation it yields indigo blue.

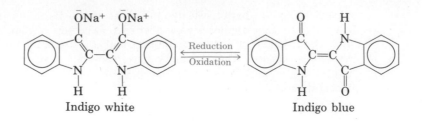

Indigo white   Indigo blue

Panel 16

ALKALOIDS

Alkaloids are loosely defined as plant-produced nitrogenous bases that have a physiological effect on animals. This definition is not all-inclusive, although it probably does describe most alkaloids. Here we present a brief survey of some important classes of alkaloids.

A. Phenylalkylamine Alkaloids

Phenylalkylamine alkaloids structurally contain a benzene ring, an alkyl group (usually ethyl), and an amino group.

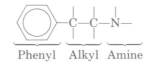

Phenyl Alkyl Amine

A number of these bases are found in the herb *Ma Huang,* which has been used medicinally in China for over 5000 years. The most important component is ephedrine, which has been used successfully to combat bronchial asthma.

To gain a concept of the physiological properties of these compounds, we might compare them to adrenalin (epinephrine), also a phenylalkylamine, which is released by the human adrenal gland in times of stress, fear, or excitement.

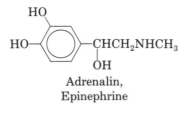

Adrenalin,
Epinephrine

1. Increased heartbeat.
2. Contraction of blood vessels; increased blood pressure.
3. Relaxation of bronchi and mucous membranes.
4. Restriction of digestive secretions.
5. Excitement, alertness.
6. Energy, release of glucose from glycogen storage.

Other phenylalkylamines have similar physiological properties. Two compounds have exaggerated effects in the fifth category. Peyote, once used in the religious rituals of Indian tribes in Mexico, is a Mexican cactus that produces the hallucinogenic drug mescaline. Amphetamine (also called dexedrine, benzedrine, dexxies, bennies, uppers, pep pills) was introduced in 1932 as a nasal decongestant (No. 3) and used in World War II to keep front-line troops alert.

[PANEL 16]

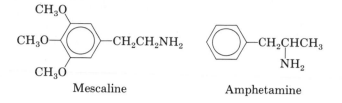

Mescaline   Amphetamine

Today, it is found in some prescription diet pills (No. 4). However, psychological dependence on amphetamines can occur and withdrawal can lead to fatigue and depression.

Many over-the-counter nasal decongestants topical and oral, contain phenylalkylamines, most commonly ephedrine, phenylephrine, and phenylpropanolamine hydrochloride (phenylpropanolamine hydrochloride is also used in non-prescription diet pills).

Ephedrine   Phenylephrine,   Phenylpropanolamine
                     neosynephrine   hydrochloride

These drugs function by contracting the arterioles within the nasal mucous membranes, thereby restricting blood flow to this area. Swelling is reduced, nasal passages open, and the ventilation and drainage of sinuses is possible. However, prolonged use of decongestants, especially topical sprays, can result in restricted nutrient flow to the area and in reduced waste removal from the sinuses, leaving the affected tissues swollen and susceptible to infection. Oral decongestants often have printed cautions to people with heart or blood pressure problems (effects 1 and 2) and to those with diabetes (effect 6). The more recently introduced long-duration nasal decongestants contain 0.05%–0.1% xylometazoline hydrochloride.

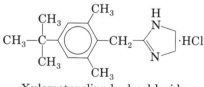

Xylometazoline hydrochloride

Many nasal decongestants and hayfever preparations contain antihistamines. Histamine is released when the body begins to experience an allergic reaction—to pollen, insect stings, or penicillin, for instance. Most symptoms of allergy are caused by histamine. Antihistamines reduce or eliminate the effects of histamine.

Antihistamines are used in nonprescription sleeping pills and sedatives. One of the most popular, methapyrilene, has been withdrawn from most preparations because it has been determined to be an animal carcinogen.

[PANEL 16]   *Antihistamines*

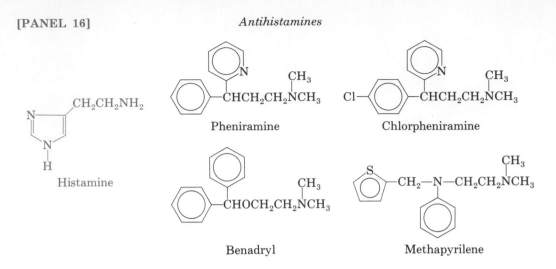

Histamine   Pheniramine   Chlorpheniramine

Benadryl   Methapyrilene

B. Pyrrolidine, Piperidine, and Pyridine Alkaloids

The alkaloids pyrrolidine, piperidine, and pyridine, as their names suggest, contain derivatives of the ring compounds of the same name.

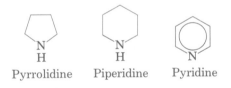

Pyrrolidine   Piperidine   Pyridine

1. Pyrrolidine Alkaloids. Pyrrolidine is found in wild carrots, and hygrine and cuscohygrine are found in the leaves of the Peruvian coca shrub. Cuscohygrine is also found in deadly nightshade (*Atropa belladonna*).

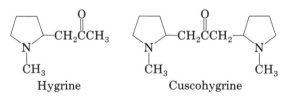

Hygrine   Cuscohygrine

2. Piperidine Alkaloids. Around 400 B.C., the Greek philosopher Socrates was executed by being forced to drink hemlock extract. The symptoms include gradual paralysis, convulsions, and finally death by paralysis of the respiratory system. Coniine is the principal alkaloid in hemlock. Other piperidine alkaloids include piperine, which occurs in black pepper (6%–11%) and lobeline, from the seeds of Indian tobacco. Lobeline is similar to nicotine in physiological action and is the basic ingredient of nonprescription cigarette-smoking deterrents.

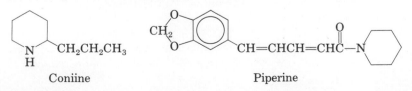

Coniine   Piperine

[PANEL 16]

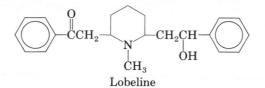

Lobeline

3. Pyridine-Pyrrolidine Alkaloids. The most familiar pyridine-pyrrolidine alkaloid is nicotine, the principal alkaloid component of tobacco (4%–6% in leaves). This substance is one of the most toxic alkaloids known; it is fatal to all forms of animal life (by respiratory paralysis) and is used as an agricultural insecticide. Nicotinic acid, niacin, is a vitamin whose lack causes pellagra in man and black tongue in dogs.

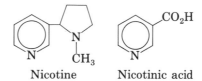

Nicotine   Nicotinic acid

4. Condensed Pyrrolidine-Piperidine (Tropane) Alkaloids. Condensed pyrrolidine-piperidine alkaloids consist of bicyclic systems incorporating both the pyrrolidine and piperidine rings. They are found in numerous species of the plant family Solanaceae, including belladonna, or deadly nightshade. The name *belladonna* arose through the ancient use of the plant extract by women to dilate the pupils of the eyes for cosmetic purposes (*belladonna* means "lovely lady"). And, at one time, the belladonna alkaloid atropine (as the sulfate salt) was used by ophthalmologists for pupil dilation.

Cocaine, from the leaves of the coca plant, has anesthetic properties and was used medicinally in the late 1800s for eye surgery and tooth extraction. Cocaine produces mental and physical stimulation, but because of its addictive properties its legal medicinal use has diminished. Scopolamine, sometimes called "truth serum," is currently used in nonprescription sedatives, calmatives, sleeping pills, and motion-sickness pills.

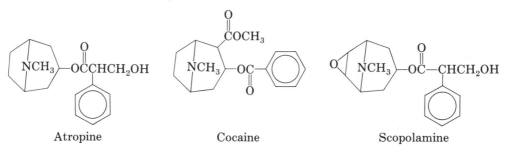

Atropine   Cocaine   Scopolamine

C. Quinoline and Isoquinoline Alkaloids

Quinoline and isoquinoline alkaloids possess the quinoline and isoquinoline ring systems.

Quinoline   Isoquinoline

[PANEL 16]   **1. Quinoline Alkaloids.** Several quinoline alkaloids are present in the bark of the cinchona tree native to the eastern slopes of the Andes. The most important is quinine, found in tonic water.

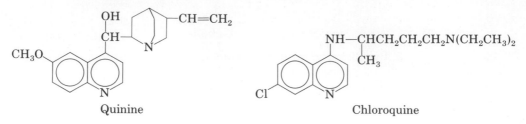

Quinine

Chloroquine

Cinchona bark was used by the Jesuits about 1600 as an antimalarial preparation. It became so much in demand that it rivaled gold in cost. By the nineteenth century new forests were planted in Java, Ceylon, and India, and trees were selectively cultivated for a high alkaloid content. When the Japanese invaded Java in 1942, the rest of the world was deprived of the bulk of the quinine supply. This led to accelerated research in the development of synthetic antimalarials. Today, quinine has been largely replaced by other drugs, such as chloroquine. Quinidine, a stereoisomer of quinine also found in cinchona bark, is used as a cardiac regulator for persons experiencing irregular heartbeat, or tachycardia.

2. Isoquinoline Alkaloids. The isoquinoline alkaloids possess the isoquinoline ring system in some form, sometimes derivatized or reduced. Tubocurarine chloride, a curare alkaloid, is a poison that African and South American tribes use. Small amounts on an arrow or blowdart can quickly kill even large animals.

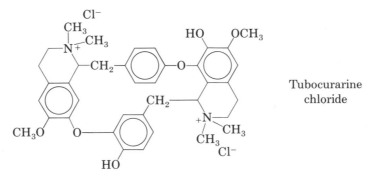

Tubocurarine
chloride

Quite a few isoquinoline alkaloids can be isolated from the opium poppy. For example, the opium alkaloid morphine and its derivatives codeine and heroin are fairly complex isoquinoline compounds.

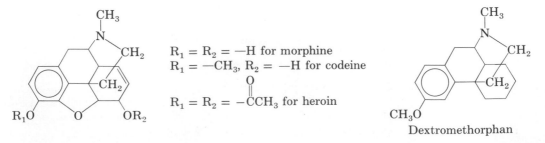

$R_1 = R_2 = $ —H for morphine
$R_1 = $ —CH$_3$, $R_2 = $ —H for codeine

$R_1 = R_2 = $ —$\overset{O}{\overset{\|}{C}}CH_3$ for heroin

Dextromethorphan

[PANEL 16] All these compounds have a pain-relieving effect and generate a feeling of well-being. Unfortunately, all three are addictive in various degrees. Codeine is a component of some cough medicines. It controls coughing by acting on the cough control center in the medulla. Although considerably less addictive than heroin and morphine, it must be used with care. A synthetic, structurally related compound, dextromethorphan, is used in nonprescription cough medicines. It is as effective as codeine and acts on the central nervous system like codeine, but is apparently nonaddictive and has minimal side effects. Methadone and meperidine (Demerol) are commercially synthesized pain relievers, but unfortunately, like morphine, they are addictive.

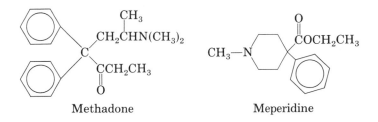

Methadone   Meperidine

D. Indole Alkaloids

Ergot alkaloids are examples of indole alkaloids. These compounds, peptide derivatives of lysergic acid and structurally related to the hallucinogenic drug LSD, were the cause of large outbreaks of ergotism during the Middle Ages. This painful and often fatal disease is characterized by gangrenous swelling of the extremities or muscular convulsions and mental damage. The alkaloids are formed by the ergot fungus, which grows on rye especially after a warm, damp summer (victims contracted ergotism by eating infected rye bread).

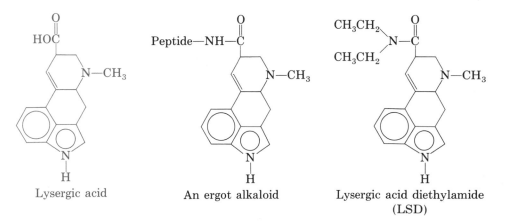

Lysergic acid   An ergot alkaloid   Lysergic acid diethylamide
                                                              (LSD)

Reserpine (from *Rauwolfia serpentina*) is a tranquilizer and a sedative, and strychnine (from *Strychnos nux vomica*), a rodenticide, is also a circulatory stimulant. Reserpine, a heavily used medication for high blood pressure, has been classified as a carcinogen by the National Cancer Institute.

[PANEL 16]

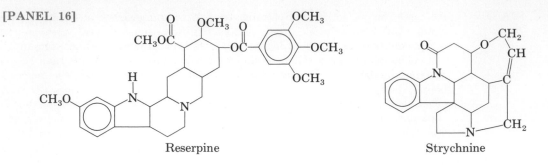

Reserpine   Strychnine

E. Purine or Xanthine Alkaloids

Caffeine, found in tea, coffee, cola, and cocoa, is a stimulant and as such is used in sleep-preventing pills and nonprescription stimulants.   Theobromine is also found in tea and cocoa-derived foods.   Like caffeine, it is a stimulant.

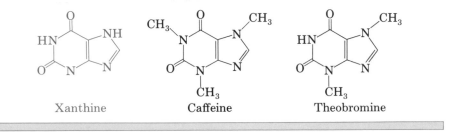

Xanthine   Caffeine   Theobromine

END-OF-CHAPTER PROBLEMS

13.13 IUPAC Nomenclature: Name by the IUPAC system of nomenclature:

(a) $CH_3(CH_2)_4CH_2NH_2$ **(b)** $CH_3CH_2\underset{\underset{NH_2}{|}}{C}HCH_3$ **(c)** $H_2N(CH_2)_8NH_2$

(d) $CH_3\underset{\underset{NH_2}{|}}{C}H(CH_2)_4\underset{\underset{O}{\|}}{C}CH_3$ **(e)** $CH_3NHCH_2CH_2\underset{\underset{O}{\|}}{C}H$ **(f)** $CH_3CH{=}CHCH_2NH_2$

(g) $CH_3\underset{\underset{N(CH_2CH_3)_2}{|}}{C}HCH_2CH_2\underset{\underset{CH_3}{|}}{\overset{\overset{CH_3}{|}}{C}}CH_3$ **(h)** $O_2N-\!\!\!\bigcirc\!\!\!-NH_2$

(i) $\bigcirc\!\!-\underset{\underset{Cl}{|}}{}N\overset{\overset{CH_2CH_3}{|}}{}CH_2CH_2CH_2CH_3$

13.14 Nomenclature: Draw the following compounds: **(a)** cycloheptanamine; **(b)** ethylpropylamine; **(c)** tributylamine; **(d)** ethylisopropylmethylamine; **(e)** N,N-dimethylaniline; **(f)** 2,4,6-trichloroaniline; **(g)** N-ethylheptanamine; **(h)** N-ethyl-N-methyl-3-propylcyclopentanamine; **(i)** 4-aminobutanoic acid; **(j)** 2-aminoethanol; **(k)** 4-N,N-diethylaminohexanal.

13.15 Physical Properties: Arrange the following in order of increasing boiling point:

(a) (I) methanamine, (II) propanamine, (III) heptanamine, (IV) decanamine
(b) (I) methanoic acid, (II) ethanamine, (III) ethanol, (IV) ethane
(c) (I) propylamine, (II) ethylmethylamine, (III) trimethylamine

13.16 Hydrogen-Bonding: Graphically illustrate hydrogen-bonding between **(a)** ammonia molecules, **(b)** 2-aminoethanol molecules, and **(c)** ethanamine and water molecules.

13.17 Basicity of Amines: Arrange the following sets of compounds in order of increasing basicity:

(a) (I) $CH_3CH_2CH_2CH_2NH_2$, (II) $CH_3\overset{\overset{\displaystyle CH_3}{|}}{N}CH_2CH_3$, (III) $CH_3NHCH_2CH_2CH_3$

(b) (I) $CH_3CH_2\overset{\overset{\displaystyle O}{||}}{C}NH_2$, (II) $CH_3CH_2CH_2NH_2$, (III) $H\overset{\overset{\displaystyle O}{||}}{C}CH_2CH_2NH_2$,

(IV) $CH_3\overset{\overset{\displaystyle O}{||}}{C}CH_2NH_2$

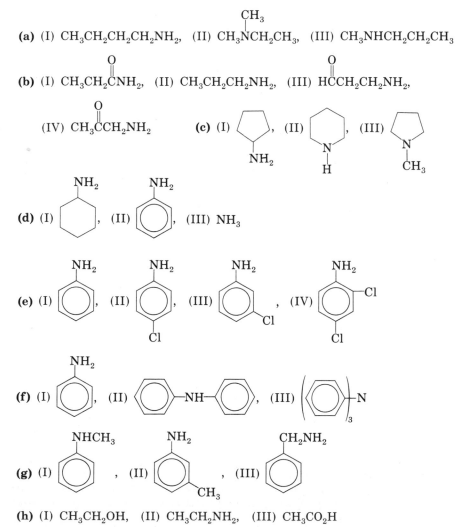

13.18 Acidity and Basicity: Aniline is much less basic than methylamine for essentially the same reasons that phenol is more acidic than methyl alcohol. Both aniline and phenol are extremely reactive toward electrophilic aromatic substitution. Explain. (See sections 13.5.C.2, 8.6.A, and 5.6.D for assistance.)

13.19 Basicity of Heterocyclic Amines: Pyridine is considerably more basic than pyrolle. Explain in terms of the π-bonding pattern and availability of the unshared electron pair. (See section 5.3.C for assistance.)

13.20 Lewis Acid–Lewis Base Reactions: Write products for the acid-base reactions between the following amines and Lewis acids:

(a) $CH_3CH_2NH_2$, HBr   **(b)** $(CH_3)_2NH$, HCl   **(c)** $2(CH_3)_3N$, H_2SO_4

(d) ⬡—NH_2, HCl   **(e)** ⬠NH, HNO_3 **(f)** $(CH_3CH_2)_3N$, BF_3

13.21 Ammonium Fertilizers: The fertilizers ammonium nitrate, ammonium sulfate, and ammonium phosphate are made by the reaction of ammonia with the appropriate mineral acids. The ammonia is produced by the Haber process in which nitrogen from the air is combined with hydrogen (from the cracking of petroleum). Write balanced equations for all of these reactions.

13.22 Basicity Constants: Write an equilibrium equation for the reaction of ammonia in water and for the reaction of trimethylamine in water. Write the basicity constant expression for both.

13.23 Reduction of Nitrogen Compounds: Write equations illustrating the reduction of the following compounds to amines:

(a) CH_3—⬡—NO_2 **(b)** $CH_3CH_2CH_2C\equiv N$ **(c)** $CH_3CH_2CH_2\overset{\displaystyle O}{\overset{\|}{C}}NH_2$

(d) $CH_3CH_2\overset{\displaystyle O}{\overset{\|}{C}}NHCH_3$ **(e)** $CH_3\overset{\displaystyle O}{\overset{\|}{C}}N(CH_3)_2$ **(f)** $CH_3\overset{\displaystyle CH_3}{\overset{\|}{C}}HCH_2CH_2NO_2$

13.24 Reactions of Amines: Write equations showing reaction of (I) propylamine, (II) ethylmethylamine, and (III) trimethylamine with each of the following:

(a) HCl   **(b)** H_2SO_4 **(c)** excess $CH_3Br(Na_2CO_3)$

(d) $CH_3\overset{\displaystyle O}{\overset{\|}{C}}Cl$ **(e)** $CH_3\overset{\displaystyle O}{\overset{\|}{C}}O\overset{\displaystyle O}{\overset{\|}{C}}CH_3$ **(f)** ⬡—$\overset{\displaystyle O}{\overset{\|}{C}}OH$, heat

(g) $CH_3CH_2\overset{\displaystyle O}{\overset{\|}{C}}OCH_3$ **(h)** ⬡—SO_2Cl **(i)** $HONO(NaNO_2/HCl)$

13.25 Hinsberg Test: Using reaction equations and words, describe how aniline, N-methylaniline, and N,N-dimethylaniline can be distinguished from one another using the Hinsberg test.

13.26 Reactions of Diazonium Salts: Write the product of the reaction between *para*-methylaniline and $NaNO_2$/HCl at 0 °C. Then show the reaction of this product with the following reagents: **(a)** CuCl; **(b)** CuBr; **(c)** KI; **(d)** CuCN; **(e)** H_2O; **(f)** HBF_4; **(g)** H_3PO_2; **(h)** phenol; **(i)** N,N-dimethylaniline.

13.27 Syntheses Using Diazonium Salts: Synthesize the following compounds using electrophilic aromatic substitution (section 5.7) and diazonium salt replacement reactions: **(a)** *p*-fluorotoluene from *p*-methylaniline; **(b)** *m*-iodobromobenzene from *m*-bromoaniline; **(c)** bromobenzene from nitrobenzene; **(d)** *m*-chlorofluorobenzene from nitrobenzene; **(e)** *m*-bromophenol from nitrobenzene; **(f)** 2,4,6-tribromoiodobenzene from aniline; **(g)** cyanobenzene from benzene; **(h)** *p*-bromoiodobenzene from benzene; **(i)** *m*-chlorofluorobenzene from benzene; **(j)** 2,4,6-tribromophenol from benzene; **(k)** 1,3,5-tribromobenzene from aniline.

13.28 Diazonium Salts—Coupling Reactions: With chemical equations, show how the following dyes can be synthesized by the diazonium coupling reaction. Start with stable available compounds. (These are the same reactions that would be used to apply these dyes by the ingrain dyeing method.)

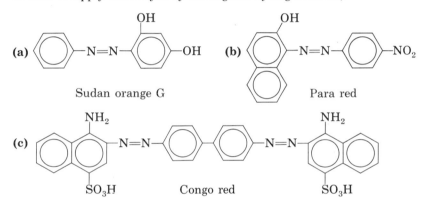

(a) Sudan orange G   **(b)** Para red

(c) Congo red

13.29 Preparations of Familiar Compounds: Write equations illustrating the following preparations:
(a) LSD: from lysergic acid (see Panel 16) and diethylamine
(b) phenacetin (pain reliever): from the reaction between *p*-ethoxyaniline and acetic anhydride
(c) hexamethylene diamine (starting material for Nylon 66): 1,4-dichlorobutane plus sodium cyanide followed by complete hydrogenation (Ni catalyst)
(d) phenylpropanolamine hydrochloride (nasal decongestant): from 1-phenyl-2-aminopropanol and hydrochloric acid
(e) amphetamine sulfate: from 1-phenyl-2-propanamine and sulfuric acid
(f) a cationic detergent: $CH_3(CH_2)_{15}CH_2NH_2 + 3CH_3Cl \longrightarrow$
(g) phenobarbital (a barbiturate):

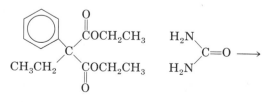

(h) sulfadiazine (a sulfa drug):

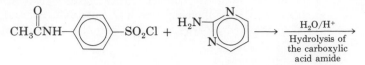

13.30 Qualitative Analysis: Using sodium hydroxide and hydrochloric acid solutions, describe how you would distinguish between benzoic acid and *p*-bromoaniline (both are white solids).

13.31 Reaction Mechanisms: Write step-by-step reaction mechanisms for each of the following reactions:
(a) S_N2 reaction between trimethylamine and methyl iodide
(b) coupling between benzene diazonium chloride and phenol by electrophilic aromatic substitution
(c) reaction between ethanoyl chloride and dimethylamine to form an amide

13.32 Consumer Chemistry: Examine the labels of nonprescription drugs in a drugstore. Find commercial products containing the following compounds. Make a chart listing brand names, types of products, and ingredients you can identify, using information in Panel 16 and your knowledge of the action of the ingredient.
(a) ephedrine   **(b)** phenylephrine   **(c)** phenylpropanolamine hydrochloride
(d) pheniramine   **(e)** chlorpheniramine   **(f)** lobeline   **(g)** scopolamine
(h) dextromethorphan   **(i)** caffeine   **(j)** xylometazoline hydrochloride

13.33 Dyes: For each of the following dyes, identify the chromophore and auxochrome groups (see section 13.7 for structures): **(a)** Acid yellow 1; **(b)** Mordant green 4; **(c)** Congo red; **(d)** Malachite green; **(e)** Methylene blue; **(f)** Alizirin.

Chapter 14

Proteins: Amino Acid Polymers

SOURCES AND FUNCTIONS

14.1   Insulin regulates the body's use of glucose. Hemoglobin carries oxygen from lungs to cells and cytochrome C transfers electrons. Hair, skin, nails, hoofs, and feathers are composed of keratin. Food is digested by enzymes, biological catalysts such as pepsin (stomach) and trypsin (intestines). Huge (macro) molecules of genetic material are synthesized with the aid of DNA polymerase. All the species just mentioned, though diverse in function, are themselves macromolecules, generally classified as proteins. The term is derived from the Greek *proteios,* meaning "first" among all substances found in living organisms in variety and most assuredly importance. They perform in every arena of living tissue and on every level of the evolutionary scale from simplest bacterium to humans. Yet these diverse functions arise from approximately 20 units all having essentially the same fundamental structure, that of the amino acid.

STRUCTURE

14.2   As the term *amino acid* suggests, the chemical structure of amino acids consists of an amine group and an acid group.

$$\underbrace{H_2N}_{\text{Amine}} - \underset{\underset{R}{|}}{\overset{\overset{H}{|}}{C}} - \underbrace{\overset{\overset{O}{\|}}{C} - OH}_{\text{Carboxylic acid}}$$

An amino acid

The side chain, R, can assume many identities. Table 14.1 lists the most common amino acids found in proteins as well as some pertinent related compounds. They are subdivided into groups dependent on the nature of the R group. Most common amino acids are also termed alpha (α) because the amine group lies on the first (α) carbon from the carboxyl. Let us consider the reactions and interactions characteristic of these two functional groups.

TABLE 14.1. THE COMMON AMINO ACIDS AT LOW pH

Basic Side Chains

$$\overset{+}{H_3}N(CH_2)_4\overset{O}{\overset{\|}{C}}HC OH$$
$$\underset{\overset{+}{NH_3}}{}$$

Lysine (Lys)[a]

$$\overset{+}{H_2}N=C-NH-(CH_2)_3\overset{O}{\overset{\|}{C}}HCOH$$
$$\underset{NH_2}{} \qquad \underset{\overset{+}{NH_3}}{}$$

Arginine (Arg)

$$HC=CCH_2\overset{O}{\overset{\|}{C}}HCOH$$
$$HN \qquad NH \quad NH_3$$
$$\overset{+}{\underset{CH}{}} \qquad \overset{+}{}$$

Histidine (His)

Acidic Side Chains

$$HO\overset{O}{\overset{\|}{C}}(CH_2)_2\overset{O}{\overset{\|}{C}}HCOH$$
$$\underset{\overset{+}{NH_3}}{}$$

Glutamic acid (Glu)

$$HO\overset{O}{\overset{\|}{C}}CH_2\overset{O}{\overset{\|}{C}}HCOH$$
$$\underset{\overset{+}{NH_3}}{}$$

Aspartic acid (Asp)

Alcohol (Polar) Side Chains

$$HOCH_2\overset{O}{\overset{\|}{C}}HCOH$$
$$\underset{\overset{+}{NH_3}}{}$$

Serine (Ser)

$$HOCH-\overset{O}{\overset{\|}{C}}HCOH$$
$$\underset{CH_3 \ \overset{+}{NH_3}}{}$$

Threonine (Thr)[a]

Sulfur-containing Side Chains

$$HSCH_2\overset{O}{\overset{\|}{C}}HCOH$$
$$\underset{\overset{+}{NH_3}}{}$$

Cysteine (CysH)[b]

$$H_3CS(CH_2)_2\overset{O}{\overset{\|}{C}}HCOH$$
$$\underset{\overset{+}{NH_3}}{}$$

Methionine (Met)[a]

Alkyl Side Chains

$$H\overset{O}{\overset{\|}{C}}HCOH$$
$$\underset{\overset{+}{NH_3}}{}$$

Glycine (Gly)

$$CH_3\overset{O}{\overset{\|}{C}}HCOH$$
$$\underset{\overset{+}{NH_3}}{}$$

Alanine (Ala)

$$CH_3CH-\overset{O}{\overset{\|}{C}}HCOH$$
$$\underset{CH_3 \ \overset{+}{NH_3}}{}$$

Valine (Val)[a]

$$CH_3CHCH_2\overset{O}{\overset{\|}{C}}HCOH$$
$$\underset{CH_3 \qquad \overset{+}{NH_3}}{}$$

Leucine (Leu)[a]

$$CH_3CH_2CH-\overset{O}{\overset{\|}{C}}HCOH$$
$$\underset{CH_3 \quad \overset{+}{NH_3}}{}$$

Isoleucine (Ile)[a]

Aromatic Side Chains

$$\bigcirc-CH_2\overset{O}{\overset{\|}{C}}HCOH$$
$$\underset{\overset{+}{NH_3}}{}$$

Phenylalanine (Phe)[a]

$$HO-\bigcirc-CH_2\overset{O}{\overset{\|}{C}}HCOH$$
$$\underset{\overset{+}{NH_3}}{}$$

Tyrosine (Tyr)

$$-CH_2\overset{O}{\overset{\|}{C}}HCOH$$
$$\underset{\overset{+}{NH_3}}{}$$

Tryptophan (Trp)[a]

Cyclic (Imino) Side Chains

$$H_2C-CH\overset{O}{\overset{\|}{C}}OH$$
$$H_2C \qquad \overset{+}{NH_2}$$
$$CH_2$$

Proline (Pro)

[a] An essential amino acid which must be provided in the diet.
[b] Often found as cystine, a dimer bonded through the sulfurs.

$$HO\overset{O}{\overset{\|}{C}}CHCH_2 \ \ CH_2\overset{O}{\overset{\|}{C}}HCOH$$
$$\underset{\overset{+}{NH_3}}{} \qquad \underset{\overset{+}{NH_3}}{}$$

A. Optical Activity

The α-carbon of an amino acid will be a chiral carbon unless the R group is another hydrogen. This means that almost all amino acids exhibit optical activity. The orientation of the amino group on the chiral α-carbon in comparison with the OH on glyceraldehyde determines whether the amino acid will be designated D or L.

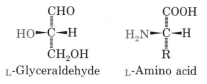

L-Glyceraldehyde   L-Amino acid

The amino acids found in proteins are of the L configuration, although some D-amino acids can be found in the cells of certain bacteria.

PROBLEM

14.1. How many optical isomers are possible for the amino acids glycine, threonine, and tyrosine?

B. Dipolar Nature of the Amino Acid

1. Ionization. The uncharged, electrically neutral picture of an amino acid so far presented is not entirely correct. Amino acids, occurring as they usually do in water solution, show activity characteristic of the acidic carboxyl group and the basic amine group. Recall that the hydrogen on a carboxylic acid group will ionize, under the appropriate pH conditions. Conversely, the lone, free pair of electrons on the primary amino group makes it a Lewis base, capable of accepting a proton. In a neutral amino acid solution (pH 7), this is just what occurs. The carboxyl loses a proton and the amino group picks one up.

$$H_2\ddot{N}-\underset{\underset{R}{|}}{\overset{\overset{H}{|}}{C}}-\overset{\overset{O}{\|}}{C}-OH \longrightarrow H_3\overset{+}{N}-\underset{\underset{R}{|}}{\overset{\overset{H}{|}}{C}}-\overset{\overset{O}{\|}}{C}-\ddot{\underset{..}{O}}:^-$$

The result is a dipolar ion, charged but electrically neutral. This species is called a *zwitterion* (German, "two ions"). Amino acids are said to be *amphoteric;* that is, their structure allows them, in reaction with the medium in which they are dissolved, to donate or accept a proton, depending on the pH of that medium.

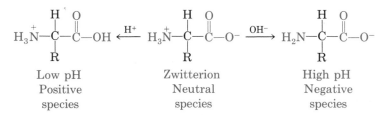

Low pH	Zwitterion	High pH
Positive	Neutral	Negative
species	species	species

The pH at which the electrically neutral species (zwitterion) predominates is dependent upon the ionization constants for the α-amino group, the carboxylic acid group, and whatever effect the side-chain group might have. If the side chain is not ionizable, the ionization picture above will exist with three ionized species. Should the side chain be acidic or basic, however, at least four ionized states are possible.   Consider aspartic acid.

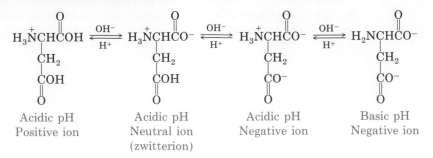

Acidic pH	Acidic pH	Acidic pH	Basic pH
Positive ion	Neutral ion	Negative ion	Negative ion
	(zwitterion)		

In all, one positive and one neutral ion and two negative ions can exist according to the pH and the K_a of the carboxyl groups and K_b of the amino group.

2. Electrophoresis.   The ionized picture of amino acids is the basis of methods used for their separation.   As you know, opposite electrical charges attract each other.   If an ionized amino acid is placed in an electric field, it will move (migrate) toward the electrode with a charge opposite to its own net charge.   For example, with our general picture of an amino acid we can see three patterns of migration possible depending upon the pH.

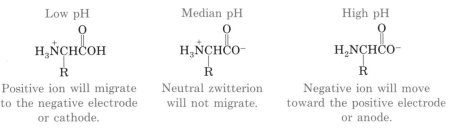

| Low pH | Median pH | High pH |

Positive ion will migrate to the negative electrode or cathode.   Neutral zwitterion will not migrate.   Negative ion will move toward the positive electrode or anode.

With a charged R group, the possibilities for variation in migration increase, depending on the net charge at a particular pH.   The pH at which an amino acid will not migrate in an electric field is known as its *isoelectric point*.   This mobility (or lack of mobility) under varying conditions of pH and electric field is used to separate, purify, and identify amino acids and their polymers.   The physical process is called *electrophoresis*.

These polar species can also be separated by means of ion exchange columns.   This is the basis for the amino acid analyzer, an invaluable instrument used by the protein research chemist.

PROBLEM

14.2. Compare the following two amino acids, lysine and serine.   **(a)** Draw out their ionized forms.   How many ionized species does each have?   **(b)** Which has the lower isoelectric pH?   **(c)** To which electrode would each migrate at pH 7?

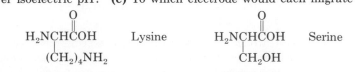

REACTIONS OF AMINO ACIDS: TOOLS FOR STUDY

14.3

The acid and amine groups of an amino acid are subject to the same chemical reactions we have described for carboxylic acids and amines. Certain reagents and reactions can be used to identify amino acids and proteins and measure their amounts.

A. The Ninhydrin Reaction

Probably the most common reagent for quantitative and qualitative amino acid analysis is ninhydrin. It is a strong oxidizing agent that causes oxidative decarboxylation of the amino acid. Each mole of amino acid reacts with a total of 2 moles of ninhydrin to produce a purple pigment that can be assayed colorimetrically at 570 nm.

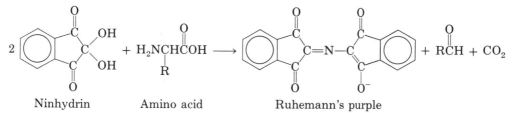

Ninhydrin   Amino acid   Ruhemann's purple

A yellow pigment instead of a purple one is produced with the two "imino" acids proline and hydroxyproline (see Table 14.1). It can be analyzed at 440 nm. The ninhydrin reagent is a very versatile one in that it can be used in a liquid analysis such as the amino acid analyzer, or it can be sprayed on an amino acid chromatogram or an electrophoretogram.

B. Sanger Reagent

The α-amino group can react with 2,4-dinitrofluorobenzene (DNFB), also known as Sanger reagent. The product, an *N*-(dinitrophenyl) amino acid (DNP-aa) is colored and can be readily identified using chromatography.

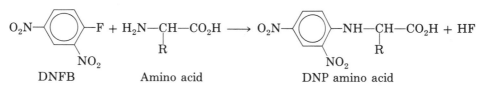

DNFB   Amino acid   DNP amino acid

The reaction is analogous to the nucleophilic substitution reaction covered in section 7.4, though not the same as that reaction.

PROBLEM

14.3. Draw the structures of the products of the reaction of dinitrofluorobenzene (DNFB) with **(a)** alanine, **(b)** tyrosine, **(c)** cysteine, and **(d)** serine.

C. Formation of an Amide Linkage;
The Peptide Bond; Proteins

The amine and carboxyl groups of an amino acid react as those typical functional groups do. This means that two amino acids can react together, eliminating the elements of water, to form an amide bond between them. In this case, it is called a *peptide bond*.

Peptide bond

The product is called a *peptide;* here it is a dipeptide (composed of two amino acids). If a peptide should contain three amino acids, it would be a tripeptide; four, a tetrapeptide; and so on. In general, they are all referred to as *polypeptides* (*poly,* "many"). As the chain increases to much larger proportions, with molecular weights in the thousands and even millions, the term *protein* is used.

A *protein,* then, is a condensation polymer, or more correctly a copolymer, with each monomer unit the same in fundamental structure but unique in its side chain. With at least 20 different monomer units, it is no wonder that the variety of proteins is potentially phenomenal.

The linear sequence of amino acids, usually written left to right from free amino to free carboxyl end, is known as the *primary structure* of a polypeptide or protein. Instead of writing out the complete chemical structure or full name of each amino acid in a polypeptide sequence, scientists find it convenient to use three-letter abbreviations in the interest of brevity. For example, bradykinin is a polypeptide (a nonapeptide) that aids in the regulation of blood pressure. Its linear sequence is expressed as

Amino end   Carboxyl end
N-terminus   Arg-Pro-Pro-Gly-Phe-Ser-Pro-Phe-Arg   *C*-terminus

The peptide bond can be broken by adding the elements of water back to the respective amino acids in the presence of acid or base (hydrolysis). The free amino acids are liberated and can be subsequently separated and identified.

In appropriately ionized forms.

PROBLEMS

14.4. How many different tetrapeptides are possible that contain the amino acids alanine, phenylalanine, valine, and arginine? Write the peptides using the abbreviations in Table 14.1.

14.5. Draw out the complete chemical structure for the polypeptide Lys-Tyr-Pro at low pH.

SEQUENCE DETERMINATION: PRIMARY STRUCTURE

14.4

One of the most important tasks of a protein chemist is to elucidate the linear sequence of amino acids in a polypeptide or protein chain—the primary structure. It is this primary structure that will ultimately determine the total three-dimensional structure and function of the macromolecule. Acid hydrolysis followed by separation on ion exchange columns and ninhydrin analysis will only give the amino acid content of a protein but will not specify how those units are arranged. How then is this accomplished?

A. End-Group Analysis

The easiest place to start a protein study is with the beginning and end of the peptide chain. The protein chemist will routinely identify the free amino and free carboxyl end groups, amino acids, of the intact polypeptide.

1. The Sanger Reagent. The reaction of free amino groups can be brought about with Sanger reagent, DNFB (Section 14.3.B), followed by acid hydrolysis, which will liberate the DNP amino acid (and all the other amino acids present as well). The DNP amino acid is yellow and can be identified using chromatography against a known sample.

$$\text{Ala-Pro-Phe-Leu} \xrightarrow{\text{DNFB}} \xrightarrow[\text{H}_2\text{O}]{\text{H}^+,} \underset{\text{Yellow}}{\text{DNP-Ala}} + \text{Pro} + \text{Phe} + \text{Leu}$$

2. Edman Degradation. The Sanger reaction only gives the *N*-terminal amino acid of a polypeptide and destroys the rest of the sequence. The Edman reaction derivatizes the *N*-terminus, and the derivatized amino acid can be removed leaving the rest of the polypeptide intact. This process can be repeated many times giving a sequence analysis of up to 20 amino acids. The whole process has been automated; one cycle of reactions and purifications takes from three to four hours.

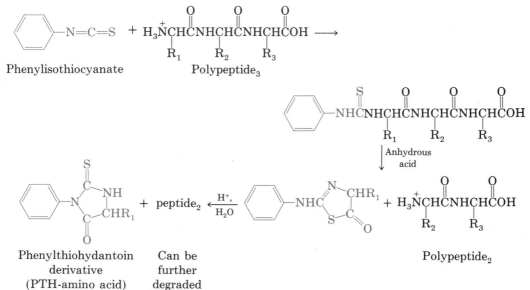

PROBLEM

14.6. Draw out the end products of three cycles of Edman degradation for the polypeptide Leu-Met-His-Ser.

---

3. Reaction with Hydrazine. The amino acid at the carboxyl end (*C*-terminal aa) of a polypeptide chain can be identified by causing the peptide to react with hydrazine. All peptide bonds will be cleaved and all the component amino acids will form hydrazides, except the *C*-terminal one (carboxyl end).

$$\underset{R_1}{H_2NCHC}\overset{O}{\underset{}{\parallel}}NHCHCOH \xrightarrow{H_2NNH_2} \underset{R_1}{H_2NCHC}\overset{O}{\underset{}{\parallel}}NHNH_2 + \underset{R_2}{H_2NCHCOH}$$

The unreacted *C*-terminal amino acid is identified by chromatography.

4. Enzyme Methods. Another method commonly used for *C*-terminal amino acids is carboxypeptidase, an enzyme that will hydrolyze the last peptide bond, liberating the final amino acid at the carboxyl end. Once the terminal amino acid is gone, the carboxypeptidase will begin to work on the penultimate, or next-to-last, unit, which has now become a new *C* terminus. Thus, this enzymatic method is one of determining relative amounts of free amino acids. The one in greatest molar amount within a specific, short time period is the *C* terminus.

B. Internal Sequences of Large Polypeptides

1. Enzymatic Fragmentation. End-group analysis is only a start at primary structure determination. The researcher must break down a very large polypeptide or protein into manageable fragments and then, using the reagents at hand, piece the puzzle back together. The most valuable tools for this detective work are themselves proteins. Those marvelous biological catalysts, the enzymes (section 14.8), are very specific in the direction of their activity. Trypsin, for example, will hydrolyze peptide bonds only at the carboxyl end of an amino acid

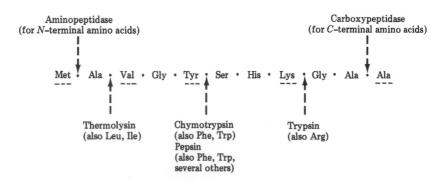

FIGURE 14.1. Polypeptide breakdown by enzymes. Most enzymes will hydrolyze the peptide bond at the carboxyl end of the amino acid residue for which they are specific. Thermolysin and carboxypeptidase, however, cleave at the amino end of the residue.

containing either a lysyl or arginyl (basic) side chain. Similarly, chymotrypsin will cleave at tyrosine, phenylalanine or tryptophan. Figure 14.1 illustrates the specificity of enzymes commonly used in sequence determination. The sequence of the smaller fragments can then be determined by the chemical methods previously mentioned. Peptide sequences can be pieced together by matching overlapping enzyme fragmentation patterns.

There are several other reagents used for peptide fragmentation and identification. We have only briefly covered a few of the common procedures.

PROBLEMS

14.7. Certain polypeptides, such as the antibiotic gramicidin, produce no free amino acids when incubated with aminopeptidase or carboxypeptidase. What does this imply?

14.8. A polypeptide, on acid hydrolysis, contained the amino acids Arg (1), Ala (1), Ile (1), Leu (2), Lys (1), Phe (2), Tyr (1). Treating the intact peptide with DNFB and subsequent hydrolysis gave DNP-Leu. Reaction with carboxypeptidase gave varying amounts of free amino acids, Phe > Leu > Ala. Digestion of the polypeptide with trypsin gave the fragments Tyr-Ile-Phe-Lys, Leu-Arg, and Ala-Leu-Phe. Chymotryptic treatment of the intact polypeptide produced Ile-Phe, Lys-Ala-Leu-Phe, and Leu-Arg-Tyr. What is the sequence of the nonapeptide?

2. Disulfide Bridges. Frequently, two cysteine residues are linked through their side chains, that is, through their sulfurs, to form a disulfide bond or bridge (see Table 14.1). This S—S bridge may occur between two cysteines in the same polypeptide chain or in two entirely different chains (Figure 14.2). To study the peptide sequence more conveniently, one must break the disulfide bridge and chemically tie up the free —SH group. Because S—S bond formation is an oxidative process, its cleavage must be effected by reduction.

The chemical treatment of hair, commonly known as a permanent, involves breaking down by reduction the interchain disulfide bonds of the hair and then physically rearranging the reduced hair fibers by curlers. This is followed by a "neutralization," when an oxidizing agent allows disulfide bonds to re-form—only they are now in different positions. The result is curl where the hair was previously straight or vice versa.

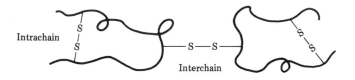

FIGURE 14.2. Disulfide bonds.

ARTIFICIAL POLYPEPTIDE SYNTHESIS

14.5   An ultimate consequence of the study of a large natural polymer like a protein, which has unique biological functions, would be to produce a synthetic polymer with the same amino acid sequence and, if possible, the same biological activity.   Ramifications of such a procedure could include altering the sequence in a predetermined way to effect a beneficial change in activity or one that would help further clarify the mode of action of a given protein or polypeptide.

Let us consider the difficulties that the relatively simple condensation of two amino acids into a dipeptide, Gly-Phe, would occasion.   For example, how many different combinations (dipeptides) could be realized from glycine and phenylalanine?

$$Gly + Phe$$
$$\downarrow$$
$$Gly\text{-}Gly + Gly\text{-}Phe + Phe\text{-}Gly + Phe\text{-}Phe$$

Four distinctly unique dipeptides are possible.   Also, how are we to prevent the higher polypeptides $(Gly)_n$ and $(Phe)_n$ and their combinations from forming?

One way to overcome some of these difficulties is to block (protect) one of the reactive ends of one of the amino acids, making it unreactive.   For example, we could block the amino end of glycine.   But this still does not preclude a polymer of phenylalanine, $(Phe)_n$, from forming.   Obviously, we must restrict the reactivity of the carboxyl group of phenylalanine to prevent phenylalanine from reacting with itself.   Another type of blocking agent is therefore used.

Finally, we can obtain the desired dipeptide by joining the appropriately protected amino acids and subsequently deblocking the *N*- and *C*-termini after peptide bond formation.

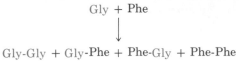

This has become the rationale used for the artificial synthesis of polypeptides.   By selectively blocking and deblocking either terminus, one can cause another amino acid to be added to the chain.   Very large polypeptides can be produced either by adding amino acids sequentially or by joining small peptide fragments.   These procedures are hampered by the same technical problems, that is, the yield of peptide at each step, separation of undesired reagents and products at each step, and purification of the desired complete polypeptide.

The chemical and technological problems of generating synthetic polypeptides were overcome by R. B. Merrifield and his associates through their development of a solid-phase method for synthesis.   As outlined in Figure 14.3, it involves chemically attaching the desired carboxy-terminal amino acid to insoluble polystyrene resin beads in such a way that the bond will resist cleavage during the sequential addition of amino acids, but can still be easily broken when the finished product has been assembled.   Each amino acid, with a protected amino group, is added and deblocked.   Since the chain is "growing" on an insoluble resin, it can be safely bathed in soluble reagents, washed, and separated by filtration.   The whole procedure has also been automated with a single cycle of amino acid linkage taking from three to four hours.   Bradykinin (section 14.3.C) and about 100 of its analogues, insulin, and numerous other biologically active polypeptides have been synthesized in this way.

THE THREE-DIMENSIONAL STRUCTURES OF POLYPEPTIDES AND PROTEINS

14.6   In protein molecules the linear sequence of amino acids, ultimately of greatest importance in the overall structure, is but a starting point for the conceptual picture of a whole polymer of molecular weight 10,000 to 4,000,000.   As a polypep-

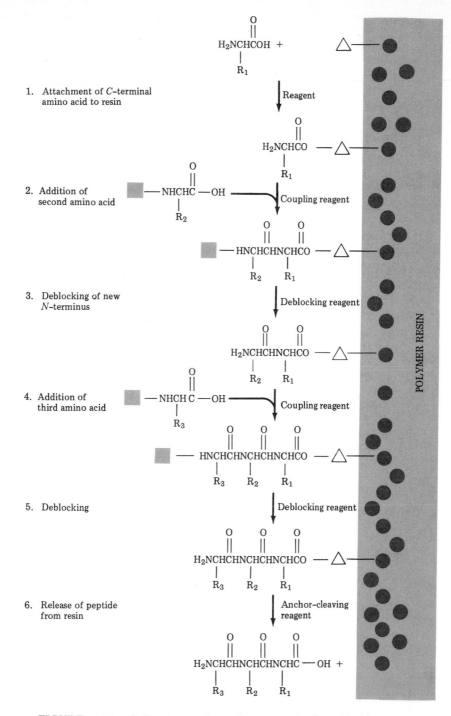

FIGURE 14.3. Solid-phase polypeptide synthesis. □ = blocking group; △ = anchor group; ○ = polystyrene beads.

tide grows, certain factors impart distinctive convolutions to the structure, eventually producing anything from an elongated muscle fiber to an amorphous (globular) hemoglobin molecule—totally different forms.   Let us consider these various contributions.

A. Primary Structure

Primary structure is but a synonym for the linear amino acid sequence.   The position of each amino acid determines where disulfide bonds will form, whether ionic (Asp, Glu, Arg, Lys) or nonpolar (Trp, Phe, Ile) side chains can come into proximity, or whether other species outside the polypeptide chain can approach or bind to it.   It is also important in studying the evolution of protein species—how molecules of similar function in the evolutionary hierarchy compare in sequence and how various amino acid modifications affect overall biological activity.

For example, Figure 14.4 shows the amino acid sequence for insulin, a glucose-regulating hormone.   Differences in amino acids at various positions in the two chains are few when comparing such diverse species as man and horse or elephant.   This similarity in structure results in a similarity in biological function so that beef insulin, for example, can be used by humans.

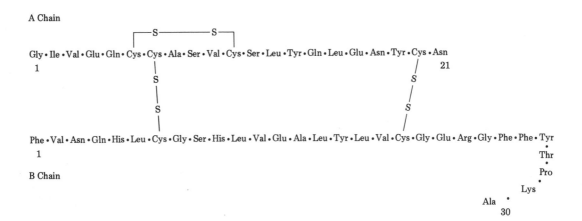

FIGURE 14.4. The primary sequence of bovine (beef) insulin. Species variations in linear sequence are minor. Human insulin contains a Thr at position 8 of the A chain and a Val at 10. In the B chain humans have a Thr at position 30.

B. Secondary Structure

The possibility of hydrogen-bonding, either intrachain or interchain, gives rise to two main three-dimensional arrangements, the alpha (α) helix and the beta (β) pleated sheet.

Recall that hydrogen-bonding can occur between hydrogens bonded to the highly electronegative elements O, N, and F and those elements.   The peptide bond contains an electronegative oxygen and an electropositive hydrogen attached to the amide nitrogen.   The normal twisting of the peptide chain at the carbon between the carbonyl and nitrogen will allow hydrogen-bonding between every four amino acids, that is, between the hydrogen attached to the nitrogen of one peptide bond and the carbonyl oxygen of the third amino acid past it (Figure 14.5a).

Linus Pauling and R. B. Corey are responsible for our knowledge of the actual structure of the peptide bond and α helix. Using X-ray diffraction studies, they postulated the α-helical arrangement after studying the relation between α-keratin and the restrictions of the peptide bond. Alpha-keratin, found in hair, feathers, horns, hoofs, wool, and scales, is composed of bundles of intertwined α helices. The individual chains are stabilized by hydrogen-bonding, and then the chains are linked covalently by disulfide bridges.

Human hair can be stretched when heated to twice its length without breaking, indicating that the α helix is not the most extended chain. But the stretched fiber no longer shows an α-helical arrangement. Rather, the X-ray diffraction pictures indicate a structure similar to that of silk fibroin, a β-keratin. The structure is called a β-pleated sheet. The chains in an α-helix bundle are arranged in the same direction; that is, all the N-termini are on the same end. In a β conformation, the strands are arranged head-to-tail (Figure 14.5b) and the hydrogen-bonding occurs between chains rather than within the same chain. The side chain

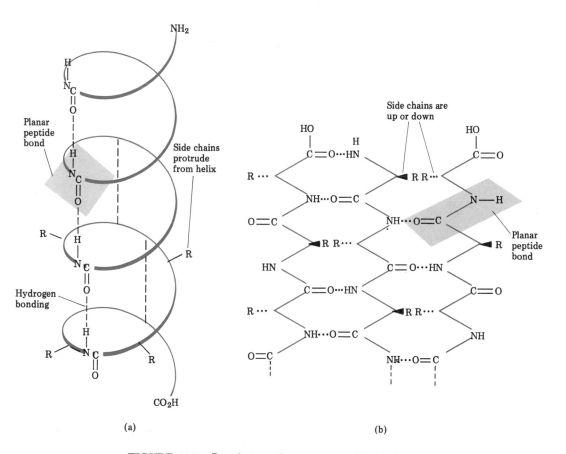

FIGURE 14.5. Protein secondary structure. (a) α-helix-intrachain hydrogen-bonding. (b) β-pleated-sheet-interchain hydrogen-bonding.

R groups stick up and down away from the pleated sheet structure. This arrangement resists stretching but allows flexibility; hence the texture of silk. The prevalent amino acids in such a conformation are found to be glycine, serine, and alanine, all with small side-chain groups that allow a stable β structure and stacking of chains.

These two variations involving hydrogen-bonding constitute what is known as *secondary structure.* Proteins do not have to show exclusively α-helical or β conformation. Rather they may show both (or neither) to varying degrees along with a more random, free orientation. The "breaking" or "bending" of an α helix will occur, for example, whenever a proline or hydroxyproline residue appears in the amino acid sequence due to their cyclic structures. In addition, various other amino acids will tend to destabilize α-helical conformations and the β structure. Hydrogen-bonding will again come into consideration in our discussion of the next level of protein superstructure.

C. Tertiary Structure

1. Classification of Tertiary Structure. Alpha-helical segments, sections arranged in β conformation, random coil—these are the constituents of the overall shape of an entire protein molecule. In general, proteins are shape-classified as either fibrous or globular, which are very descriptive terms designating either long, tubular structures (mainly α-helix or β-pleated sheet) or almost spherical ones (mixtures). There are various forces holding a protein in its predominant native (natural) conformation. These include hydrogen-bonding not only for secondary structures such as the α helix and β pleated sheet but also hydrogen-bonding between sections of the polypeptide chain that may be separated widely in linear amino acid sequence but close physically because of the folding of the long chain.

2. Interactions Contributing to Tertiary Structure. In addition to hydrogen-bonding between various portions of the amino acid chain, ionic interactions can occur between charged side chains (salt bridges) in different segments of the molecule or between side chains and a polar medium in which the protein may be dispersed. The contrary situation to these salt-type interactions also is evident. That is, nonpolar side chains such as tryptophan, valine, and leucine may form hydrophobic interactions, eliminating any contact with a polar solvent and becoming part of a nonpolar core. (Figure 14.6 summarizes these contributions to the tertiary structure or conformation of a protein molecule.) Added to these are the disulfide bridges that occur between cysteine residues and covalently hold the protein in its 3-dimensional shape.

How do we know that all these interactions can occur within such a large and complicated structure? It would be convenient if protein chemists had microscopes powerful enough to see an actual individual amino acid side-chain residue, but nothing like that exists—yet. A type of "picture" can be taken, though, by X-ray diffraction, and the thousands of diffraction spots can be mathematically analyzed to give a fairly good picture of the convoluted protein molecule. Myoglobin, which stores oxygen in muscle cells, was the first protein studied in this way. Figure 14.7 illustrates the structural model drawn from the X-ray diffraction pattern analysis. Ribonuclease (an enzyme that hydrolyzes ribonucleic acid), lysozyme (an enzyme, found in mucus, that hydrolyzes bacterial cell walls), and α-chymotrypsin and trypsin (intestinal enzymes) are among the macromole-

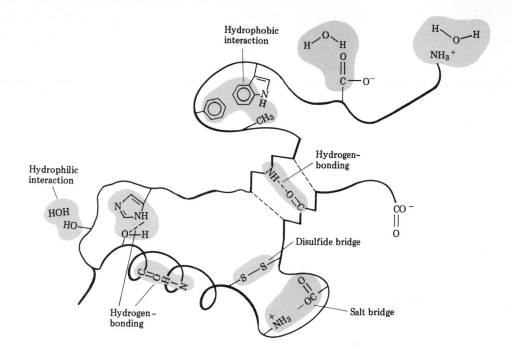

FIGURE 14.6. Forces stabilizing the tertiary structure or conformation of a protein molecule.

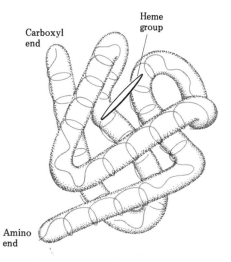

FIGURE 14.7. Myoglobin containing 153 amino acid residues and noncovalently attached heme group. Hypothetical structure based on X-ray data.

cules so far analyzed. These X-ray pictures have been confirmed by chemical evidence studying the various environments within such a large molecule.

3. Disruption of Tertiary Structure; Denaturation. Because of the varied factors stabilizing a protein's biologically active (native) tertiary structure, there are, in turn, a variety of outside influences that can and will disrupt the conformation. The process is called *denaturation* and results in temporary or permanent loss of biological activity. The most common method of denaturation is heating. An increase in temperature will increase the kinetic energy of molecules, eventually causing the vibrational disruption of stabilizing interactions. This type of denaturation is usually irreversible, as when one boils an egg. The white of the egg, containing the protein albumin, coagulates, changing from a clear, gelatinous material to an opaque white solid. This is not to say that all proteins are subject to heat denaturation. Certain bacteria have been found to thrive in the almost boiling waters of hot springs. Denatured protein materials are easier to digest, and so high-protein foods, such as meats, are cooked before eating.

Proteins show native conformations within a fairly narrow pH range. Extremes of acidity or alkalinity will affect the ionization of charged amino acid side chains, disrupting ionic interactions. Likewise, agents that show strong hydrogen-bonding, such as ethyl alcohol or urea $\left(\begin{array}{c} H_2NCNH_2 \\ \parallel \\ O \end{array} \right)$, or nonpolar tendencies will compete with various appropriate side chains and destabilize the functional protein structure.

PROBLEM
_____

14.9. Explain the following phenomena in terms of protein denaturation: **(a)** microwave ovens; **(b)** whipped egg whites; **(c)** curdling in milk caused by lactic acid-producing bacteria; **(d)** isopropyl alcohol (rubbing alcohol) used as an antibacterial agent.

D. Quaternary Structure

There is yet one higher level of organization in protein structure—the noncovalent assembly of individual polypeptide chains, each subunit lacking biological activity but together forming a viable, active species. This aggregation of subunits is referred to as *quaternary structure*. Hemoglobin is a prevalent example of a protein consisting of four polypeptide chains, two α chains and two β chains, each resembling myoglobin in structure. The total weight of *tetra*meric (four subunits) hemoglobin is about 64,000, with each subunit weighing approximately 16,000. There are other oligomeric (different subunits) proteins, but as might be expected, their structures are much more difficult to elucidate than those with only one polypeptide chain.

It must not be forgotten that many proteins, regarded as having only one subunit, are actually composed of several polypeptide chains covalently linked through disulfide bonds. Insulin, chymotrypsin, and antibodies are examples.

CONJUGATED PROTEINS

14.7   As you can see, not only does the existence of 20 or more different amino acids give infinite variety to the number of proteins and polypeptides possible but also the differences in conformation and assembly (tertiary and quaternary structure) can impart even more diversity. Now add to the roster of variation the fact that polypeptides may combine either covalently or noncovalently with nonprotein factors to form conjugated proteins. The nonprotein portion is called a *prosthetic group* or *cofactor* (*coenzyme* if specifically working with an enzyme). The prosthetic group may be a carbohydrate, lipid (fat), metal ion, nucleic acid, vitamin, or combination of one of these with something else such as the iron protoporphyrin (heme) group in the hemoproteins hemoglobin, (section 14.9.A), myoglobin, and cytochrome C. It may or may not be attached covalently to its protein portion. Thiamine, or vitamin B, is an example of a water-soluble cofactor which aids enzymes in their catalytic role.

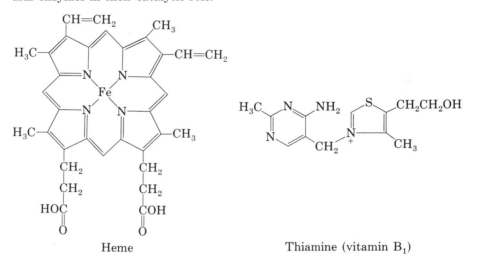

Heme   Thiamine (vitamin B_1)

ENZYMES

14.8   The catalysis of biological reactions is probably the most important task that nature has assigned the protein. Recall that a catalyst is a substance that alters the rate of a chemical reaction without itself being altered overall during the reaction. We have covered numerous examples of metal, acid, base, free-radical, and light-catalyzed reactions. Enzymes are proteins because of their component amino acids. They aid in producing the most profound expressions of human spirit—natural (cholinesterase in nerve transmission) and bottled (sucrase and zymase in fermentation).

A. Types of Enzymes

In general, enzymes are identified by the suffix *-ase*. Some older known enzymes were named for their source or first-discovered function and the common name has persevered. For example, the enzymes trypsin, chymotrypsin, and pepsin derive their names from Greek roots meaning "to wear down," "juice," and "digestion," respectively. All three are involved in the gastrointestinal digestion of proteins. Recently, enzymes have been given names suitable to their function. Very often an enzyme will have both common and systematic names.

The classification and systematic naming of enzymes embraces six main categories pertaining to their specific mode of action. Table 14.2 summarizes the system of categorizing enzymes. As you can see, these biological catalysts carry out and influence a phenomenal variety of reactions.

TABLE 14.2. ENZYME CLASSIFICATION BY INTERNATIONAL ENZYME COMMISSION: A SUMMARY

Class 1 *Oxido-reductases* carry out and influence oxidation-reduction reactions with alcohols, carbonyls, carbon-carbon double bonds, amines, etc.

Class 2 *Transferases* facilitate the transfer of certain functional groups such as carbonyl, acyl, sugar, alkyl, and phosphate groups.

Class 3 *Hydrolases* catalyze the hydrolysis of esters, ethers, peptide bonds, glycosidic bonds, halides, acid anhydrides, and more.

Class 4 *Lyases* allow addition reactions with carbon-carbon double bonds, carbonyls, etc., or form such bonds themselves.

Class 5 *Isomerases* promote isomerization, optical and geometric, and also catalyze various intramolecular reactions, resulting in skeletal isomerization.

Class 6 *Ligases* aid in bond formation between carbon and sulfur, oxygen, nitrogen, and another carbon.

B. Mode of Action

To focus on another facet of these concepts, each enzyme is specific—first, for a certain type of chemical reaction; and second, for particular reagents, called *substrates*. What is more, when two reactions, differing only in catalyst used—one with a conventional catalyst, such as a metal, and the other with a biological catalyst, or enzyme—are compared in rate, the enzyme-catalyzed experiment is faster by a factor of 10^3 to 10^7. So the enzyme chemist has primarily focused on two questions: What makes an enzyme so specific? Why is the catalysis so efficient?

The specificity of an enzyme depends both directly and indirectly on its primary structure, that is, its linear amino acid sequence. In most enzymes, specific amino acids are involved in binding the substrate molecule and in catalyzing its particular chemical reaction. For example, trypsin cleaves peptide bonds at the carboxyl end of a lysyl or arginyl residue.

Trypsin

Phe-Val-Lys—Ala-Ala-Gly

The enzyme has to "recognize" and then grab onto the polypeptide at the lysyl (or arginyl) group, a basic group, before it can proceed to catalyze the hydrolysis of the bond. There must be, then, a binding site on the enzyme consisting of amino acid side chains that would attract and interact with the basic side chains of the substrate (see Figure 14.8). For the hydrolysis of the peptide bond to proceed, an amino acid or several amino acids are needed that will react with a group—in this case, the carbonyl group of the bond. The hydroxyl oxygen of serine, or the thiol group of cysteine, is electron-rich, or nucleophilic, and will react with the electrophilic carbonyl carbon in a peptide bond. Other amino acids can donate or accept

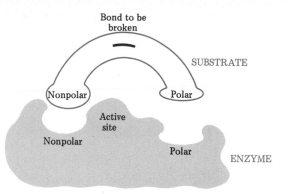

FIGURE 14.8. Simplified picture of a possible enzyme-substrate interaction in a cleavage-type reaction.

protons to help catalyze the hydrolysis of the amide linkage. This whole process, while very complex, is much more efficient than an ordinary chemical reaction, since the participating molecules are effectively embracing each other, and all the elements necessary for the reaction to occur are close to the action.

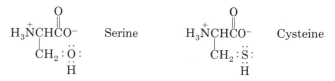

While some enzymes have a very small and selective number of substrates, others are effective over a much broader range. Again the tertiary structure (as dictated by the primary sequence) will closely regulate the binding sites and other interacting factors such as the physical size of various sites and the stereospecificity of the sites. Most enzymes are highly specific toward the optical isomer they will accept as a substrate. Because amino acids are optically active themselves, it follows that as components of enzyme macromolecules, they impart the effects of optical activity to the polymer (Figure 14.9).

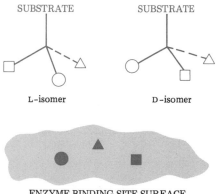

FIGURE 14.9. Enzyme stereospecificity. Only the D-isomer can bind to this "enzyme."

The specificity of an enzyme's action can also be its own demise. Poisons are usually elements, ions, or compounds that can bind or react irreversibly with key enzymes, blocking their life-sustaining catalysis. Pesticides (such as Malathion, Diazinon, or Systox; see section 7.5.B) are often formulated with a cholinesterase inhibitor, which mimics the natural substrate or ties up the enzyme, preventing its normal function in the transmission of nerve impulses. Mercury(II) ion, Hg^{2+}, present in antiseptics and fumigants and fungicides, will react with —SH groups of amino acid side chains. This reaction results in the inactivation of enzymes in which the sulfhydryl is part of the active site, and it can actually precipitate various proteins out of cellular solution. Arsenic, lead, and silver can also act to inhibit sulfur enzymes in this way. Nerve gases, such as diisopropylfluorophosphate, will form a stable enzyme complex with an active-site serine or cysteine, thereby inactivating two classes of biological catalysts. Antidotes to various enzyme-specific poisons are often designed in turn to imitate the site of attack. Penicillamine, for example, and its derivatives have been used to alleviate poisoning by mercury salts. Notice the —SH (thiol) group which can take the place of a cysteine and react with the poison.

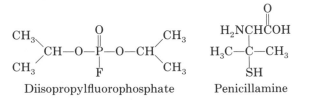

Diisopropylfluorophosphate   Penicillamine

PERTINENT PROTEINS AND POLYPEPTIDES OF DIVERSE FUNCTION

14.9   An infinite variety of structures is afforded by the polymerization of 20 or so amino acid units. Knowing the primary functions proteins undertake, we can deal in detail with a few proteinaceous materials that hold key metabolic positions. All are yet subject to intense research for full understanding.

A. Hemoglobin

Hemoglobin is but one of many proteins present in the blood of vertebrates. It is specifically located in the red blood cells, or erythrocytes, and has the principal function of transporting molecular oxygen, O_2, from the lungs to cells throughout the body. Normal adult hemoglobin is represented as HbA and is a conjugated tetrameric protein (section 14.7). A hemoglobin molecule is globular or roughly spherical, and consists of four chains—two identical α chains each with 141 amino acids and two identical β chains each with 146 amino acids. The four chains or subunits are not covalently attached and have a total molecular weight of about 64,500. The nonprotein portion of HbA is a protoporphyrin group also known as heme. Iron, in the form of Fe^{2+}, is bound in the center of the heme, and this whole ferro-protoporphyrin group is nestled in a hole, or crevasse, formed in each HbA subunit. Molecular oxygen can complex in a reversible reaction with the Fe^{2+}. Therefore, a total of four molecules of O_2 can be bound and transported by each hemoglobin molecule. Not only is O_2 bound by HbA but carbon monoxide, CO, and hydrogen cyanide, HCN, are also capable of binding. In fact,

CO binds 200 times better than O_2 does. Hydrogen cyanide, though even more tightly bound than CO to Hb, has its most detrimental effect in the electron transport system, where it more quickly binds to the cytochromes, electron transport agents that are also iron porphyrin–protein complexes. This essentially stops metabolism—a fatal situation.

A most dramatic deviation in conformation is the cause of a fatal genetic disease, sickle-cell anemia. Originating in the populations of Central Africa and Southern India, sickle-cell anemia affects more than 50,000 Americans, mostly of black ancestry. A person may carry the gene that could produce a mutant hemoglobin, HbS, in which *one* amino acid in each of the β chains, glutamic acid in position 6, is replaced with a valine. Someone with only one such gene is said to have sickle-cell trait and produces normal HbA as well as HbS. However, should two persons with sickle-cell trait have an offspring with two deviant genes, sickle-cell anemia will develop. The amino acid replacement in HbS has little effect when the blood flows through areas of high oxygen content (pressure), such as the arteries and lungs. The blood cells assume their normal disc-like shape (Figure 14.10a). However, as blood flows into the extremities, where oxygen pressure is low, the HbS loses its solubility and shape, and literally precipitates in the red cells. The entire cell collapses, resulting in a permanently sickled, rigid appearance (Figure 14.10b) and clogging the capillaries and arterioles. Clinically observed symptoms include heart damage because of increased strain trying to pump the plugged-up blood, gangrene, excessive excretion of sickled blood cells, which are highly susceptible to rupture and destruction by enzymes, and kidney and brain damage due to overall poor circulation. Persons with the sickle-cell trait are more resistant to the fatal effects of a type of cerebral malaria, and the gene most likely mutated as an adaptation to the prevalence of malaria in the original populations.

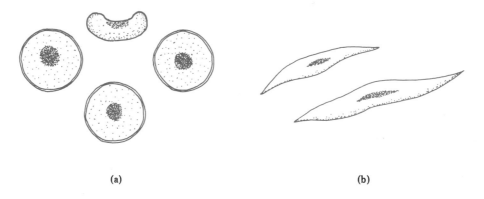

(a)   (b)

FIGURE 14.10. Blood cells. (a) Normal cells, disc-shaped. (b) Sickled cells.

Approximately 300 variations of hemoglobin have been discovered. Some, like HbS, have fatal symptoms and others are harmless anomalies. Much research is going on to reverse the deleterious conformational effects of HbS. It has been termed a disease to be cured by chemistry because it is caused by the abnormal chemistry of hereditary (nucleic acids) and protein materials.

PROBLEM

14.10. Why could the substitution of valine for glutamic acid in hemoglobin cause the observed decrease in solubility?

B. Polypeptide Antibiotics

One of the most significant occurrences in modern drug research was the discovery of penicillin and numerous other antibiotics. The term *antibiotic* strictly means "against life," and as a class, these compounds either severely restrict or stop altogether the growth of bacteria that cannot be otherwise interrupted by the body's own defense system. While antibiotics come in many chemical structures, some are polypeptides or analogues of peptides. Can you detect, for example, the two amino acid analogues present in penicillin?* (*Hint:* Use the dotted lines to separate the components.)

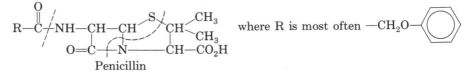

Penicillin

Certain bacteria produce polypeptide antibiotics with some unique features. D-amino acids and cyclic polypeptide chains introduce a unique character to their structures. Figures 14.11 and 14.12 illustrate a few of these interesting compounds.

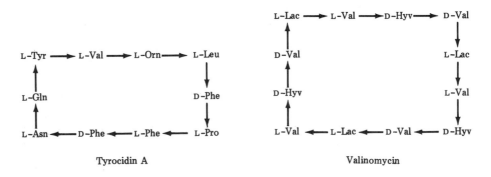

Tyrocidin A   Valinomycin

FIGURE 14.11. Polypeptide antibiotics. These polypeptides, synthesized by bacteria, show the inclusion of D-amino acids, some amino acid analogues, and cyclic structures. Lac = lactic acid. Hyv = α-hydroxy valeric acid.

---

* Penicillin exerts its effect by inhibiting the formation of cross-linkages in bacterial cell-wall molecules. Although penicillin was discovered by Dr. Alexander Fleming in 1929 in England, its use was not fully developed until the 1940s because of difficulty in isolating it from mold in sufficient quantities. In 1942 this antibiotic was given its first human test on a woman whose streptococcal infection following a miscarriage had resisted all possible treatment. As a last resort she was given injections of the then experimental penicillin, and within 12 hours her body temperature went from over 105 °F to normal. Now that the substance had been tested under fire, the chief problem became producing increased quantities. A moldy cantaloupe in a Peoria, Illinois, market was the serendipitous find of a researcher. It proved to be the source of a new penicillin strain of much higher antibiotic yield than Fleming's original mold.

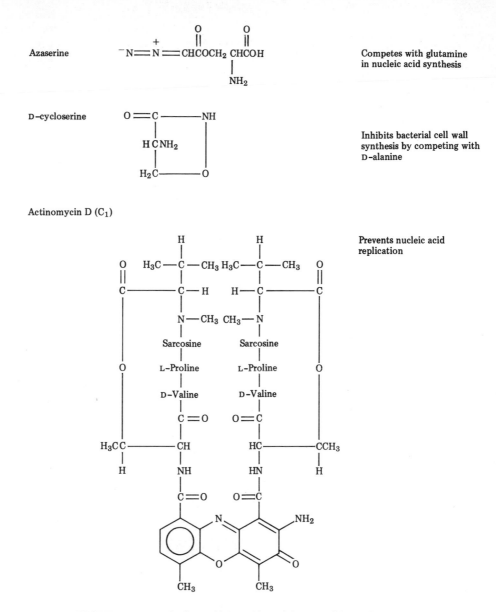

FIGURE 14.12. Amino acid (peptide) analogues with antibiotic activity.

Antibiotics often function by inhibiting the synthesis of bacterial cell walls. In many cases, this is because the antibiotic mimics the normal substrate of a synthesizing enzyme; this happens with D-cycloserine (see Figure 14.12), which is structurally similar to the D-alanine existing in cell walls. Other antibiotics may prevent correct reading of the genetic code (Chapter 15), interfere with the electron transport system, or in some way interrupt key metabolic reactions.

Other types of structures also prevail for antibiotics.

C. Protein and Peptide Hormones

Hormones are substances secreted by endocrine glands that regulate the various metabolic functions of the body. The exact methods or mechanisms by which they accomplish this inconceivably complex task is still being studied. The structures of numerous hormones and the effects of hormonal activity or lack of activity are known, however. The most familiar of these substances are the steroid hormones, specifically the sex hormones (section 12.7.C). Here, let us briefly consider some polypeptide hormones, especially insulin, epinephrine, and oxytocin and its sister, vasopressin.

1. Insulin. Insulin regulates the metabolism of glucose and indirectly also affects the metabolism of fats and proteins. Until several years ago, it was believed that lack of insulin was the chief cause of diabetes, and that artificially increased insulin levels either by injection of insulin itself or by stimulated secretion of insulin with drugs was the way to treat the symptoms of the disease most effectively. Recently this simple theory has been modified drastically by the discoveries that (a) diabetes is not always due to a lack of insulin and (b) at least two other peptide hormones, glucagon and somatostatin, are also involved in glucose metabolism. Somatostatin appears to be a regulator of insulin and glucagon activity. There is also growing evidence that what is called "juvenile-onset" diabetes may be caused by a virus that affects only those who are genetically predisposed. More about diabetes can be found in Panel 14 in Chapter 10.

Insulin is secreted by the β cells of the pancreas in an inactive form, called proinsulin, which contains 81 amino acids. Various enzymes cleave the single-chained molecule, excising a large (30 amino acid) segment and leaving what is known as active insulin (Figure 14.4). This active molecule is released into the blood when glucose is present in high concentration (along with other factors). In some way, the insulin enhances the movement of glucose across the cell membrane. The action of insulin is not limited to this one function. It also stimulates the activity of certain enzymes while inhibiting others in a complex metabolic interplay.

PROBLEM

14.11. The pancreas contains only about 10 mg of insulin, and the amount secreted daily into the blood is 1 to 2 mg. If the molecular weight of insulin is 5500, how many moles and how many molecules are in the blood? The total volume of blood is about 6 liters. What is the molarity of insulin in the blood?

2. Glucagon. Glucagon, with 29 amino acid residues (Figure 14.13), is also secreted by the pancreas (α cells) in an inactive form (37 amino acids). This hormone has an action opposed to that of insulin. It acts on the liver to break down glycogen to glucose whenever blood sugar levels get below normal, and so restores blood sugar to normal amounts (about 65–95 mg/100 ml).

PROBLEM

14.12. From the structure of glucagon would you expect it to migrate to the anode (−) or cathode (+) during electrophoresis at pH 2? Explain your answer.

His·Ser·Glu·Gly·Thr·Phe·Thr·Ser·Asp·Tyr·Ser·Lys·Tyr·Leu·Asp·Ser·Arg·Arg·Ala·Glu·Asp·Phe·Val·Glu·Trp·Leu·Met·Asn·Thr

     10   20

FIGURE 14.13. The structure of glucagon.

3. Epinephrine. Epinephrine is an amino acid analogue hormone that also has effects antagonistic to those of insulin. It is secreted in the adrenal medulla, the central portion of the adrenal gland (hence the common name *adrenalin*), and has a relatively simple structure.

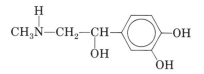

The L-isomer causes the breakdown of glycogen to glucose (as glucagon does) in the liver and to lactate in muscle tissue. In addition, the classic responses include an increase in blood pressure and heart rate and the contraction and relaxation of various muscles. The sequence of reactions that end in these symptoms is complex.

4. Oxytocin and Vasopressin. Oxytocin and vasopressin are two polypeptide hormones (Figure 14.14) of extremely similar structure, secreted by the pituitary gland located at the base of the brain. Oxytocin regulates uterine contraction and lactation and is often used to induce delivery. Vasopressin controls the excretion of water by the kidneys and also the elevation of blood pressure. In the disease *diabetes insipidus* there exists a deficiency of vasopressin, resulting in the excretion of too much urine. The hormone can be injected to relieve these symptoms.

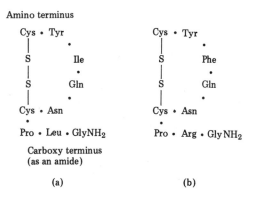

FIGURE 14.14. The amino acid composition of (a) bovine oxytocin and (b) vasopressin.

There exist a great number of polypeptides with hormonal or hormone-regulatory activity. They range in size from 3 amino acids (thyroid-hormone-releasing factor) to almost 200 (191 amino acids for prolactin which regulates milk production in the mammary glands).

PROBLEM

14.13. From the structures of oxytocin and vasopressin in Figure 14.14, is there any way to distinguish between the two dependent upon their structures.

CONCLUSION

14.10   Protein chemistry is a complex and continuing area of research that requires the talents and coordination of almost every aspect of the physical and medical sciences. The field of immunochemistry, that is, antigen-antibody structures and interactions, is much too large to be synopsized. We hope the background given here will encourage the interested student to read further in biochemistry.

END-OF-CHAPTER PROBLEMS

14.14 Terms: Define and illustrate the following terms relating to amino acids: **(a)** α-amino acid; **(b)** L-amino acid; **(c)** zwitterion; **(d)** isoelectric pH; **(e)** basic amino acid; **(f)** nonpolar amino acid.

14.15 Structure: Draw the ionized structures of the amino acids arginine and glutamic acid under varying conditions of pH. Indicate the zwitterion structure of each.

14.16 Structure: Identify the amino acids having the following characteristics: **(a)** optical inactivity; **(b)** a phenolic group; **(c)** involvement in interchain and intrachain bridging; **(d)** two structural isomers; **(e)** responsibility for bending a peptide chain and "breaking" a helical structure; **(f)** hydrogen-bonding through an R side-chain group.

14.17 Structure: How will glutamine (Gln) differ in isoelectric point from glutamic acid?

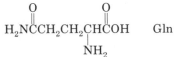

Gln

14.18 Reactions: Write the principal products of the following reactions, elaborating the structure:

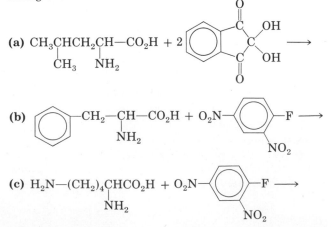

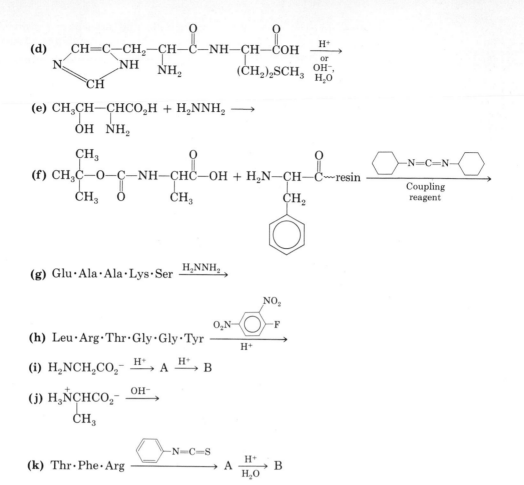

(d)

(e) $CH_3CH-CHCO_2H + H_2NNH_2 \longrightarrow$
       $\quad\quad$ OH $\;$ NH$_2$

(f)

(g) Glu·Ala·Ala·Lys·Ser $\xrightarrow{H_2NNH_2}$

(h) Leu·Arg·Thr·Gly·Gly·Tyr $\xrightarrow{H^+}$

(i) $H_2NCH_2CO_2^- \xrightarrow{H^+} A \xrightarrow{H^+} B$

(j) $H_3\overset{+}{N}CHCO_2^- \xrightarrow{OH^-}$
       $\quad\quad$ CH$_3$

(k) Thr·Phe·Arg $\longrightarrow$ A $\xrightarrow[H_2O]{H^+}$ B

14.19 Structure: α-Amanitine is a polypeptide analogue that is the deadly component of a type of poisonous mushroom, *Amanita phalloides*. From its structure, try to identify the component amino acids and any novel linkage (besides the α-amino-carboxyl peptide bond).

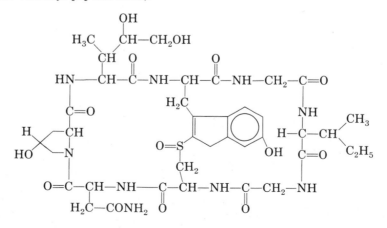

14.20 Protein Structure: Where do the following terms fit in a protein's hierarchy of structure (as 1°, 2°, 3°, or 4°): **(a)** the α and β subunits of hemoglobin; **(b)** Phe-Val interactions; **(c)** intrachain hydrogen-bonding; **(d)** linear sequence of amino acids; **(e)** salt-type bridges; **(f)** disulfide bridges?

14.21 Structure: What kinds of interactions would the following pairs of amino acids be involved with through their side chains: **(a)** Trp-Ile; **(b)** Ser-Tyr; **(c)** Phe-Gly; **(d)** Val-Thr; **(e)** Asp-Arg; **(f)** Glu-Water?

14.22 Protein   Interactions: Urea, $H_2N-\overset{\overset{\displaystyle O}{\|}}{C}-NH_2$,   and   guanidine, $H_2N-\overset{\overset{\displaystyle NH}{\|}}{C}-NH_2$, are used to denature proteins temporarily for chemical studies. How could these two substances distort a protein's native conformation?

14.23 Peptide Sequence: A peptide with the amino acid composition presented herewith (alphabetically) was analyzed as indicated. What is the linear sequence of the peptide?

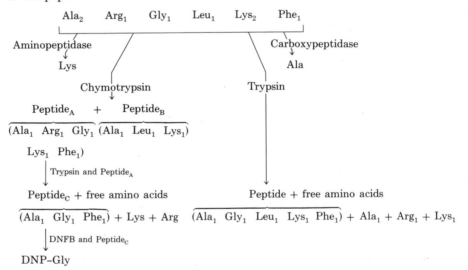

14.24 Enzymes: What is an enzyme? How would extremes of temperature and pH affect enzyme activity?

14.25 Peptide Synthesis: Show with equations how the peptide Gly-Ala-Phe can be prepared by the Merrifield synthesis.

Chapter 15

Nucleic Acids

15.1   Carbohydrates supply the energy necessary for running the complex machinery of a living organism while also acting as part of the support structure needed to contain that machine. Proteins are the cogs and wheels that make the machine perform its assigned tasks. Fats give more structure and energy and added resiliency. But what constitutes the computer needed to direct the intricacies of the living organism? What will determine whether raw material will be used as energy, wheel, or structure, and in what part of the sequence the substance should be located? Why is a blood cell just that and not muscle tissue? Could a yeast bacterium reproduce the cellular structure of a green plant? These queries lead us to a fourth class of biologically imperative compounds, the nucleic acids.

   The term *nucleic* refers to the nuclei of cells, which is where these substances were first found by Miescher in 1869. Not until the 1940s, though, did scientists begin to uncover the significance of these polymeric materials. Since that time, research on them has progressed at an unbelievable pace, with awe-inspiring ramifications for the future of not only the scientific world but philosophy, economics, and politics. Nucleic acids direct the synthesis of every substance in the body as well as hold the blueprints for the reproduction of themselves.

CHEMICAL STRUCTURE

15.2   The nucleic acids contain the elements carbon, oxygen, nitrogen, hydrogen, and phosphorus. They are biopolymers, macromolecules with molecular weights in the millions, tens of millions, and hundreds of millions. Yet nucleic acids, like polysaccharides and proteins, have a specific, repetitive structure that can be analyzed.

A. General Structure

All nucleic acids can undergo stepwise degradation to three fundamental units: a basic heterocycle (a cyclic compound which contains elements other than carbon in the ring), a carbohydrate (a sugar), and phosphoric acid. Partial decomposition gives a mixture of nucleotides. These, in turn, will give phosphate and nucleosides; and nucleosides are degradable to a heterocycle and a sugar. Rebuilding these units will give us an idea of the copolymeric nature of the huge biomolecule.

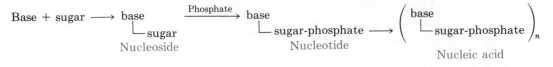

The obvious questions now are what the structures of the base and sugar are and how the monomer (nucleotide) units are linked. This information will depend on the kind of nucleic acid we are discussing. There are two principal types of nucleic acid—deoxyribonucleic acid, or DNA, and ribonucleic acid, or RNA.

B. DNA

Deoxyribonucleic and ribonucleic acids are named for the characteristic sugar unit found in each. Deoxyribonucleic acid contains the pentose sugar 2-deoxy-D-ribose shown here in both the Fischer and Haworth forms. Note the absence of a hydroxy group at the second carbon (hence the term *deoxy*).

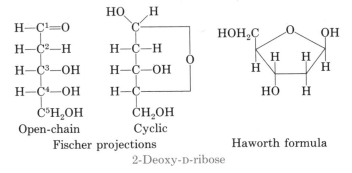

The bases found in DNA are four, two of which are structurally derived from the basic heterocycle purine and two from pyrimidine.

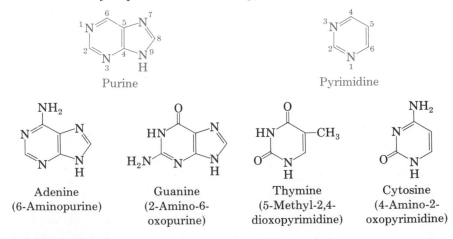

PROBLEMS

15.1. What makes purine, pyrimidine, and their DNA derivatives basic?

15.2. What type of secondary interaction might the DNA bases be capable of due to their basic nature?

Recall that a sugar and a base combine to form a nucleoside. This linkage occurs between the 9 position of the purine base or the 1 position of the pyrimidine and the C_1 of the deoxyribose unit (the anomeric or hemiacetal carbon). So that the numbering of the bases can be distinguished from the numbering of the sugars, the carbons of the sugars are primed—1′, 2′, 3′, etc. The bond is therefore to $C_1′$ of the sugar.

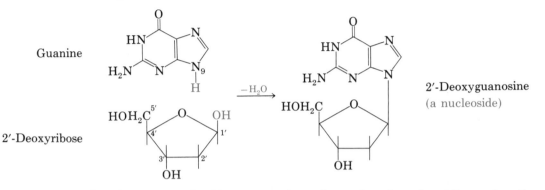

Guanine

2′-Deoxyribose

2′-Deoxyguanosine
(a nucleoside)

The other DNA nucleosides are named as 2′-deoxyadenosine, -thymidine, and -cytidine.

PROBLEMS

15.3. What is the configuration at the anomeric (1′) carbon of the monosaccharide unit in 2-deoxyguanosine (α or β)?

15.4. Draw the structures of 2′-deoxycytidine and 2′-deoxyadenosine.

Progressing from nucleoside to nucleotide involves the addition of phosphate to the 5′ end of the sugar.

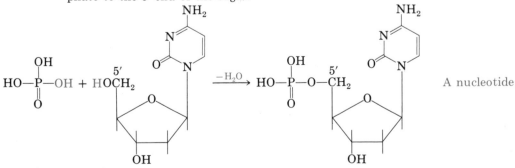

A nucleotide

These DNA nucleotides are named as 2′-deoxyribonucleoside 5′-monophosphates.

PROBLEM

15.5. Draw the structure of 2'-deoxythymidine 5'-monophosphate.

The covalent union of nucleotide units, forming polynucleotides and eventually nucleic acids, occurs through the formation of a phosphate ester between the phosphate group of one nucleotide and the 3' —OH group of another.

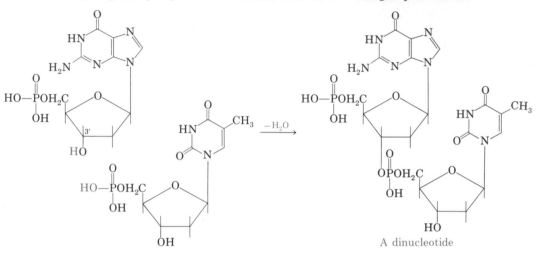

A dinucleotide

A polynucleotide then can be represented schematically as in Figure 15.1. Most often, nucleotide sequences are written left to right in the 5' to 3' direction.

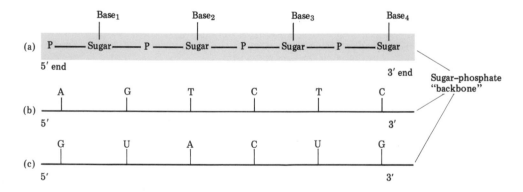

FIGURE 15.1. Representations of a polynucleotide sequence. (a) A general scheme holding for both DNA and RNA. (b) A DNA sequence. (c) An RNA sequence. The sugar in DNA is 2'-deoxyribose, and that in RNA is ribose.

C. RNA

Ribonucleic acid contains, as you might conjecture, D-ribose as its sugar unit. Three of its four component bases are the same as those found in DNA—adenine, guanine, and cytosine. However, RNA contains uracil instead of thymine. In addition, RNA exhibits a small percentage of minor bases, such as 2-

methyladenine and 2-thiouracil. Otherwise, RNA has the same basic structure as DNA (see Figure 15.1).

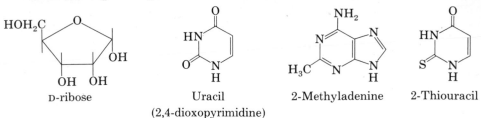

| D-ribose | Uracil (2,4-dioxopyrimidine) | 2-Methyladenine | 2-Thiouracil |

PROBLEMS

15.6. Describe two similarities and two differences between a polyribonucleotide and a polydeoxyribonucleotide.

15.7. Draw the structure of the nucleotide uridine 5′-monophosphate.

Both DNA and RNA can be selectively hydrolyzed at various positions in their chains by specific enzymes. Some of these enzymes are found in snake venoms and are responsible for part of the venom's toxicity.

PROBLEM

15.8. What would be the products of the complete hydrolysis of the following polynucleotides (molar amounts)?

(a)   A A U G C G   (b)   C T T A G C

D. Nucleoproteins

Nucleic acids do not exist alone in the cell but can be associated with other materials, basic proteins called *histones*. The histones are composed of a high percentage of lysine and arginine, which are basic amino acids; the histones are electropositive and strongly associated with the electronegative nucleic acids by electrostatic interactions.

E. Mononucleotides of Specialized Function

Not all nucleotides are polymerized into nucleic acids. Several nucleotide species are extremely important in metabolism as cofactors, intermediates, and energy transferral units.

1. ATP, Adenosine Triphosphate.

Adenosine triphosphate is the main agent involved in the transfer of phosphate for several important metabolic reactions, releasing energy in the process.

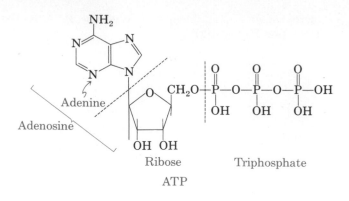

Its formation, conversely, uses energy. The last two phosphate ester bonds, especially the last, are high-energy bonds; that is, on hydrolysis, energy is released.

The phosphate released in the hydrolysis of ATP can be transferred to another species, such as glucose, imparting part, but not all, of the energy to the new bond. The "energized" molecule can then readily participate in some phase of metabolism, eventually expending the energy. The release and consumption of energy in the ATP-ADP reactions, and those similar to it, constitute cycles that allow an organism to operate.

2. Other Energized Nucleotide Triphosphates. ATP is not the only high-energy workhorse in metabolism, although it is the most prevalent. For example, uridine triphosphate, UTP, helps out in the metabolism of carbohydrates; guanosine triphosphate, GTP, participates in protein and carbohydrate cycles; and cytidine triphosphate, CTP, works in lipid metabolism. All four nucleotides combine, of course, to form RNA.

PROBLEM
_____

15.9. Draw out the structures for UTP, GTP, and CTP.
_____

3. Nucleotide Vitamins. There are several other heterocyclic-base nucleotides that participate in enzymatic reactions as cofactors. They are also known as water-soluble vitamins; that is, they are organic compounds essential to life, which are water-soluble, not synthesized or accumulated within the body, and must be obtained by food intake. Following are their structures and functions in brief.

Riboflavin, vitamin B_2, consists of a flavin base and the sugar alcohol ribitol. It is found in meats, dairy products, and also green, leafy vegetables. The riboflavin moiety is joined either to phosphate or to adenosine diphosphate to form the active cofactors flavin mononucleotide (FMN) or flavinadenine dinucleotide (FAD), respectively. These cofactors aid in the biotransfer of electrons by directly undergoing reduction by the addition of H_2 and oxidation by H_2 removal. A deficiency of B_2 results in dermatitis of the face, inflamed tongue, and eye disorders.

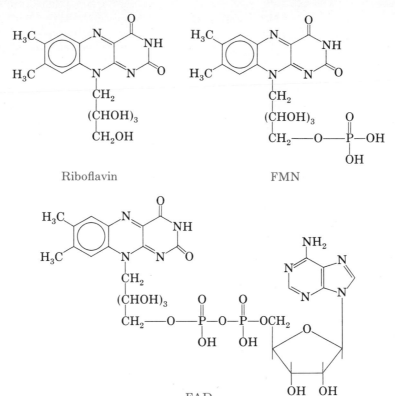

Riboflavin   FMN

FAD

Niacin or nicotinic acid in the form of its amide (nicotinamide) will join with an adenine nucleotide to make the hydrogen transferral coenzyme, nicotinamide adenine dinucleotide (NAD$^+$). The 2′ —OH group of the ribose attached to the adenine can also be phosphorylated for another active cofactor, NADP$^+$.

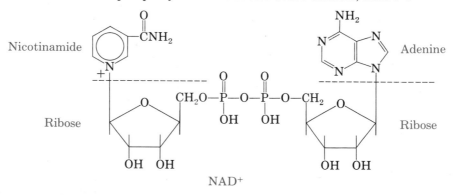

NAD$^+$

Niacin and nicotinamide are found in yeast, meats, and wheat germ   Their absence from the diet results in pellagra, a fatal disease with diarrhea, indigestion, and dermititis as symptoms.

There are numerous other water-soluble vitamins that are not of a nucleotide structure.

4. cAMP, Cyclic Adenosine Monophosphate.   Although much remains unknown about the ways in which hormonal secretions affect cells, several aspects of the transfer of hormone "messages" have been elucidated.   One of these is the role of the "second messenger," cyclic AMP or cAMP.   Cell membranes contain specific receptors for certain hormones such as glucagon (Section 14.9.C.2), which is produced, for instance, during exercise, and an enzyme system called adenyl cyclase.   Adenyl cyclase causes the cyclization of ATP within the cell with the removal of inorganic pyrophosphate.   This is an energy-releasing reaction.

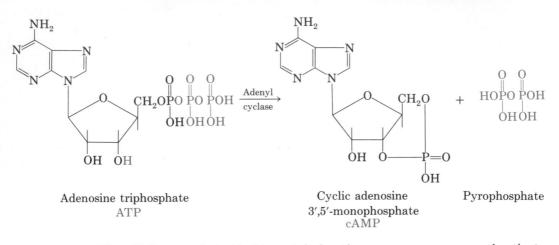

Adenosine triphosphate
ATP

Cyclic adenosine
3′,5′-monophosphate
cAMP

Pyrophosphate

The cAMP proceeds to bind to certain inactive enzyme precursors and activate them, eventually resulting in greatly increased metabolic activity.   Another enzyme, a phosphodiesterase, hydrolyzes cAMP to AMP, adenosine-5′-monophosphate, thereby stopping its activity.   It has been found that caffeine and theobromine (found in tea), both methylxanthine derivatives, inhibit the diesterase, hence the stimulatory effects of coffee and tea.   (See Panel 16 in Chapter 13 for structures.)

THE THREE-DIMENSIONAL STRUCTURE OF DNA

15.3   The actual physical orientation of the DNA macromolecule was not known until 1953, when Watson, Crick, and Wilkins postulated a novel and fascinating picture of the giant biopolymer that fitted X-ray and chemical analysis.

A. The Double Helix

It had been known that the molar ratio of adenine to thymine and guanine to cytosine was always 1, no matter what the source of the DNA was.   The number of the individual bases varied but those ratios always remained the same.   The reason for this depends on a recurring phenomenon in organic chemistry, hydrogen-bonding.   Looking at the structures for the bases, we can see that the opportunity for hydrogen-bonding exists since there are electronegative oxygens of the carbonyl groups or ring nitrogens and electropositive hydrogens of the amine or imine groups.

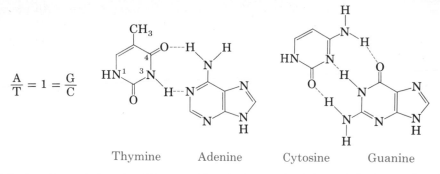

$$\frac{A}{T} = 1 = \frac{G}{C}$$

Thymine   Adenine   Cytosine   Guanine

Adenine (A) and thymine (T) are said to be base-paired through hydrogen-bonding, as guanine (G) and cytosine (C) are. These are often written as A≡≡≡T or A–T pairs and G≡≡≡C or G–C pairs. Maximum base pairing occurs when A and T are joined by two hydrogen bonds and G and C by three. Although other combinations are certainly possible, nature deems that the maximum number prevail. One can see then why the A/T and G/C ratios might be 1.

The question then is, how does the DNA molecule allow this base pairing to occur? Did the long biopolymer bend and twist back on itself like a protein? The Watson-Crick hypothesis was that DNA is double-stranded with base pairing occurring between bases on different strands. The two polynucleotide chains are twisted around each other in a double helix. The sugar-phosphate backbone of each polynucleotide strand makes up the helical arrangement, with the bases arranged toward the inside of the helix for base pairing, as Figure 15.2 shows. There are 10 nucleotides for every complete turn of the helix (360°) covering a 34-Å distance (angstrom Å = 10^{-8} cm). This hypothesis won for Watson, Crick, and Wilkins a Nobel Prize in 1962, and it is the basis for the chemical and structural correlations used today for DNA.

The protein portions associated with DNA are believed to nestle in the groove present between the helical strands. The two strands of the helix are complementary and also antiparallel, that is, they run in directions opposite to each other (head-to-tail) in order to accommodate maximum hydrogen-bonding.

The hydrophilic sugar-phosphate backbone is exposed to the aqueous cell environment, while the requirements for the bases are met with their hydrogen-bonding and hydrophobic interior. Notice that the base pairs occur between a purine and pyrimidine, allowing approximately equivalent spacing between the polynucleotide chains. Other helical arrangements for DNA have been found to exist, but not in large amounts.

B. Superstructures of DNA

Deoxyribonucleic acid from different sources can also show interesting variations of the double helical structure. Many DNA's are circular or supercoiled or both; some can be noncovalently linked as dimers. Such forms can be pictured by thinking of two strings tied in a circle and then twisted or two pairs of string circles

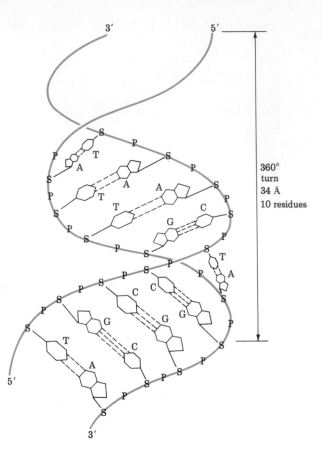

FIGURE 15.2.   The DNA double helix of Watson and Crick. S = sugar.
P = phosphate.

tied after being looped.   The structures are not merely interesting but allow DNA
to function as a continuously flowing (circular) synthetic blueprint.

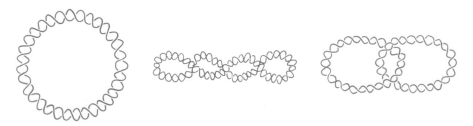

C. RNA Structure

Ribonucleic acid molecules can be helical or take some other structure in
either single or double strands, although single-strand RNA predominates in
higher organisms.   Because RNA has the same general base composition as DNA,
except in the substitution of uracil for thymine, it can still form hydrogen bonds.

The variety of functions RNA fills guides the three-dimensional structure of the polynucleotide.

THE FUNCTIONS OF NUCLEIC ACIDS

15.4   Deoxyribonucleic acid has two main purposes: to reproduce itself and to direct the reproduction of essentially everything else within the organism. The nucleus of the human cell carries its genetic information, that is, its essence, in its 46 chromosomes. In turn, the chromosomes contain smaller segments called genes; these are the blueprints for every protein in the body. Genes are made up of DNA, the ultimate hereditary information.

A. Replication of DNA

The process by which DNA reproduces itself is called *replication*. It is a simple sequence in its overall concept but complicated in its actual execution within an organism. As you might expect, knowing the structure of DNA, replication depends on the hydrogen-bonding of the heterocyclic polynucleotide bases. The parent DNA helix unravels in certain segments to form two separate chains. Free nucleotides in the medium then base-pair with the exposed strands. Short sections of nucleotides are joined by enzymes, and these polynucleotide segments are finally biosynthesized into larger strands. Thereby, both chains of the double helix are reproduced simultaneously, with each strand of the parent DNA molecule intertwining with one of the new daughter strands. These associations form two identical DNA's where only one existed before (Figure 15.3). Studies per-

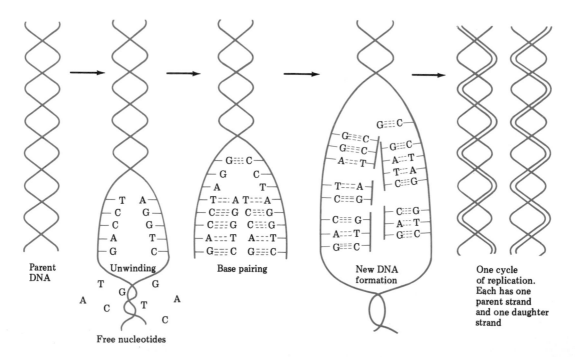

FIGURE 15.3. Replication of DNA. Note that each strand replicates its complement.

formed using ^{15}N radioactive free nucleotides showed that the radioactivity was distributed in one strand of each DNA molecule after replication. This fact corresponds to known data on the transmission of genetic traits.

B. Transcription and Translation of DNA

The second function for which DNA is responsible is the awesome feat of directing the synthesis of every protein in the organism. It is in this process that RNA exhibits its multifaceted capacities.

1. Triplet Code. If DNA contains the blueprint for proteins, there must exist a code that can be used to direct the relations among the four bases of DNA and the 20 common amino acids. Allowing each base to correlate to one amino acid is plainly insufficient. Two nucleotide bases per amino acid would still only give 16 (4^2) combinations. A base triplet, however, would provide more than

TABLE 15.1. RNA TRIPLET CODE FOR AMINO ACIDS

Amino Acid	Type	Triplet Code (Codons)[a]
Lysine	Basic	AAA, AAG
Arginine		AGA, AGG, CGA, CGG, CGC, CGU
Histidine		CAU, CAC
Aspartic Acid	Acidic	GAU, GAC
Glutamic Acid		GAA, GAG
Asparagine	Amide	AAU, AAC
Glutamine		CAA, CAG
Glycine	Aliphatic	GGU, GGC, GGA, GGG
Alanine		GCU, GCC, GCA, GCG
Valine		GUU, GUC, GUA, GUG
Leucine		UUA, UUG, CUU, CUC, CUA, CUG
Isoleucine		AUU, AUC, AUA
Phenylalanine	Aromatic	UUU, UUC
Tyrosine		UAU, UAC
Tryptophan		UGG
Serine	Alcoholic	UCU, UCC, UCA, UCG, AGU, AGC
Threonine		ACU, ACC, ACA, ACG
Cysteine	Sulfur-	UGU, UGC
Methionine	containing	AUG
Proline	Cyclic	CCU, CCC, CCA, CCG
End of code Termination of polypeptide chain		UAA, UAG, UGA
Initiation of code		AUG

[a]A = adenine; G = guanine; C = cytosine; U = uracil.

enough "codewords"—4^3, or 64—for the 20 amino acids. The excess triplets could be used to signal various "start-stop" activities in the synthesis and could also represent a "degeneracy" in the code; that is, there could be more than one base code for most amino acids. And this is the case. (See Table 15.1.) Certain "nonsense" codons also exist, which stand for no amino acid.

2. Messenger RNA: Transcription of the Code. So far we have not discussed the function of RNA or its structure in any detail. As you will see now, this is because RNA plays three roles in the conversion of the DNA template, or blueprint, to finished protein. The first task that RNA assumes is to become a messenger entrusted with the code prescribed by DNA. Once again hydrogen-bonded base pairing is responsible for a faithful reproduction of each triplet sequence. As in the replication of DNA, where each strand of the double helix came in contact with free deoxyribonucleotides and a new DNA strand was produced complementary to the original, so too, free ribonucleotides will base-pair with one strand of the DNA. Uracil instead of thymine will pair with adenine but in an equivalent manner.

Only one strand of the DNA double helix is used to produce a complementary RNA, though more than one RNA may be coded on a single DNA strand (Figure 15.4). Each three-base complementary codeword on a messenger RNA (mRNA) molecule is called a *codon*. The codons for bacterial systems have been deciphered (Table 15.1) experimentally. It has also been determined that the "code" is universal, that is, it holds for all species from yeast to human.

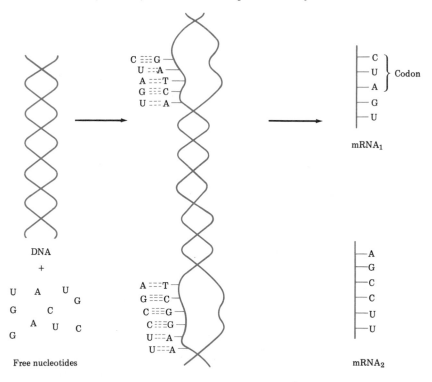

FIGURE 15.4. Formation of messenger RNA (mRNA) from a DNA template. Each step is catalyzed by enzymes. Only one strand of DNA is used as a model.

3. Ribosomal RNA.   The mRNA needs a place to interact with amino acids in order to produce proteins.   Another type of RNA called ribosomal RNA, or rRNA, combines with protein material to form large granules termed *ribosomes*. The mRNA will complex with rRNA and provide this "place" for the rest of the process to occur.   Many ribosomes can complex along an mRNA strand to form a polysome.   The ribosomes will move along the mRNA strand allowing the biosynthesis of a complete protein on one ribosome complex.   See Figure 15.6.

4. Transfer RNA.   The last link in the transcription must be a translation of the code to amino acids.   This is accomplished via transfer RNA (tRNA). Transfer RNA must contain two important parts: an end capable of reading the codon and one able to carry an amino acid.   The general structure of tRNA has been determined and it indeed does possess two such ends (Figure 15.5).   The three ribonucleotide bases responsible for the recognition of the code are called the *anticodon* and are specific for the amino acid with which the particular tRNA will react.   The anticodon will base-pair with the mRNA codon, thereby bringing the amino acid with it to the ribosomal site of protein synthesis.

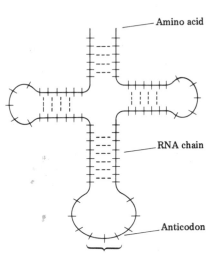

FIGURE 15.5.   General structure of transfer RNA (tRNA).

A tRNA must also have nucleotide sequences corresponding to attachment sites on rRNA and ones specific for the enzymes necessary to effect biosynthesis of the protein at that point.

Once the tRNA is paired to the mRNA-ribosome complex, synthesis of the protein can begin.   The ribosome will move along the mRNA strand, facilitating the necessary chemical reactions.   In this way, many polypeptide chains can undergo synthesis from the same mRNA at the same time.   When a termination codon is reached, the synthesis is stopped and the complete protein released. Figure 15.6 illustrates the synthesis of a protein in a schematic, stepwise manner.

The total process is much more complex than this; it would have to be, in organisms as specialized and complicated as dogs, monkeys, and humans.   But this should give one an appreciation of the inspiring feats performed within living

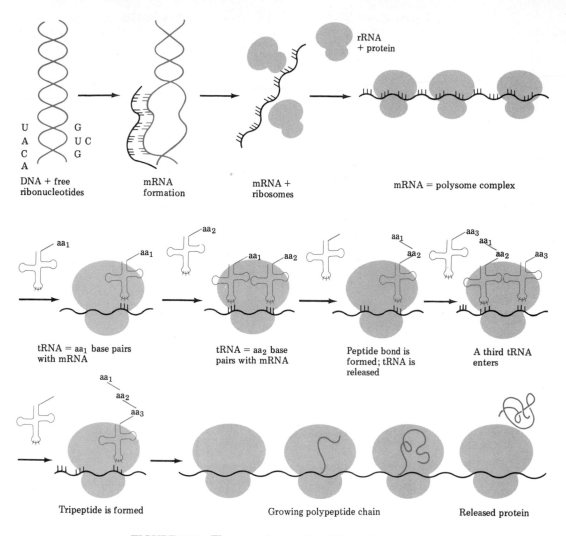

U
A
C
A

G
U C
G

DNA + free
ribonucleotides

mRNA
formation

mRNA +
ribosomes

rRNA
+ protein

mRNA = polysome complex

aa₁

tRNA = aa₁ base pairs
with mRNA

aa₂

aa₁

tRNA = aa₂ base
pairs with mRNA

aa₁
aa₂

Peptide bond is
formed; tRNA is
released

aa₃
aa₁
aa₂
aa₃

A third tRNA
enters

aa₁
aa₂
aa₃

Tripeptide is formed

Growing polypeptide chain

Released protein

FIGURE 15.6. The transcription of the DNA triplet code for the biosynthesis of protein via mRNA, rRNA, and tRNA.

organisms and the equally inspired experimentation that has elucidated the mechanisms.

THE GENETIC CODE: ITS FUNCTION AND MALFUNCTION

15.5

The true pictures of DNA and RNA are far from simple. Basic concepts concerning them can nevertheless be used to theorize about transferral of genetic traits, the mutation of cells by outside agents, and similar matters. The triplet code is remarkably accurate and its translation is inconceivably fast. It has been estimated that should the total number of DNA molecules within the human body be stretched out and attached end to end, the chain would reach to the sun and back 400 times. The entire α chain of rabbit hemoglobin (146 amino acid residues) can be synthesized within three minutes. Of course, natural mistakes can occur. A break in the DNA template chain, the insertion of a stray nucleotide, the misreading of an mRNA code, chemical or radiation-induced structural change—all can influence the pattern carried by the nucleic acids and ultimately alter the structure of a protein. If the mutation is inconsequential, nothing will happen. In fact, a change could be beneficial to a species. However, there are those that are intrinsically fatal to the particular organism in which they occur or perhaps to a species as a whole.

A. Congenital Errors

Several of the genetic diseases we have already considered are known to involve only subtle differences in a protein's amino acid sequence and yet yield dramatic, deadly consequences. Sickle-cell anemia is a pertinent example in that only one amino acid is changed in the β chain of hemoglobin. A valine is located where a glutamic acid normally occurs. Although these two amino acids each have several codons, it is interesting to note the similarities in two.

Valine	GUG, GUA
Glutamic Acid	GAG, GAA

How the original DNA template becomes altered is unknown, but the coding errors are a helpful lead.

Other kinds of genetic diseases involve the malfunction of various enzymes, which are proteins. Phenylketonuria and galactosemia are examples of inborn enzyme deficiencies; in the first, the amino acid phenylalanine cannot be metabolized and in the second, it is the monosaccharide galactose. Both diseases can be fatal, but when they are detected in early life, they can be alleviated by controlling the infant's diet.

Tay-Sachs disease, another fatal hereditary disorder which is the result of an enzyme deficiency in lipid metabolism, can, like many others, be controlled through genetic counseling, because the malfunctioning gene can be detected experimentally. Samples of the amniotic fluid can be taken during pregnancy in order to detect hereditary defects. The future hopes of the scientific world include the repair of mutant genes either before or during pregnancy.

B. Environmentally Induced Mutations

With such a complex and interrelated system as the DNA–RNA scheme, we are just beginning to find out the effects of modern technology and natural environment on its delicate balance. Some general results are known. For example, high-energy radiations such as X rays or gamma rays can break covalent bonds, thereby destroying the integrity of a DNA code sequence or messenger RNA. Such radiation can also convert the heterocyclic bases of DNA and RNA into less stable tautomeric forms, resulting in an alteration of hydrogen-bonding.

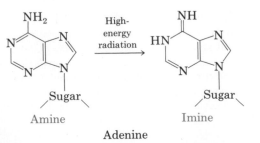

Adenine

Ultraviolet (UV) radiation, though not quite as energetic as X rays, can produce dimerization of adjacent thymine residues. This mutation stops replication temporarily, but the dimer can be enzymatically excised and repaired by the organism.

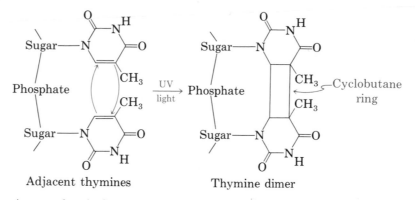

Adjacent thymines   Thymine dimer

Among chemical agents that are believed to affect the genetic material of living cells is nitrous acid, HNO_2, and its salt, sodium nitrite, $NaNO_2$. Sodium nitrite is used in preparing cured meats such as bacon and sausage to prevent botulism and impart a red color. In the acidic conditions of the stomach, sodium nitrite becomes HNO_2, which can theoretically react with amine groups of amino acids to form nitrosoamines (which are carcinogenic), and with the amine group substituents on the nucleotide bases to form alcohols. These alcohols are tautomeric forms of ketones, and when they tautomerize to the more stable ketone, the hydrogen-bonding pattern of the nucleic acid sequence will be altered.

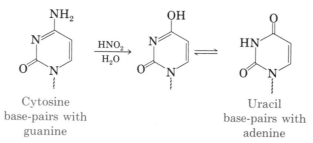

Cytosine   Uracil
base-pairs with   base-pairs with
guanine   adenine

This will obviously upset the replication and transcription process. Adenine and guanine can also undergo conversion by this reaction, with accompanying alterations in base pairing.

Other chemicals such as acridine dyes and hydroxylamine are mutagens whose actions are directly documented. Of course, many substances—saccharin, caffeine, LSD—are highly suspect because of their mutagenic effects on bacteria (Aames test), though the actual alteration in the chemical structure of genetic material is still under investigation.

PROBLEM

15.10. How does the radiation-induced tautomeric change in adenine alter its hydrogen-bonding abilities?

VIRUSES

15.6   The virus is a unique organism consisting of nucleic acids enclosed in a protein coat. Plant viruses contain RNA exclusively, whereas animal viruses can have either DNA or RNA. Viruses reproduce by harnessing the biological machinery of a host cell, closing down the cell's own functions. It is extremely difficult to overcome the action of a virus, as anyone with a cold or influenza can testify. The body must counteract the viral attack with its own immune system (a protein system), and do that before various bacterial side infections fatally weaken the defenses. Immunization involves introducing small doses of virus into the body to trigger the protective response without using too much virus and causing illness. The Sabin-Salk polio vaccine is a prominent, dramatic example of the effectiveness of proper immunization. Before the Salk vaccine was introduced to the public in 1954, there were almost 40,000 new cases of polio annually in the United States and close to 15,000 deaths. By 1960, the new cases were in the hundreds and today poliomyelitis is very rare. Smallpox is another disease that is almost nonexistent today. However, studies in viral research are cautiously approached, and in many cases they are quickly discontinued because of the epidemic possibilities. A deadly virus imported from Africa called Lassa fever has made headlines because of its virulency. As a result of its unexpected spread even with the proper precautions, the U.S. National Institutes of Health has banned research on it.

Panel 17

RECOMBINANT DNA—SCIENCE FICTION TODAY?

One of the most exciting topics today, both scientifically and philosophically, is the transfer of genetic material from one type of organism to another. A strain of bacteria of the *E. coli* family, different from that found in the human intestinal tract, has been used to biomanufacture polypeptides and proteins from human DNA. Human insulin, growth hormone, and somatostatin have been produced by bacterial cells. The technique is being commercially developed to provide these vital hormones to the public.

How can human genes be put into bacteria? The answer lies in the simpler substructure of the *E. coli* cell. Besides chromosomal DNA, bacteria contain DNA in separate "packages" known as plasmids. Plasmids serve various functions such as coding for proteins that act as cell protection. Specific enzymes can be used to cut out a portion of plasmid DNA and insert a piece of human DNA or, for that matter, DNA from any other source. The modified plasmid DNA will now serve as a template for the production of human protein as well as bacterial protein. Because bacteria reproduce at a much greater rate than the cells of higher organisms, the recombined DNA is replicated rapidly and the polypeptide or protein should be produced in yields far exceeding the supplies currently available from conventional sources. This technique also addresses the problem of incompatibility and immune rejection. For example, although humans can use insulin from other species such as cattle or sheep, immune rejection can occur after a period of time. This would be avoided if human insulin were used instead.

[PANEL 17]Many substances like growth hormone must be of human origin* or may be needed in therapeutic dosages so large as to be prohibitive.

Another area of research involves a substance known as interferon. It is produced naturally in the body, in minute amounts, as a response to viral infection. It protects the organism against attack from another virus. The protective "interference" has led researchers to believe that interferon may be a potent anticancer agent. Recombinant DNA techniques have led to the production of human interferon by bacteria. This has aided immeasurably in the study of interferon and its effects. Interferon is species-specific and the quantities produced in the human body are so small as to make it impossible to isolate it in a pure form from a human source.

A stunning Supreme Court decision in June 1980 granted patent rights on the products of recombinant DNA techniques—the mutant bacteria—to their developers. This landmark case involved the production of a crossbred strain of Pseudomonas bacteria which consume oil and may prove invaluable in the cleaning up of oil spills. The hybrid bacteria were produced by recombining plasmid material from three oil-eating bacteria with a fourth strain. New life forms, the products of human ingenuity, are now patentable.

For the science fiction fan all of this brings to mind repeated tales of futuristic ecstasy and horror—from superhumans to mutated ants. These very real concerns prompted scientists in 1974 to declare a voluntary moratorium on such research until guidelines could be drawn up to avoid any lethal situations. The National Institute of Health guidelines now govern any government funded projects in this field, and commercial development is burgeoning. However, the debate continues, increasingly involving the local citizenry in final decisions. What are the limits to this experimentation, physical and moral? Other nations do not have the governmentally imposed restrictions the United States has, nor do privately endowed facilities. What are the potentials for the development of toxic bacteria or new and better strains of wheat and corn to feed a starving world? Here the moral issues facing the research scientist and those in power politically and economically must be met with caution and good judgment. To further unveil the complexities of the bioorganic world is necessary to the betterment of the human race on physical and intellectual levels. To temper the uncontrolled fervor of research with logical constraint will advance humanity.

---

*It takes 50 pituitary glands from human cadavers to supply the growth hormone needed to treat one child for a year.

END-OF-CHAPTER PROBLEMS

15.11 Structure: Draw and name an example of each of the following: **(a)** purine base; **(b)** pyrimidine base; **(c)** ribonucleoside; **(d)** deoxyribonucleotide; **(e)** trinucleotide-GUA.

15.12 Nucleic Acid Structure: Give three important structural differences between DNA and RNA.

15.13 Genetic Code: A DNA sequence contains the code G-T-A-A-C-G-T-G-C. What are the appropriate codon-anticodon sequences? Which amino acids will be called upon for biosynthesis?

15.14 Nucleic Acid Structure: There are 10 nucleotide bases per 360° turn of the DNA strands. This corresponds to a linear distance of about 34 Å (1 Å = 1 angstrom = 10^{-8} cm). How long, in meters, would a DNA molecule be if it contained 1 million nucleotide bases?

15.15 Energied Cofactors: The complete hydrolysis of ATP, NAD, FAD will yield what products?

15.16 Base Pairing: Show the hydrogen-bonding that uracil encounters when it base pairs with adenine.

15.17 Genetic Code: Two variations discovered in the amino acid sequence for hemoglobin involve substituting a lysine for a glutamic acid in the β chain (hemoglobin E), and substituting a tyrosine for histidine in the β chain (hemoglobin M, Boston). How could you explain this in terms of the genetic code?

15.18 Genetic Code: Glucagon is a peptide hormone that is derived from an inactive form, proglucagon, with 37 amino acid residues. How many nucleotides would be needed to code for this inactive form, including initiation and termination sequences?

Chapter 16

Organic Chemical Instrumentation

In reading the daily newspaper or weekly news magazines, we frequently find scientific reports on topics such as the search for life on Mars and other planets, the chemical composition of the moon, criminal investigations, or federal investigations of drugs and consumer products. These articles often enumerate the various analytical chemical techniques used to arrive at the conclusion that furnished the headline: infrared analysis, gas chromatography, mass spectrometry, amino acid analysis, ultraviolet spectroscopy. These are all essential laboratory tools, and to us their application is of particular interest.

SPECTROSCOPY

16.1

A. Absorption of Electromagnetic Radiation

All chemical substances interact with electromagnetic radiation in some way. It is evident then that measuring this interaction can provide valuable information about the substance. When a molecule absorbs energy, a transformation or perturbation occurs that may be either temporary or permanent. Low-energy radiation may merely cause a molecular rotation or a bond vibration. Higher-energy radiation may effect the promotion of electrons to higher energy levels; and radiation of even greater energy can result in bond cleavage and permanent disruption of the molecule.

Energy can be visualized as traveling in waves (Figure 16.1). The distance between waves is the *wavelength,* and the number of waves that pass by a point in a given time or the number of waves per unit distance is the *frequency,* expressed in cycles per second, or by the newer designation hertz (Hz). High-energy radiation is characterized by short wavelengths and high frequency, and low-energy radiation by long wavelengths and low frequency. Radiation of a particular wavelength has a definite, constant amount of energy associated with it. Figure 16.2 shows the spectrum of electromagnetic energy varying in wavelengths from a fraction of an angstrom to thousands of kilometers.

As we have indicated, the interaction of a molecule with electromagnetic radiation causes molecular transformations. Whether the transformation in-

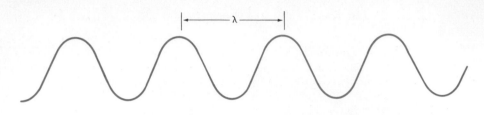

FIGURE 16.1. Electromagnetic radiation travels in waves characterized by a wavelength λ and frequency ν.

volves molecular rotation, bond vibration, or electronic transition, the molecule absorbs only the wavelength of radiation with exactly the energy necessary for the transition. Wavelengths of lower energy are not accumulated until the total energy of transition is attained. Neither can the energy of transition be extracted from higher-energy radiation. The situation is analogous to a vending machine that takes only dimes. You can obtain your item only if you have a dime. Trying to insert ten pennies or two nickels is useless. Likewise, quarters or half-dollars will not be accepted.

Since the absorption of wavelengths of radiation is selective for the particular transition and this transition depends on molecular structure, spectroscopy is invaluable both qualitatively and quantitatively. By measuring the absorption spectra of known compounds, we can correlate the wavelengths of energy absorbed with characteristic structural features. This information is then used to identify structural units in unknowns.

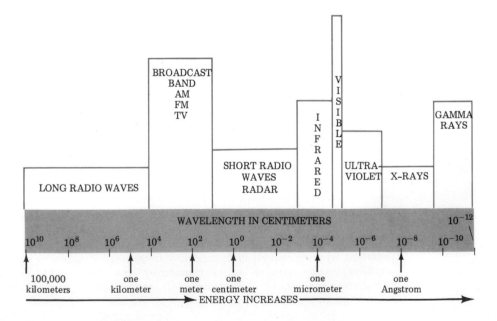

FIGURE 16.2. The spectrum of electromagnetic radiation.

B. Spectrophotometers

A spectrophotometer measures the absorption of energy by a chemical compound (see Figure 16.3). Its basic components are radiation source, intensity control, wavelength selector, sample holder, detector, and recorder. The first three components determine the intensity and wavelength of light presented to the sample. The general area of the electromagnetic spectrum used is dependent on the radiation source. A wavelength selector separates and selects wavelengths to be presented individually from the source to the sample. Each wavelength is then divided into two beams, a reference beam that is transmitted directly to the detector and a sample beam, which passes through the sample before striking the detector. If the sample interacts with a particular wavelength, it will absorb it and the intensity of the sample beam will be diminished. The detector compares the intensity of the reference and sample beams. The percentage of the sample beam transmitted is recorded as a peak or band by the recorder.

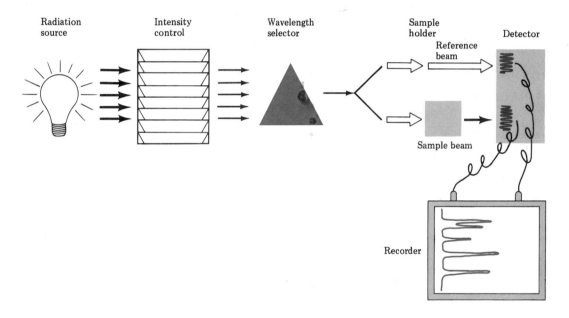

FIGURE 16.3. A schematic diagram of a spectrophotometer.

INFRARED SPECTROSCOPY

16.2

An infrared spectrometer subjects a sample compound to infrared radiation in the 2–15-micrometer* (μm) wavelength range. This region is also described in terms of wavenumber (frequency), 5000 cm^{-1}–670 cm^{-1}, which is essentially the number of cycles or waves in a distance of one centimeter. Although this radiation is weak and unable to inflict permanent alteration on a molecule, it does

---

*Micrometer (μm) is the currently accepted term for 10^{-6} meter in the metric system In infrared spectroscopy, *micron* (μ) has been used for some time and has the same meaning as *micrometer*.

supply sufficient energy for bonds in the molecule to vibrate by stretching, scissoring, bending, rocking, twisting, or wagging (Figure 16.4). The atoms of a molecule can be conceived as linked by springs that are set in motion by the application of energy. As the molecule is subjected to the individual wavelengths in the 2–15 μm range, it absorbs only those possessing exactly the energy required to cause a particular vibration. Energy absorptions are recorded as bands on chart paper.

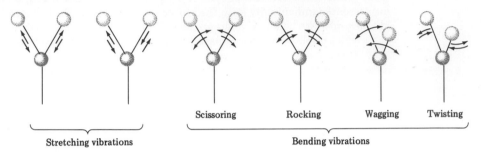

FIGURE 16.4. Molecular vibrations caused by infrared radiation.

Since different bonds and functional groups absorb at different wavelengths, an infrared spectrum is usually applicable in qualitative analysis, that is, determining what types of groups are in a molecule. For example, a carbon-carbon triple bond is stronger than a double bond and requires a shorter wavelength (greater energy) to stretch. The same considerations apply to carbon-oxygen and carbon-nitrogen bonds.

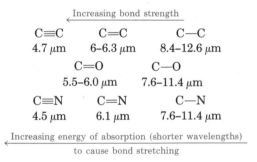

As is evident, from the position of an absorption peak one can identify the group that caused it. Figure 16.5 illustrates the general area in which various bonds absorb in the infrared.

An infrared spectrum is usually studied in two sections. The area from 2 μm to 7 or 8 μm is the functional group area. The bands in this region are particularly useful in determining the types of groups—alkene, alkyne, aldehyde, ketone, alcohol, acid—present in the molecule. The remainder of the spectrum is called the fingerprint region. A peak-by-peak match of an unknown spectrum with the spectrum of the suspected compound in this region can be used, much like a fingerprint, to confirm its identity. Table 16.1 summarizes some infrared assignments useful in functional group analysis and Figure 16.6 contains some sample spectra.

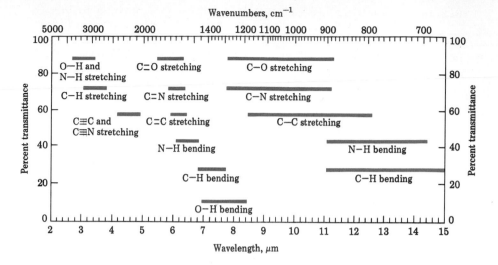

FIGURE 16.5. Areas of absorption of infrared radiation by various bonds. The lower scale is the wavelength in micrometers (μm). The upper scale is frequency expressed in wavenumbers (the number of waves in 1 centimeter). The vertical scale describes percentage of transmittance of the sample beam.

ULTRAVIOLET AND VISIBLE SPECTROSCOPY

16.3   In ultraviolet-visible spectroscopy, the 200–750-nm (200–750 nanometers*) region of the electromagnetic spectrum is used. This includes both the visible, 400–750 nm, and near ultraviolet, 200–400 nm. Radiation of these wavelengths is sufficiently energetic to cause the promotion of loosely held electrons, such as nonbonding electrons or electrons involved in a π bond to higher energy levels. For absorption in this particular region of the ultraviolet, however, there must be conjugation of double bonds. An alternating system of double and single bonds lowers the energy of transition of an electron moving to a higher energy level. If the conjugation is extensive, the molecule may absorb in the visible region and show color (section 13.7, Table 13.2).

In general, ultraviolet and visible spectroscopy are not used for functional-group analysis as extensively as infrared analysis. Rather, they show the presence of conjugated unsaturated systems such as the ones illustrated.

Compounds that absorb in this area have characteristic wavelengths of absorption. Thus, their presence and concentration in a solution can be detected and measured. This is useful in identifying product ratios and reaction rates and also in determining other quantitative data.

---

*A nanometer is 10^{-9} meter in the metric system.

TABLE 16.1. INFRARED ABSORPTION ASSIGNMENTS

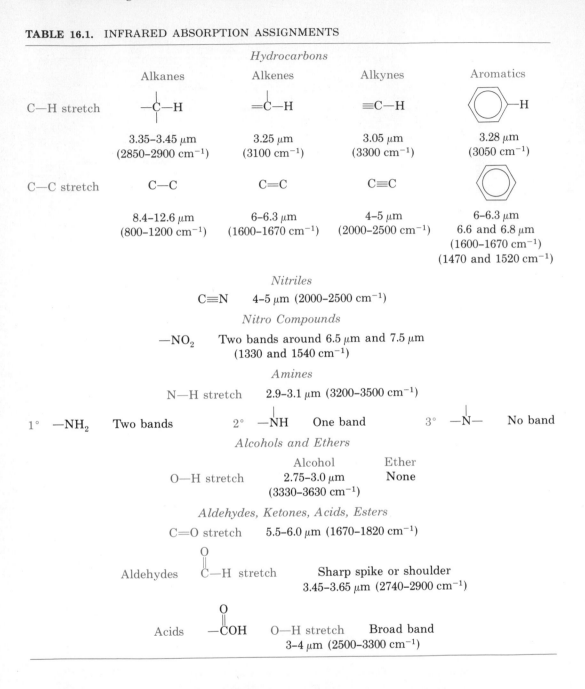

Hydrocarbons

	Alkanes	Alkenes	Alkynes	Aromatics

C—H stretch:

Alkanes: 3.35–3.45 μm (2850–2900 cm⁻¹)

Alkenes: 3.25 μm (3100 cm⁻¹)

Alkynes: 3.05 μm (3300 cm⁻¹)

Aromatics: 3.28 μm (3050 cm⁻¹)

C—C stretch:

Alkanes: C—C 8.4–12.6 μm (800–1200 cm⁻¹)

Alkenes: C=C 6–6.3 μm (1600–1670 cm⁻¹)

Alkynes: C≡C 4–5 μm (2000–2500 cm⁻¹)

Aromatics: 6–6.3 μm, 6.6 and 6.8 μm (1600–1670 cm⁻¹) (1470 and 1520 cm⁻¹)

Nitriles

C≡N   4–5 μm (2000–2500 cm⁻¹)

Nitro Compounds

—NO₂   Two bands around 6.5 μm and 7.5 μm (1330 and 1540 cm⁻¹)

Amines

N—H stretch   2.9–3.1 μm (3200–3500 cm⁻¹)

1° —NH₂   Two bands   2° —NH   One band   3° —N—   No band

Alcohols and Ethers

O—H stretch   Alcohol 2.75–3.0 μm (3330–3630 cm⁻¹)   Ether None

Aldehydes, Ketones, Acids, Esters

C=O stretch   5.5–6.0 μm (1670–1820 cm⁻¹)

Aldehydes $\overset{O}{\overset{\|}{C}}$—H stretch   Sharp spike or shoulder 3.45–3.65 μm (2740–2900 cm⁻¹)

Acids $-\overset{O}{\overset{\|}{C}}OH$ O—H stretch   Broad band 3–4 μm (2500–3300 cm⁻¹)

FIGURE 16.6 (next two pages). Infrared absorption spectra. Compare bands in the 2–8μm region with assignments for each functional group in Table 16.1. Infrared spectra, except 3-butyn-2-one, courtesy Sadtler Research Laboratories. 3-Butyn-2-one courtesy of Aldrich Chemical Company Inc. (Charles J. Pouchert, author), Aldrich Library of Infrared Spectra.

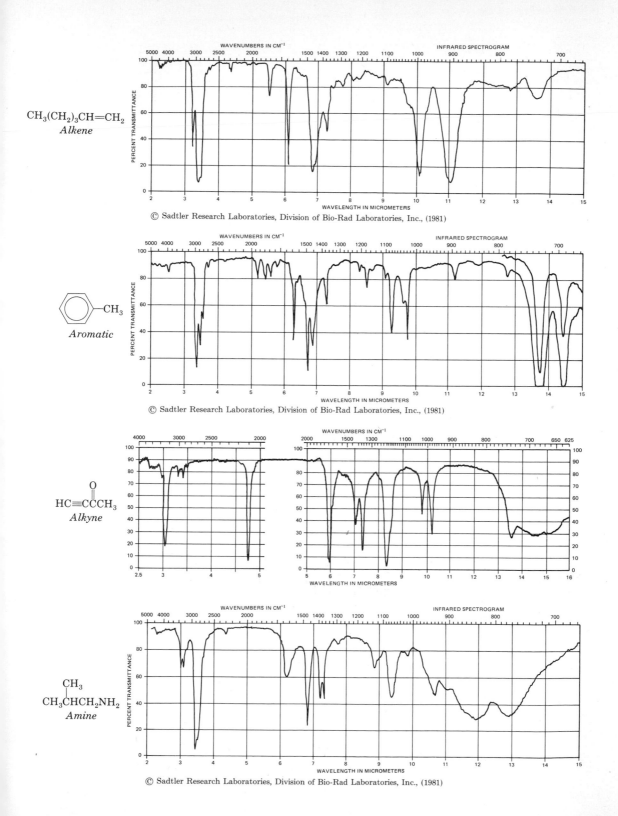

WAVENUMBERS IN CM⁻¹

INFRARED SPECTROGRAM

$CH_3(CH_2)_3CH{=}CH_2$
Alkene

© Sadtler Research Laboratories, Division of Bio-Rad Laboratories, Inc., (1981)

—CH_3

Aromatic

© Sadtler Research Laboratories, Division of Bio-Rad Laboratories, Inc., (1981)

$HC{\equiv}C\overset{\overset{\displaystyle O}{\|}}{C}CH_3$
Alkyne

$\underset{|}{CH_3}$
$CH_3CHCH_2NH_2$
Amine

© Sadtler Research Laboratories, Division of Bio-Rad Laboratories, Inc., (1981)

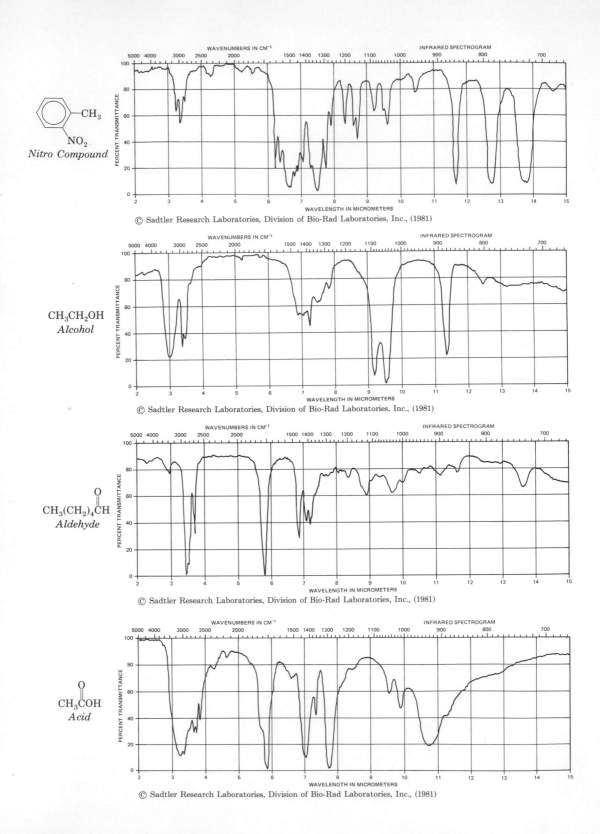

WAVENUMBERS IN CM⁻¹   INFRARED SPECTROGRAM

CH₃
NO₂
Nitro Compound

© Sadtler Research Laboratories, Division of Bio-Rad Laboratories, Inc., (1981)

CH₃CH₂OH
Alcohol

© Sadtler Research Laboratories, Division of Bio-Rad Laboratories, Inc., (1981)

$$CH_3(CH_2)_4\overset{O}{\overset{\|}{C}}H$$
Aldehyde

© Sadtler Research Laboratories, Division of Bio-Rad Laboratories, Inc., (1981)

$$CH_3\overset{O}{\overset{\|}{C}}OH$$
Acid

© Sadtler Research Laboratories, Division of Bio-Rad Laboratories, Inc., (1981)

NUCLEAR MAGNETIC RESONANCE

16.4   The nuclei of some atoms are thought to spin. In doing so, they generate a magnetic moment along their axis of spin, acting as tiny bar magnets. The nucleus of the hydrogen atom exhibits this property and is the one most often analyzed by nuclear magnetic resonance (nmr) spectroscopy. If a hydrogen atom is placed in an external magnetic field, its nucleus can align with the field (the more stable arrangement) or against the field (a more energetic, less stable state) (Figure 16.7).

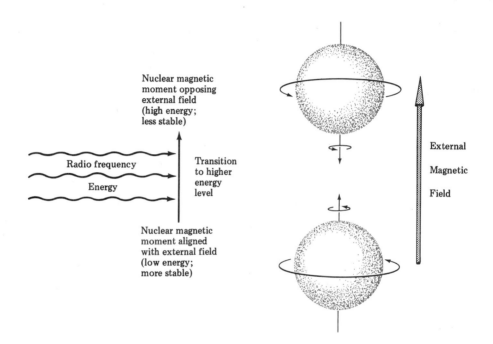

FIGURE 16.7. The spinning nucleus of the hydrogen atom acts like a tiny magnet that can go into alignment or nonalignment with an externally applied magnetic field. Applying radiofrequency energy can flip protons in the more stable "aligned" state to nonalignment.

To make a nucleus flip from alignment to nonalignment, energy in the radiofrequency range must be applied. For example, a hydrogen nucleus in an external field of 14,092 gauss requires a frequency of 60 million hertz (cycles per second) for the transition. When this exact frequency is applied, it is absorbed, and the absorption recorded on chart paper. In practice, either the magnetic field can be held constant and the radiofrequency varied, or more commonly, the radiofrequency can be held constant and the magnetic field varied.

A. Chemical Shift

If nmr's main feat were to detect the presence of hydrogen in a molecule, it would not be worth discussing here. Nuclear magnetic resonance spectroscopy can, however, distinguish between hydrogens in different chemical environments within the molecule. Hydrogens on a benzene ring, or on a carbon bearing a chlorine, or on a carbon adjacent to a carbonyl group absorb radiofrequency energies at different applied magnetic fields and appear at different locations on the recording paper. Furthermore, the position of absorption is relatively constant for hydrogens in a particular chemical or structural environment. Hence, the number of signals recorded on the nmr chart paper indicates the number of different types of hydrogens in a molecule. The position of the peak can give information about the molecular structure in the vicinity of the hydrogens.

To understand fully the value of nmr then, we must gain a concept of equivalent and nonequivalent hydrogens. Equivalent hydrogens are positioned in structurally and chemically equivalent areas in the molecule. For example, consider the following molecules and convince yourself of the different types of hydrogens shown. The first compound has two methyl groups connected to the same oxygen. The hydrogens on these carbons are chemically equivalent. However, in the second example, bromoethane, the —CH_2— group is bonded to a carbon and a bromine and the CH_3— is bonded to just a carbon. The hydrogens on these two carbons are in significantly different chemical environments and are nonequivalent. In the third and fourth examples, note that the methyl groups are equivalent but in the fifth example the two methyl groups are in two different chemical environments and are nonequivalent.

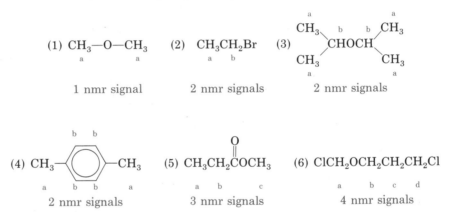

Nuclear magnetic resonance chart paper is rectangular with a linear scale of so-called δ units across the bottom. Most chart papers have scales from $\delta = 0$ to $\delta = 8$ or 9, although peaks at higher δ values can be easily recorded. To every sample a small amount of tetramethylsilane (TMS) is added as a reference (the TMS signal is at $\delta = 0$).

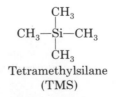

Tetramethylsilane
(TMS)

All other signals are relative to TMS. Table 16.2 summarizes the characteristic chemical shifts (from TMS) of different types of hydrogens.

Now let us consider a specific example, benzyl alcohol (nmr in Figure 16.8(a)).

TABLE 16.2. CHEMICAL SHIFTS

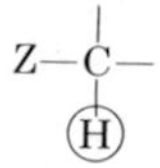

Z—C—(H)

Z Groups[a]

Group	Structure	Actual δ Value	Chemical Shift Relative to Alkyl H
Z = alkyl	R—C—(H)	0.9–1.6	0
Z = C	O=C—C—(H)	~2–2.5	1
Z = aromatic	⬡—C—(H)	2.3–2.9	1.3–1.5
Z = alkene	C=C—C—(H)	1.8–2.8	0.9–1.2
Z = O	O—C—(H)	3.3–5	2.4–3.4
Z = Cl	Cl—C—(H)	3.2–4	2.3–2.6
Z = Br	Br—C—(H)	2.7–3.8	1.8–2.2
Z = amines	N—C—(H)	2.2–3	1.3–1.4
Z = NO₂	O₂N—C—(H)	4.4–4.6	3–3.5

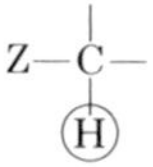

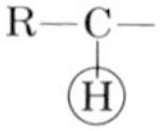

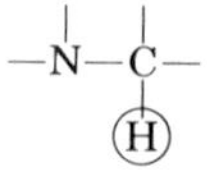

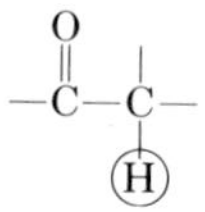

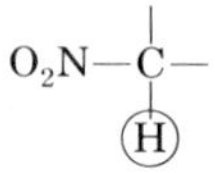

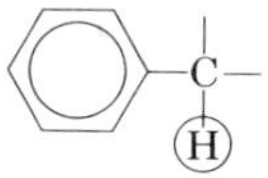

Other Groups

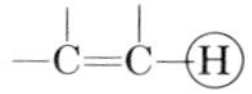

Group	Structure	Actual δ Value	Chemical Shift Relative to Alkyl H
Vinyl hydrogens	C=C—(H)	4.5–6	—
Aldehydes, acid	O=C—(H), —CO—(H)	9–12	—
Tetramethylsilane, TMS reference		0	—
Aromatic hydrogens	⬡—(H)	7–8	—
Alcohol, phenols, amines	C—O—(H), ⬡—O—(H), N—(H)	variable	—

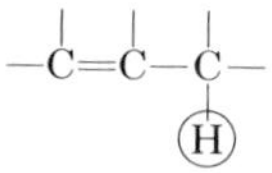

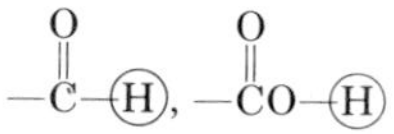

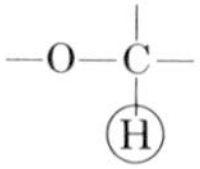

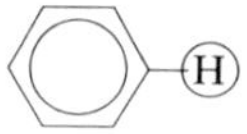

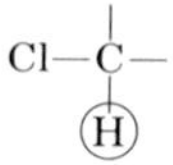

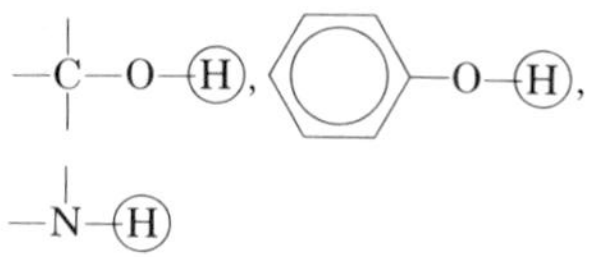

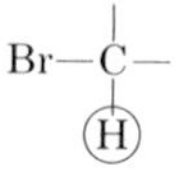

[a]The table indicates the shift of the circled hydrogen under the influence of the Z group.

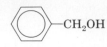 —CH$_2$OH   Benzyl alcohol

The nmr has three distinct peaks at $\delta = 2.4$, $\delta = 4.6$, and $\delta = 7.3$ ($\delta = 0$ is the TMS reference). Examination of the molecule confirms three types of hydrogen present: one hydrogen bonded to an oxygen, two hydrogens bonded equivalently to a carbon, and five essentially equivalent hydrogens attached to the benzene ring. Using Table 16.2, we can now assign each hydrogen type to a signal. Aromatic hydrogens occur between $\delta = 7$ and 8, while hydroxy hydrogens have variable δ values. So the $\delta = 7.3$ must belong to the five benzene hydrogens. The third column in Table 16.2 indicates that the methylene hydrogens, if attached to an alkyl group, R—CH$_2$—, would generate a signal at 0.9–1.6. However, the fourth column specifies the chemical shift due to adjacent groups. Since a benzene ring (Z = aromatic) and an oxygen (Z = O) are in those positions, their influences will shift the methylene hydrogen's signal 1.4 and 2.8 units, respectively. Added to the normal signal ($\delta = 1.0$), the sum is $\delta = 5.2$, very close to the 4.6 δ value. By elimination, the $\delta = 2.4$ signal must be due to the hydroxy hydrogen.

As another example, let us predict the chemical shifts of the three types of hydrogens in the following molecule.

a. aromatic $\delta = 7$–8

b. 1.0 normal δ value
1.0 adjacent C=O
2.0 adjacent Br

1.4 adjacent—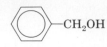

$\delta = 5.4$ total

c. Z = C $\delta = 2$–2.5

(a)

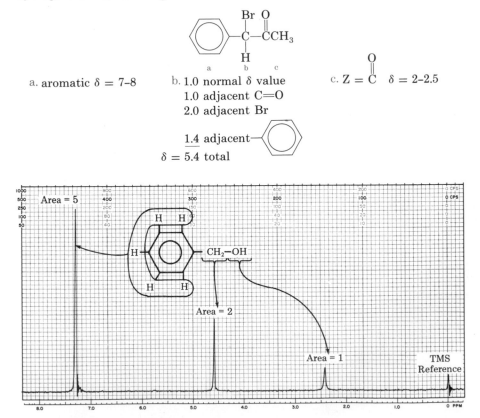

FIGURE 16.8. (see next page). Some nmr spectra. (a) Benzyl alcohol. (b) Ethyl alcohol. (c) Isopropyl benzene. (d) Sample problem (see Section 16.4. E). NMR spectra courtesy of Varian Associates.

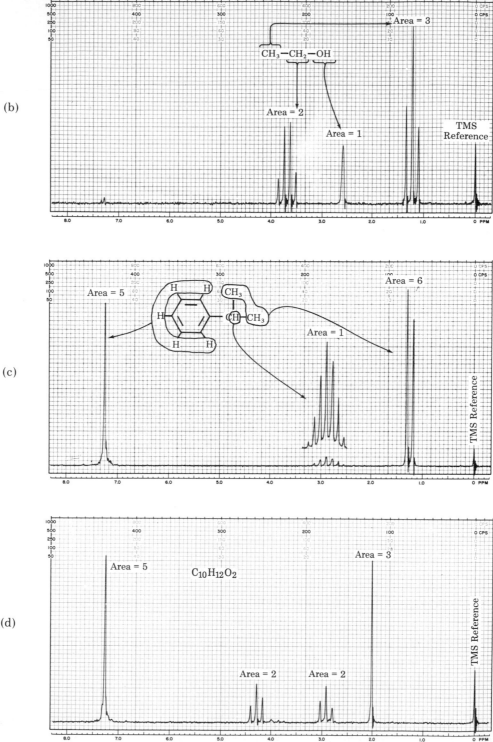

(b)

CH₃—CH₂—OH
Area = 3
Area = 2
Area = 1
TMS Reference

(c)

Area = 5
Area = 6
Area = 1
TMS Reference

(d)

Area = 5
$C_{10}H_{12}O_2$
Area = 3
Area = 2
Area = 2
TMS Reference

B. Integration

The relative areas under the various peaks of an nmr spectrum are in proportion to the number of hydrogens contributing to each signal. These areas can be electronically integrated by an nmr spectrometer. Comparison of the areas provides the ratio among the various kinds of hydrogens in the molecule. Consider the nmr spectrum of benzyl alcohol (Figure 16.8(a)), for example. The hydrogens in the molecule are in a 1:2:5 ratio, like the corresponding peak areas in the spectrum.

C. Peak Splitting

Hydrogens on adjacent carbons, each with a different chemical shift, can influence the signal of the other. This influence appears as peak splitting. We can generalize the phenomenon by saying that the number of peaks into which a particular hydrogen's signal is split equals one more than the total number of hydrogens on directly adjacent carbons. Assuming that each of the following types of hydrogens is nonequivalent, we should obtain the indicated splitting patterns.

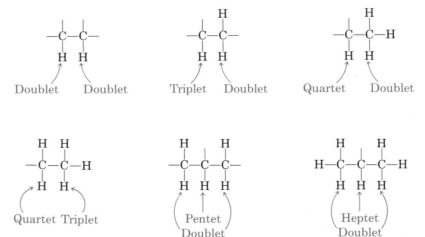

In Figure 16.8(b) (ethyl alcohol), note that the ethyl group is indicated by a quartet and a triplet and that the isopropyl group in Figure 16.8(c) (isopropyl benzene) shows as a heptet and a doublet. The split peaks are not of equal height. The number and relative heights of the peaks follow the mathematical triangle that appears here.

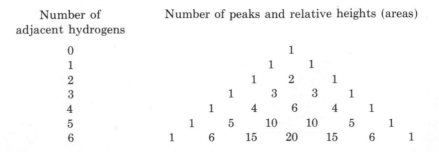

Number of adjacent hydrogens			Number of peaks and relative heights (areas)						
0						1			
1					1		1		
2					1	2	1		
3				1	3		3	1	
4			1	4		6		4	1
5		1	5	10		10	5		1
6	1	6	15	20		15	6		1

D. Summary of NMR

The following aspects of nmr provide information.

Chemical Shift. The number of signals corresponds to the number of different types of hydrogens in the molecule. The positions of each signal give information about the structural environment of the hydrogens.

Integration. The relative areas under the signals give the ratio of the numbers of each hydrogen type in the molecule. If the molecular formula is known, the actual number of each type of hydrogen can be determined.

Splitting. The number of peaks into which a signal is split is one more than the total number of hydrogens on directly adjacent carbons.

E. Sample NMR Problem

To conclude our discussion of nmr, let us identify an unknown compound. Consider Figure 16.8(d), the nmr spectrum of a compound with the formula $C_{10}H_{12}O_2$. The four signals indicate four different types of hydrogens. At $\delta = 7.3$, there are five hydrogens in the aromatic region—probably a monosubstituted benzene ring. The simplest way of expressing two equivalent hydrogens as in the signal at $\delta = 2.9$ and also at $\delta = 4.3$ is with methylene, CH_2, groups. Finally, the signal at $\delta = 2$ worth three hydrogens is most simply expressed as a methyl group. Remaining in the formula are a carbon and two oxygens; these are most simply expressed as —CO—. Although obviously there are other arrangements of

$$\underset{\overset{\|}{O}}{-CO-}$$

all the groups mentioned, these are the simplest expressions and should be considered first. The pieces of the puzzle are

$$\delta = 7.3 \quad \bigcirc\!\!\!\!\!\bigcirc\!\!- \qquad \delta = 4.3 \quad -CH_2- \qquad \delta = 2.9 \quad -CH_2-$$

$$\delta = 2.0 \quad -CH_3 \qquad \qquad \underset{\overset{\|}{O}}{-CO-}$$

The spectrum shows that the two CH_2 groups split each other and thus must be adjacent to each other (—CH_2CH_2—). In this arrangement, each CH_2 splits the other into a triplet (one more peak than the number of hydrogens). Now the puzzle has fewer pieces.

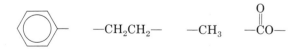

Since one of the CH_2 groups has a chemical shift of $\delta = 4.3$, it must be connected to an oxygen (see Table 16.2). We are now down to three segments.

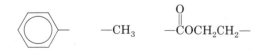

The methyl group cannot be bonded to the benzene ring (even though the chemical shift is right) since this would make a complete molecule (toluene) and we should not yet have used all the pieces. It cannot be bonded to the other CH_2 group because the chemical shift is wrong, the methyl would have appeared as a triplet if so attached and the methylene group would have been even more heavily split. Bonding it to the carbonyl (C=O) is consistent with the chemical shift and lack of splitting. Attaching the benzene ring to the remaining methylene is also consistent with the spectrum. The unknown then is

MASS SPECTROMETRY

16.5   By mass spectral analysis, it is possible to determine the molecular weight and molecular formula of a compound. The structure of the compound is determined by breaking the molecule into smaller, identifiable fragments and then piecing them back together, like a puzzle.

Mass spectral analysis is initiated by bombarding a vaporized sample with an electron beam. This can cause an electron to be dislodged from the molecule, producing a positive molecular ion. If the electron beam is sufficiently energetic, it may cause the molecule to rupture into a variety of positive fragments.

$$\text{ABC Molecule} \xrightarrow[\text{beam}]{\text{Electron}} \begin{matrix} ABC^+ & & \\ AB^+ & C^+ & \\ A^+ & BC^+ & \end{matrix}$$

Molecular ion

Fragmentation ions

The ions are then subjected to magnetic and electric fields. Since most of them have a single positive charge, they are separated according to mass (actually, mass to charge ratio, m/e), and the separation is recorded on chart paper. The intensity of the peak describes the relative abundance of a particular ion. Usually the spectrum is then recorded in tabular form, correlating the mass and relative abundance of each ion. For example, the mass spectrum of carbon dioxide is as shown in Table 16.3.

TABLE 16.3. MASS SPECTRUM OF
CARBON DIOXIDE

Mass (m/e)	Relative Abundance, %
28	20
29	0.2
44	100
45	1.11

The most intense peak in the mass spectrum is called the base peak and is assigned a value of 100%. The peak formed by the loss of one electron from the molecule is called the molecular ion M. In CO_2, the base peak and molecular ion peak are the same. Any peaks of less mass than the molecular ion are called fragment peaks.

A. Molecular Formula Determination

The atomic weights of common elements are averages of the weights of naturally occurring isotopes. For example, the atomic weight of chlorine is 35.5, since there are two abundant isotopes of chlorine in nature—Cl^{35}, 75%; and Cl^{37}, 25%. The mass spectrometer detects each isotope separately, and for chlorine there would be a peak at $m/e = 35$ and a peak at $m/e = 37$, one-third (25/75) as high. By considering such isotopic abundances, one can often determine the elemental composition of a compound.

Carbon. Most natural carbon is C^{12} but about 1.1% is C^{13}. For every carbon in the molecular ion M, the next higher peak M + 1 is 1.1% of the M peak. Note in Table 16.3, the M + 1 peak of CO_2 is 1.11% of the M peak.

$$\frac{\text{Number of}}{\text{carbons in M peak}} = \frac{\text{rel. abund. M} + 1}{0.011 \times \text{rel. abund. M}}$$

C_5H_{12}				$C_{10}H_{24}$			
$C_5^{12}H_{12}$	M	72	100%	$C_{10}^{12}H_{24}$	M	144	100%
$C_4^{12}C^{13}H_{12}$	M + 1	73	5.5%	$C_9^{12}C^{13}H_{24}$	M + 1	145	11%

Chlorine. In compounds containing chlorine, the M + 2 peak (two mass units heavier than molecular ion) is about 33% of the molecular ion for each chlorine.

CH_3Cl				CH_2Cl_2			
CH_3Cl^{35}	M	50	100%	$CH_2Cl_2^{35}$	M	84	100%
CH_3Cl^{37}	M + 2	52	33%	$CH_2Cl^{35}Cl^{37}$	M + 2	86	66%

Bromine. Naturally occurring bromine is almost equally abundant in Br^{79} and Br^{81}. So for every bromine in a molecule, the M + 2 peak is approximately 100% of the M peak.

CH_3Br				CH_2Br_2			
CH_3Br^{79}	M	94	100%	$CH_2Br_2^{79}$	M	172	100%
CH_3Br^{81}	M + 2	96	99%	$CH_2Br^{79}Br^{81}$	M + 2	174	198%

Sulfur. For compounds containing sulfur, the M + 2 peak is 4.5% of the M peak for each sulfur, owing to the small isotopic abundance of S^{34} compared with S^{32}.

Nitrogen. If the molecular ion has an odd mass number, there are an odd number of nitrogens in the compound.

Hydrogen, Oxygen. Hydrogen, oxygen, and other common elements must be deduced by elimination after the other elemental components have been determined.

B. Fragmentation Patterns

Using the M, M + 1, and M + 2 peaks, we can obtain the molecular mass and either a partial or a complete molecular formula. How do we obtain a structural formula? This is accomplished by analyzing the fragment peaks. In mass spectrometry, we take a large molecule whose structure is unknown and break it down with a beam of electrons into smaller, more easily identifiable fragments. The fragments are then pieced back together to obtain the structure of the unknown molecule.

In a mass spectrometer, a molecule can undergo almost any possible cleavage to form all imaginable fragment ions. Fortunately, not all fragment ions form with equal ease. In general, we can say that the probability of fragmentation depends on: (1) bond strengths—almost all important fragmentations in organic molecules are at single bonds rather than at stronger double and triple bonds; and (2) carbocation stability—the fragments are positive and the more stable the fragment, the greater the ease of formation. Table 16.4 summarizes some of the main cleavage patterns for the types of compounds we have studied in this text. The table also summarizes the numerical sequences of mass numbers associated with the particular fragment types. For example, consider the masses of alkyl groups. The simplest methyl, CH_3^+, has a mass of 15. Ethyl, $CH_3CH_2^+$, is 29, 14 more, and each of the subsequent alkyl fragments has masses 14 units more than the one before.

Let us take a specific example, methyl ethyl ketone. From Table 16.4, we see that ketones fragment predominantly on either side of the carbonyl group, giving four principal peaks, in this case at $m/e = 15$, 29, 43, and 57.

$$CH_3 \!+\!\! \overset{\displaystyle O}{\overset{\|}{C}} \!+\! CH_2CH_3 \longrightarrow \underset{15}{CH_3^+} \quad \underset{29}{CH_3CH_2^+} \quad \underset{43}{CH_3\overset{\displaystyle O}{\overset{\|}{C}}{}^+} \quad \underset{57}{CH_3CH_2\overset{\displaystyle O}{\overset{\|}{C}}{}^+}$$

Methyl ethyl
ketone

It is evident that both alkyl (R—) and acyl ($R\overset{\displaystyle O}{\overset{\|}{C}}$—) groups have the same numerical sequence. However, if we were to piece this together, we should assign the lowest mass number to the smaller alkyl group and the largest to the larger acyl group.

Now let us identify an unbranched ketone with the formula $C_7H_{14}O$ and principal peaks at 29, 43, 57, and 71. The smallest fragment, 29, corresponds to an ethyl group $CH_3CH_2^+$ and the largest, 71, to the acyl group $CH_3CH_2CH_2\overset{\displaystyle O}{\overset{\|}{C}}{}^+$. The compound is ethyl propyl ketone.

$$CH_3CH_2\overset{\displaystyle O}{\overset{\|}{C}}CH_2CH_2CH_3 \longrightarrow \underset{29}{CH_3CH_2^+} \quad \underset{43}{CH_3CH_2CH_2^+} \quad \underset{57}{CH_3CH_2\overset{\displaystyle O}{\overset{\|}{C}}{}^+} \quad \underset{71}{CH_3CH_2CH_2\overset{\displaystyle O}{\overset{\|}{C}}{}^+}$$

Ethyl propyl ketone

Another approach to structure determination is to identify the group that must have fallen off the molecule to produce an abundant fragment. To do this, we determine the mass difference between the molecular ion and the important fragment ions. For example, from Table 16.4 we see that alcohols preferentially cleave at the carbon bearing the hydroxy group.

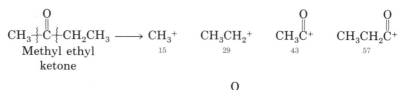

TABLE 16.4.   MAIN CLEAVAGE PATTERNS IN MASS SPECTROMETRY

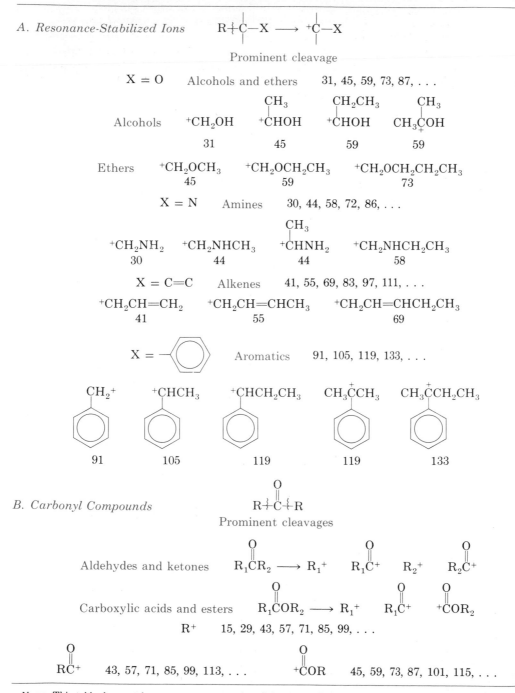

A. Resonance-Stabilized Ions
$$R \!\!+\!\! \overset{|}{\underset{|}{C}} \!\!-\!\! X \longrightarrow {}^{+}\overset{|}{\underset{|}{C}} \!\!-\!\! X$$

Prominent cleavage

X = O   Alcohols and ethers   31, 45, 59, 73, 87, . . .

Alcohols ${}^{+}CH_2OH$ ${}^{+}\overset{CH_3}{\underset{}{CHOH}}$ ${}^{+}\overset{CH_2CH_3}{\underset{}{CHOH}}$ $CH_3\overset{CH_3}{\underset{+}{COH}}$
                    31   45   59   59

Ethers ${}^{+}CH_2OCH_3$ ${}^{+}CH_2OCH_2CH_3$ ${}^{+}CH_2OCH_2CH_2CH_3$
                 45   59   73

X = N   Amines   30, 44, 58, 72, 86, . . .

${}^{+}CH_2NH_2$ ${}^{+}CH_2NHCH_3$ ${}^{+}\overset{CH_3}{\underset{}{CHNH_2}}$ ${}^{+}CH_2NHCH_2CH_3$
   30   44   44   58

X = C=C   Alkenes   41, 55, 69, 83, 97, 111, . . .

${}^{+}CH_2CH{=}CH_2$ ${}^{+}CH_2CH{=}CHCH_3$ ${}^{+}CH_2CH{=}CHCH_2CH_3$
      41   55   69

X = ⬡   Aromatics   91, 105, 119, 133, . . .

$CH_2{}^{+}$ ⬡ ${}^{+}CHCH_3$ ⬡ ${}^{+}CHCH_2CH_3$ ⬡ $CH_3\overset{+}{C}CH_3$ ⬡ $CH_3\overset{+}{C}CH_2CH_3$ ⬡
   91   105   119   119   133

B. Carbonyl Compounds
$$R \!\!+\!\! \overset{O}{\overset{\|}{C}} \!\!+\!\! R$$

Prominent cleavages

Aldehydes and ketones $R_1\overset{O}{\overset{\|}{C}}R_2 \longrightarrow R_1{}^{+}$ $R_1\overset{O}{\overset{\|}{C}}{}^{+}$ $R_2{}^{+}$ $R_2\overset{O}{\overset{\|}{C}}{}^{+}$

Carboxylic acids and esters $R_1\overset{O}{\overset{\|}{C}}OR_2 \longrightarrow R_1{}^{+}$ $R_1\overset{O}{\overset{\|}{C}}{}^{+}$ ${}^{+}\overset{O}{\overset{\|}{C}}OR_2$

R^{+} 15, 29, 43, 57, 71, 85, 99, . . .

$R\overset{O}{\overset{\|}{C}}{}^{+}$ 43, 57, 71, 85, 99, 113, . . . ${}^{+}\overset{O}{\overset{\|}{C}}OR$ 45, 59, 73, 87, 101, 115, . . .

NOTE: This table does not by any means summarize all important fragmentation patterns; it is a good start, however.

Suppose we have an unknown alcohol that gives a mass spectrum with a molecular ion of 116 and principal peaks at 101, 87, and 73. Now let us determine the difference between the molecular ion and fragment peaks, M-101, M-87, and M-73. The peak M-101 is 15. One of the R groups then must have a mass of 15 and be a methyl. M-87 is 29, and must represent an ethyl group. Finally, M-73 is 43, either a propyl or an isopropyl group. A possible structure then for the unknown is one in which $R_1 = $ —CH_3, $R_2 = $ —CH_2CH_3, and $R_3 = $ —$CH_2CH_2CH_3$.

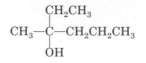

$$
\begin{array}{c}
CH_2CH_3 \\
| \\
CH_3\text{—}C\text{—}CH_2CH_2CH_3 \\
| \\
OH
\end{array}
$$

END-OF-CHAPTER PROBLEMS

16.1 Infrared Spectroscopy: How could one distinguish between the members of the following sets of compounds by infrared spectroscopy? Give the wavelength or wavenumber of an easily identifiable absorption band that would appear in one molecule but not the other. Identify the bond responsible for the absorption.

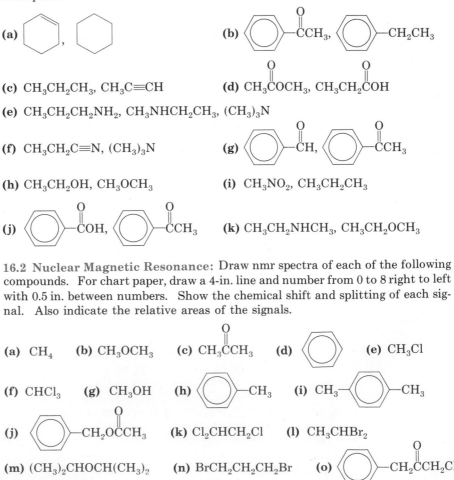

(a)

(b)

(c) $CH_3CH_2CH_3$, $CH_3C{\equiv}CH$

(d) $CH_3\overset{O}{\overset{||}{C}}OCH_3$, $CH_3CH_2\overset{O}{\overset{||}{C}}OH$

(e) $CH_3CH_2CH_2NH_2$, $CH_3NHCH_2CH_3$, $(CH_3)_3N$

(f) $CH_3CH_2C{\equiv}N$, $(CH_3)_3N$

(g)

(h) CH_3CH_2OH, CH_3OCH_3

(i) CH_3NO_2, $CH_3CH_2CH_3$

(j)

(k) $CH_3CH_2NHCH_3$, $CH_3CH_2OCH_3$

16.2 Nuclear Magnetic Resonance: Draw nmr spectra of each of the following compounds. For chart paper, draw a 4-in. line and number from 0 to 8 right to left with 0.5 in. between numbers. Show the chemical shift and splitting of each signal. Also indicate the relative areas of the signals.

(a) CH_4 **(b)** CH_3OCH_3 **(c)** $CH_3\overset{O}{\overset{||}{C}}CH_3$ **(d)**   **(e)** CH_3Cl

(f) $CHCl_3$ **(g)** CH_3OH **(h)** —CH_3 **(i)** CH_3—⟨⟩—CH_3

(j) —$CH_2O\overset{O}{\overset{||}{C}}CH_3$ **(k)** Cl_2CHCH_2Cl **(l)** CH_3CHBr_2

(m) $(CH_3)_2CHOCH(CH_3)_2$ **(n)** $BrCH_2CH_2CH_2Br$ **(o)** —$CH_2\overset{O}{\overset{||}{C}}CH_2Cl$

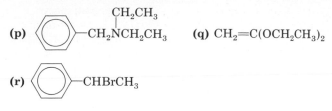

(p) <chemical structure> —CH$_2$NCH$_2$CH$_3$ with CH$_2$CH$_3$ group

(q) CH$_2$=C(OCH$_2$CH$_3$)$_2$

(r) <chemical structure> —CHBrCH$_3$

16.3 Nuclear Magnetic Resonance: In each of the following problems, an nmr spectrum is described (chemical shift, splitting, ratio of hydrogens), and two or three isomeric compounds are given. Pick the compound whose spectrum is described and explain your choice.

(a) $\delta = 2.2$ singlet: CH$_3$CH$_2$CH (with O double bond) or CH$_3$CCH$_3$ (with O double bond)

(b) $\delta = 3.8$ singlet (3), $\delta = 7.5$ (5):

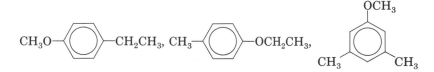

(c) $\delta = 1.2$ doublet (6), $\delta = 3.5$ singlet (3), $\delta = 4.0$ heptet (1):

CH$_3$CH$_2$OCH$_2$CH$_3$, CH$_3$OCHCH$_3$ (with CH$_3$ branch)

(d) $\delta = 1.3$ triplet (3), $\delta = 2.7$ quartet (2), $\delta = 7.2$ singlet (2):

<chemical structures: C(CH$_3$)$_3$-benzene, CH$_3$CH$_2$-benzene-CH$_2$CH$_3$, and tetramethylbenzene>

(e) $\delta = 1.2$ triplet (3), $\delta = 2.5$ quartet (2), $\delta = 3.7$ singlet (3), $\delta = 7.2$ (4):

<chemical structures: CH$_3$O-benzene-CH$_2$CH$_3$, CH$_3$-benzene-OCH$_2$CH$_3$, and dimethyl-methoxybenzene>

16.4 Nuclear Magnetic Resonance: In each of the following problems, a molecular formula is given and the nmr spectrum described (chemical shift, splitting, ratio of hydrogens). Draw a structural formula consistent with the molecular formula and the nmr spectrum.

(a) C$_3$H$_6$O$_2$: $\delta = 2.0$, singlet (1); $\delta = 4.0$, singlet (1)
(b) C$_6$H$_{12}$O$_2$: $\delta = 1.4$, singlet (3); $\delta = 2.1$, singlet (1)
(c) C$_2$H$_6$O: $\delta = 1.2$, triplet (3); $\delta = 2.6$, singlet (1); $\delta = 3.7$, quartet (2)
(d) C$_4$H$_8$O: $\delta = 1.1$, triplet (3); $\delta = 2.1$, singlet (3); $\delta = 2.4$, quartet (2)
(e) C$_3$H$_7$Br: $\delta = 1.7$, doublet (6); $\delta = 4.3$, heptet (1)
(f) C$_2$H$_4$O$_2$: $\delta = 2.0$, singlet (3); $\delta = 11.4$, singlet (1)
(g) C$_2$H$_3$Cl$_3$: $\delta = 3.9$, doublet (2); $\delta = 5.8$, triplet (1)
(h) C$_3$H$_6$O: $\delta = 2.7$, pentet (1); $\delta = 4.7$, triplet (2)
(i) C$_3$H$_8$S: $\delta = 1.3$, triplet (3); $\delta = 2.1$, singlet (3); $\delta = 2.5$, quartet (2)
(j) C$_7$H$_8$: $\delta = 2.3$, singlet (3); $\delta = 7.2$, singlet (5)

(k) $C_{13}H_{11}Cl$: $\delta = 6.1$, singlet (1); $\delta = 7.3$, singlet (10)
(l) $C_{15}H_{14}O$: $\delta = 2.2$, singlet (3); $\delta = 5.1$, singlet (1); $\delta = 7.3$, singlet (10)
(m) $C_3H_5ClO_2$: $\delta = 1.8$, doublet (3); $\delta = 4.5$, quartet (1); $\delta = 11.2$, singlet (1)
(n) $C_4H_6Cl_2O_2$: $\delta = 1.4$, triplet (3); $\delta = 4.3$, quartet (2); $\delta = 6.9$, singlet (1)
(o) $C_7H_{12}O_4$: $\delta = 1.3$, triplet (3); $\delta = 3.4$, singlet (1); $\delta = 4.2$, quartet (2)
(p) C_8H_{10}: $\delta = 1.3$, triplet (3); $\delta = 2.7$, quartet (2); $\delta = 7.2$, singlet (5)

16.5 Mass Spectrometry: In each of the following problems the M, M + 1, and M + 2 peaks of the mass spectra of the unknown compounds are given. Calculate a molecular formula.
(a) 114 = 100%, 115 = 8.8%, 116 = 0.1%
(b) 64 = 100%, 65 = 2.2%, 66 = 33%
(c) 136 = 40%, 137 = 1.3%, 138 = 39%
(d) 48 = 100%, 49 = 1.1%, 50 ≃ 4.5%
(e) 96 = 80%, 97 = 1.8%, 98 = 54%
(f) 234 = 50%, 235 = 3.3%, 236 = 99%
(g) 59 = 75%, 60 = 2.5%, 61 = 0.05%
(h) 140 = 30%, 141 = 2.3%, 142 = 10%
(i) 92 = 70%, 93 = 1.5%, 94 = 6.3%

16.6 Mass Spectrometry: In each of the following problems, a partial description of an unknown compound is presented along with the important peaks of its mass spectrum. Write a structure consistent with the data for the unknown.
(a) A straight-chained ketone with the formula $C_8H_{16}O$ (M = 128) and major m/e peaks at 29, 57, 71, and 99.
(b) A straight-chained alkene with the formula C_7H_{14} (M = 98) and major m/e peaks at 69 and 83.
(c) A monosubstituted benzene with the formula $C_{12}H_{18}$ (M = 162) and major m/e peaks at 133 and 147.
(d) A straight-chained secondary alcohol with the formula $C_7H_{16}O$ (M = 116) and major m/e peaks at 59 and 87.
(e) A straight-chained secondary amine with the formula $C_5H_{13}N$ (M = 87) and major m/e peaks at 58 and 72.
(f) A straight-chained ester with the formula $C_{10}H_{20}O_2$ (M = 172) and major m/e peaks at 57, 85, and 115.

16.7 Mass Spectrometry: Using the concepts in Table 16.4, predict the major peaks (one to five peaks) in the mass spectra of the following compounds:

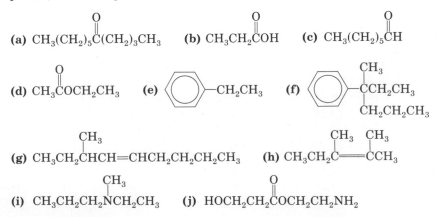

(a) $CH_3(CH_2)_5\overset{\text{O}}{\overset{\|}{C}}(CH_2)_3CH_3$ **(b)** $CH_3CH_2\overset{\text{O}}{\overset{\|}{C}}OH$ **(c)** $CH_3(CH_2)_5\overset{\text{O}}{\overset{\|}{C}}H$

(d) $CH_3\overset{\text{O}}{\overset{\|}{C}}OCH_2CH_3$ **(e)** $\langle\text{C}_6\text{H}_5\rangle-CH_2CH_3$ **(f)** $\langle\text{C}_6\text{H}_5\rangle-\overset{CH_3}{\underset{CH_2CH_2CH_3}{C}}CH_2CH_3$

(g) $CH_3CH_2\overset{CH_3}{\underset{|}{C}}HCH=CHCH_2CH_2CH_2CH_3$ **(h)** $CH_3CH_2\overset{CH_3}{\underset{|}{C}}=\overset{CH_3}{\underset{|}{C}}CH_3$

(i) $CH_3CH_2CH_2\overset{CH_3}{\underset{|}{N}}CH_2CH_3$ **(j)** $HOCH_2CH_2\overset{\text{O}}{\overset{\|}{C}}OCH_2CH_2NH_2$

Appendix

Summary of IUPAC Nomenclature

In this appendix the nomenclature of organic compounds presented in the text is compiled and summarized. More detailed presentations of the various aspects of organic nomenclature and examples can be found in the following sections.

	Section
Nonaromatic Hydrocarbons	3.2
Aromatic Compounds	5.4
Halogenated Compounds	7.1
Alcohols, Phenols, and Ethers	8.1
Aldehydes and Ketones	9.2
Carboxylic Acids and Derivatives	11.1
Amines	13.2

PART I: NOMENCLATURE OF NONAROMATIC COMPOUNDS

Nomenclature Rule 1: Naming the Carbon Chain

The base of the name of an organic compound is derived from the Greek name for the number of carbon atoms present in the longest continuous chain of carbons (Table A.1). A cyclic chain is designated by the prefix *cyclo-*. Side-chain alkyl and alkoxy groups are named as shown in Table A.2.

TABLE A.1. CONTINUOUS-CHAIN HYDROCARBONS

First Ten Hydrocarbons

CH_4	Methane	$CH_3(CH_2)_4CH_3$	Hexane
CH_3CH_3	Ethane	$CH_3(CH_2)_5CH_3$	Heptane
$CH_3CH_2CH_3$	Propane	$CH_3(CH_2)_6CH_3$	Octane
$CH_3(CH_2)_2CH_3$	Butane	$CH_3(CH_2)_7CH_3$	Nonane
$CH_3(CH_2)_3CH_3$	Pentane	$CH_3(CH_2)_8CH_3$	Decane

Higher Hydrocarbons

$C_{11}H_{24}$	Undecane	$C_{21}H_{44}$	Heneicosane	$C_{42}H_{86}$	Dotetracontane
$C_{12}H_{26}$	Dodecane	$C_{22}H_{46}$	Docosane	$C_{53}H_{108}$	Tripentacontane
$C_{13}H_{28}$	Tridecane	$C_{23}H_{48}$	Tricosane	$C_{64}H_{130}$	Tetrahexacontane
$C_{15}H_{32}$	Pentadecane	$C_{30}H_{62}$	Triacontane	$C_{99}H_{200}$	Nonanonacontane
$C_{20}H_{42}$	Eicosane	$C_{31}H_{64}$	Hentriacontane	$C_{100}H_{202}$	Hectane

TABLE A.2. ALKYL AND ALKOXY GROUPS

CH_3—	Methyl	CH_3O—	Methoxy
CH_3CH_2—	Ethyl	CH_3CH_2O—	Ethoxy
$CH_3CH_2CH_2$—	Propyl	$CH_3CH_2CH_2O$—	Propoxy
$CH_3(CH_2)_2CH_2$—	Butyl	$CH_3(CH_2)_2CH_2O$—	Butoxy
$CH_3(CH_2)_3CH_2$—	Pentyl	$CH_3(CH_2)_3CH_2O$—	Pentoxy

Branched alkyl groups

		CH_3	CH_3
CH_3CHCH_3	$CH_3CHCH_2CH_3$	CH_3CHCH_2—	CH_3CCH_3
Isopropyl	Secondary butyl	Isobutyl	Tertiary butyl
	(*sec-*)		(*tert-*, or *t-*)

Nomenclature Rule 2: Describing Carbon-Carbon Bonds

If all carbon-carbon bonds in the longest continuous carbon chain are single bonds, this is indicated by the suffix -*ane*. Double bonds are described by the suffix -*ene*, and triple bonds by the suffix -*yne*. See Table A.3a.

TABLE A3. GROUP NOMENCLATURE: PREFIXES AND SUFFIXES

Class	Functional Group	Prefix	Suffix
a. Groups Indicated by Suffix Only			
Alkanes	C—C	. . .	-*ane*
Alkenes	C=C	. . .	-*ene*
Alkynes	C≡C	. . .	-*yne*
b. Groups Indicated by Prefix or Suffix			
* Carboxylic acids	—COOH, $-\overset{O}{\overset{\|}{C}}OH$, —$CO_2H$	*Carboxy-*	-*oic acid*
Aldehydes	—CHO, $-\overset{O}{\overset{\|}{C}}-H$	*Carboxaldo-*	-*al*
Ketones	$-\overset{O}{\overset{\|}{C}}-$	*Keto-*	-*one*
Alcohols	—OH	*Hydroxy-*	-*ol*
Amines	—NH_2	*Amino-*	-*amine*
c. Groups Indicated by Prefix Only			
Halogenated compounds	—F	*Fluoro-*	. . .
	—Cl	*Chloro-*	. . .
	—Br	*Bromo-*	. . .
	—I	*Iodo-*	. . .
Nitrated compounds	—NO_2	*Nitro-*	. . .
Alkylated compounds	—R	*Alkyl-*	. . .
Ethers	—OR	*Alkoxy-*	. . .

* Nomenclature of carboxylic acid derivatives: carboxylic acid $R\overset{O}{\overset{\|}{C}}OH$, suffix -*oic acid*

Acid anhydrides	Acid chlorides	Esters, salts	Amides
$R\overset{O}{\overset{\|}{C}}O\overset{O}{\overset{\|}{C}}R$	$R\overset{O}{\overset{\|}{C}}Cl$	$R\overset{O}{\overset{\|}{C}}OR'(M^+)$	$R\overset{O}{\overset{\|}{C}}NH_2$
change *acid* to *anhydride*	change -*ic acid* to -*yl chloride*	change -*ic acid* to -*ate* preceded by name of alkyl, R', or cation, M^+	change -*oic acid* to *amide*

Nomenclature Rule 3: Naming Functional Groups

The functional groups in Table A.3b are designated by a suffix when only one such group is present. If more than one such group is present, the group highest in the table is allocated the suffix and the rest are indicated by prefixes.

Nomenclature Rule 4: Naming Substituents

The groups in Table A.3c are named only by prefixes.

Nomenclature Rule 5: Numbering the Carbon Chain

In numbering a carbon chain, the lowest numbers are given preferentially to (1) groups in Table A.3b named by suffixes, followed by (2) carbon-carbon double bonds, (3) carbon-carbon triple bonds, and last (4) groups named by prefixes.

Procedure for Naming Organic Compounds

1. Determine and name the longest continuous chain of carbons (Rule 1, Table A.1).
2. Name carbon-carbon double and triple bonds with suffixes (Rule 2, Table A.3a).
3. Name the group highest in Table A.3b with a suffix (Rule 3, Table A.3b).
4. Number the carbon chain (Rule 5). Indicate in the name the positions of groups named by suffixes.
5. Name (and number) all other groups with prefixes. (Rule 4, Table A.3b and c).

PART II: NOMENCLATURE OF SUBSTITUTED AMINES

Nomenclature Rule 6: Locating Substituents on Amines

If one or more hydrogens on the nitrogen of the —NH_2 group are replaced by substituents, the positions of these substituents are indicated by a capital N—.

PART III: NOMENCLATURE OF CARBOXYLIC ACID DERIVATIVES

Nomenclature Rule 7: Suffix Endings of Acid Derivatives

Carboxylic acid derivatives are named by modifying the suffix ending on the name of the parent acid as shown in the footnote of Table A.3.

PART IV: NOMENCLATURE OF AROMATIC HYDROCARBONS

Nomenclature Rule 8: Parent Aromatic Ring Systems

Following are common aromatic ring systems and their names:

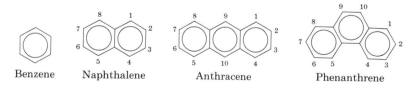

Benzene   Naphthalene   Anthracene   Phenanthrene

Nomenclature Rule 9: Monosubstituted Aromatics

Monsubstituted aromatic compounds are named as derivatives of the parent ring system (such as chlorobenzene). Some monosubstituted benzenes are frequently referred to by common names.

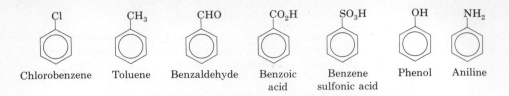

Chlorobenzene   Toluene   Benzaldehyde   Benzoic acid   Benzene sulfonic acid   Phenol   Aniline

Nomenclature Rule 10: Polysubstituted Benzenes

The positions of groups on disubstituted benzenes can be designated by numbers or *o-*, *m-*, *p-*; *ortho-* or 1,2; *meta-* or 1,3; and *para-* or 1,4. On more highly substituted benzenes, numbers only are used to indicate the relative positions of groups.

Nomenclature Rule 11: Aromatic Rings as Prefixes

Aromatic rings can be named as prefixes when this will simplify the overall name.

Phenyl   Benzyl

Index

Abietic acid, 99
ABS detergents, 302
ABS polymers, 101
Acetaldehyde, 210
Acetals, 220–21
  in carbohydrates, 242
Acetate rayon, 251–52
Acetic acid, 257, 260
Acetoacetic ester synthesis, 291
Acetoaminophen, 283
Acetohexamide, 253
Acetone, 210
Acetylene, 17, 103
Acetylsalicylic acid, 282–84
Acid anhydrides
  condensation reactions,
      271–72
  nomenclature, 262
  structure, 261
Acid chlorides
  condensation reactions,
      270–71
  nomenclature, 262
  structure, 261
Acidity
  α-hydrogens, 222–24
  amino acids, 347
  carboxylic acids, 265–69
  phenols, 191–92
Acidity constants, 268
Acids
  carboxylic, 265–69
  Lewis concept, 22
  phenols, 191–92
  sulfonic, 111
Acrilan, 86
Actinomyein D, 367
Addison's disease, 308
1,4 Addition, 87–90
Addition polymers, 81–92, 162
  commercial examples, 84–87
  mechanism of formation,
      81–83
  rubber, 87–92
Addition reactions
  aldehydes and ketones, 212–23
  alkenes and alkynes, 74–81

general equation, 58
1,4 addition, 87–90
Adenine, 374
Adenosine diphosphate, 243, 378
Adenosine triphosphate, 243,
    377–78
Adenyl cyclase, 380
Adipic acid, 276
Adipose tissue, 305
ADP, 243, 378
Adrenalin, 154, 334, 369
Aerobic metabolism, 216
Aerosol propellants, 163
Agar agar, 250
Alanine, 346
Alcohol, 185–87
Alcoholic beverages, 186–87
Alcohols, 32, 178–205
  Grignard synthesis, 182
  nomenclature, 178–81
  physical properties, 182–85
  preparations, 181–82
    Grignard synthesis, 217–20
    hydrogenation, 214
    LiAlH$_4$ reduction, 214–15
  1°, 2°, 3°; 178
  reactions, 191–97
    dehydration, 194
    with hydrogen halides,
        194–96
    oxidation, 196–97
    with PX$_3$, 196
    with sodium, 193
    with thionyl chloride, 196
  structure, 178
  sulfur analogues, 198–99
  uses, 188
Aldehydes, 32, 206–34
  addition reactions
    acetal formation, 220–21
    Grignard addition, 217–20
    hydrogenation, 214
    hydrogen cyanide, 213–14
    imine formation, 221–22
    LiAlH$_4$ reduction, 214–17
    mechanism, 213
    water, hydrogen halides, 212

aldol condensation, 225–27
α-hydrogens, 222–24
haloform reaction, 224–25
nomenclature, 207–10
oxidation, 210–12
physical properties, 207
preparations, 210
reactions, 210–27
reaction summary, 223
reduction, 182
structure, 206
Aldol condensation, 225–27
Aldoses, 235, 237
Aldrin, 129, 172
Alizarin, 332
Alkaloids, 334–40
Alkanes, 28, 32, 49
  halogenation, 60–63
  nomenclature, 50–53
  physical properties, 55–58
Alkenes, 32, 49
  addition reactions, 74–81
    halogenation, 75, 78
    halohydroxylation, 75, 78
    hydration, 75, 78
    hydroboration, 92–93
    hydrogenation, 75
    hydrohalogenation, 75, 78
    mechanism, 76–80
    orientation of addition, 79–80
    polymerization, 81–92
  nomenclature, 54–55
  oxidation
    KMnO$_4$, 93
    ozonolysis, 93–94
  physical properties, 56
  preparations, 63–65, 169–70
  stability, 65
Alkylation of amines, 127,
    320–21
Alkyl benzene sulfonates, 302
Alkyl groups, 51
Alkyl halides, 159–73
  nomenclature, 160
  physical properties, 160
  preparations, 161
  1°, 2°, 3°; 159

Alkynes, 32, 49
  acidity of terminal alkynes, 94
  addition reactions, 74, 76, 80
    halogenation, 76
    hydration, 76
    hydrogenation, 76
    hydrohalogenation, 76
  nomenclature, 54–55
  physical properties, 56
  preparation, 63–65, 164,
    169–70
Allergy, 335
Allose, 237
Allylic free radical, 87–90
Allylic group, 159
Allylic systems
  carbanion, carbocation, free
    radical, 89–90
α and β Configurations, 237–38
Alpha helix, 356–57
Alpha hydrogens, 222–24
Altrose, 237
Aluminum hydroxide, 284
Aluminum trichloride, 22
α-Amanitine, 371
Ames test, 131
Amides
  hydrolysis, 273
  LiAlH$_4$ reduction, 320
  nomenclature, 263–64
  structure, 261
Amines, 32, 313–44
  nomenclature, 314–15
  physical properties, 315–16
  preparations, 319–20
  1°, 2°, 3°; 313
  reactions, 317–28
    alkylation, 320–21
    amide formation, 321
    basicity, 317–19
    diazonium salts, 324–28
    nitrous acid, 323–25
    sulfonamide formation,
      322–23
Amino acid analyzer, 348
Amino acids, 345–50
  reactions
    dipolar ion, 347–48
    ninhydrin, 349
    peptide formation, 350
    Sanger reagent, 349
  table of, 346
p-Aminobenzoic acid, 323
Ammonia, 22, 313
Ammonium nitrate, 22, 189, 317
Ammonium phosphate, 23
Ammonium sulfate, 22
Amobarbital, 284
cAMP, 380
Amphetamine, 334–35
α-Amylase, 155, 243
Amylopectin, 250

Amylose, 250
Amytal, 284
Anaerobic metabolism, 216
Androgens, 309
Anesthetics
  barbiturates, 284–85
  ether, 188
  local, 285
  phenols, 188–89
Angle strain, 38
Aniline, 111, 313
  tribromination of, 319
Anionic detergents, 302
Anionic polymerization, 83
Anions, 6
Anomers, 238
Antacids, 283–84
Anthracene, 111
Anthracite, 66
Anthraquinone dyes, 332
Antibiotics, 366–67
Antifreeze, 185, 187, 197
Antihistamines, 335–36
Antimalarial drugs, 338
Antioxidants, 190, 297, 298, 311
Antiseptics, 185, 188, 302
APC tablets, 283
Apricot oil, 262
Arabinose, 237
Arginine, 346
Aromatic compounds, 32,
    104–36
  carcinogens, 130
  characteristic properties, 108
  electrophilic substitution,
    114–23
    activating, deactivating
      groups, 122
    Friedel-Crafts reaction, 115,
      117
    halogenation, 115, 117
    nitration, 115, 118
    orientation, 118–22
    o, m, p directors, 119
    reaction mechanism, 116–18
    sulfonation, 115, 118
  IUPAC nomenclature, 111–13
  oxidation of alkylbenzenes,
    123
  physical properties, 110
  structural and bonding re-
    quirements, 108–10
  uses, 114
Aromaticity, 108–10
Aromatization, 128
Arsenic poisoning, 364
Arthritis, 308
Aspartame, 254
Aspartic acid, 346
Asphalt, 67
Aspirin, 114, 282–84, 305
Atherosclerosis, 307

Atom, 1
Atomic orbitals, 2–3
ATP, 243, 377–78
Atropine, 337
Aufbau principle, 3
Auxochrome groups, 330
Axial hydrogens, 41
Azaserine, 367
Azine dyes, 332
Azine group, 329
Azo compounds, 327
Azo dyes, 331
Azo group, 329
Azoxy group, 329

Baekeland, Leo, 228
Baeyer test, 93
Bakelite, 83, 228–30
Baking soda, 284
Banana oil, 261
Barbiturates, 284–85
Base peak, 408
Bases, Lewis concept, 22
Basicity
  amines, 317–19
  amino acids, 347
Basicity constants, 317
Basil, 95–96
Bayberry wax, 95–96
Bayer Company, 283
Baygon, 173
Beer, 186
Beeswax, 293
Belladonna alkaloids, 337
Benadryl, 336
Benedict's test, 211, 241
Benzaldehyde, 111
Benzalkonium chlorides, 302
Benzedrine, 334
Benzene, 104
  bonding, 106–8
  bond lengths, 106
  as a carcinogen, 129
  resonance, 106
  resonance energy, 105
Benzene diazonium chloride,
    324–27
Benzene hexachloride, 177
Benzene sulfonic acid, 111
Benzenesulfonyl chloride, 322
Benzocaine, 285
Benzoic acid, 111
Benzpyrene, 129–30
Benzylic group, 113, 159
Beta-pleated sheet, 356–58
Beverage alcohol, 185–87
BHA, 190
BHC, 177
BHT, 190, 297
Bicarbonate of soda, 284
Bile acids, 308

Biodegradable detergents, 303
Biolipids, 303–11
Biological resolution of enan-
    tiomers, 153
Biomass, 69
Birth control pills, 310
Bituminous coal, 66
Blood buffer, 24
Blood sugar levels, 211, 368
Boat form of cyclohexane, 40
Bond angles
    acetylene, 17
    benzene, 108
    ethene, 15
    methane, 14
    summary, 20
Bond energy, 20
Bonding, 1–21
    covalent, 9–11
    ionic, 6
    single, double, triple bonds,
        10–11
Bond lengths
    benzene, 106
    carboxylate anion, 267
    single, double, triple bonds, 20
Bond rotation, 34–35
Bond strengths, 20
Boron trifluoride, 22
Bradykinin, 350
Bromination, 60–63
Butane, 28
Butter, 257, 294–96
Butylated hydroxy anisole, 190
Butylated hydroxy toluene, 190,
    297
Butyl rubber, 102
Butyraldehyde, 210
Butyric acid, 257, 260
BXT, 126

Cadeverine, 313
Caffeine, 283, 340
Cahn-Ingold-Prelog convention,
    148–53
Cahn, R.S., 148
Calcium carbonate, 284
Calcium cyclamate, 253–54
Calcium propionate, 266
Calcium undecylate, 266
Camphene, 96
Camphor, 46, 96
Cancer, 128–31
Cannabinols, 191
Capric acid, 257
Caproic acid, 257, 260
Caprolactam, 278
Caprylic acid, 257, 260
Carbamate insecticides, 173–74
Carbanions, 59, 223

Carbaryl, 173
Carbocations, 59
    definition, 79
    in dehydration, 64
    in electrophilic addition, 77–80
    in electrophilic substitution,
        116–18
    in nucleophilic substitution,
        168
    primary, secondary, tertiary,
        79–80
    stability, 79–80
Carbofuran, 173
Carbohydrates, 235–56
    disaccharides, 243–48
    monosaccharides, 235–43
    polysaccharides, 248–52
Carbon dioxide, 48, 66
Carbonium ions, 59
Carbon monoxide, 129
    poisoning, 364–65
Carbon tetrachloride, 61, 160
Carbonyl group, 206
Carboxylic acids, 32
    acidity, 265–69
    derivatives, 261–64
    nomenclature, 258–60
    preparations, 220, 265
    reactions of derivatives, 270–
        276
    salt formation, 265–66
    structure, 257
Carboxylic acid salts
    formation and uses, 265–66
    nomenclature, 263
Carboxypeptidase, 352
Carcinogens, 128–131
Carnauba wax, 293
β-Carotene, 98, 330
Carrots, 330
    pigment, 98
Carvone, 98
Cation, 6
Cationic detergents, 302
Cationic polymerization, 81
Cedar oil, 97
Cedrol, 97
Celcon, 228
Celery, oil of, 97
Cell membrane, 306
Cellobiose, 244–45, 248–49
Cellophane, 252
Cellosolves, 198
Celluloid, 251
Cellulose, 155, 245, 248–52
Cellulose acetate, 252
Cellulose nitrate, 189, 251
Cellulose xanthate, 251–52
Cephalin, 303
Ceramide, 304
Cerebroside, 304
Cetylpyridinium chloride, 302

Chain reactions
    free-radical addition poly-
        merization, 82, 87–89
    freons and ozone shield, 162–
        164
    halogenation of alkanes, 62
Chair conformation
    in cyclohexane, 40
    in diamond, 43
    in glucose, 237
Chemical instrumentation,
    393–414
Chemical shift, 401–4
Chiral carbons, 141, 147
    R,S designation, 148–52
Chiral compound, 147
Chloral hydrate, 233
Chlordane, 129
Chlorination
    of alkanes, 60–63
    of aromatics, 115, 117
Chlorofluorocarbons, 162–64
Chloroform, 61, 129, 160, 162
p-Chlorophenoxyacetic acid, 258
Chloroprene, 101–2
Chloroquine, 338
Chlorothymol, 189
Chloroxylenol, 189
Chlorpheniramine, 336
Chlorpropamide, 253
Cholecalciferol, 311
Cholesterol, 39, 97, 307–8
Cholinesterase inhibitors, 173,
    364
Chromophore groups, 329
Chromosomes, 383
Chymotrypsin, 353, 358
Cigarette smoke, 129
Cinnamaldehyde, 104, 227
Cinnamon, oil of, 104
Circular DNA, 382
Cis-trans isomerism
    cyclic compounds, 42
    double bonds, 36–38
Citral, 95–96, 200, 208
Citric acid, 257–58
Citronella, oil of, 98
Citrus fruits, 257
Claisen condensation, 280–81
Cloves, oil of, 99, 104
Coal, 66–68
    gasification, liquefaction, 68
Cocaine, 337
Coconut oil, 295, 299
Codeine, 338
Cod liver oil, 295
Codons, 385
Collodion, 251
Color, 328–30
Combination pain relievers, 283
Combustion, 66
Complementary colors, 329

Complex carbohydrate, 247
Compounds, 1
Condensation polymers, 276–79
Condensation reactions of car-
    boxylic acids and deriva-
    tives, 270–76
Conformational isomerism,
    34–36
    cyclohexane, 40–42
Conformations eclipsed and
    staggered, 34
Congeners, 186
Congo red, 331, 343
Coniine, 336
Conjugated proteins, 361
Contact lenses, 86
Contraceptives, 310
Copolymers, 83
Corey, R. B., 357
Cork, 99
Corn oil, 295
Cortisol, 308
Cortisone, 308
Cotton, 250–51
Cottonseed oil, 295
Cough medicine, 162, 338–39
Covalences, 9–10, 27
Covalent bonding, 9–11
    bonding and nonbonding
        electrons, 11
    polar bonds, 19–20
    valences, 9–10
Cracking, petroleum, 67, 127
Crick, Francis, 380
Crude oil, 66
    enhanced recovery, 69
CTP, 378
Cuscohygrine, 336
Cyanide poisoning, 364–65
Cyanohydrins, 213–14
Cyclamates, 129, 131, 253–54
Cycles per second, 393
Cyclic adenosine menophos-
    phate, 380
Cyclic AMP, 380
Cyclic compounds, 33–34
Cycloalkanes, 38–42
    nomenclature, 50
Cyclohexane
    axial and equatorial positions,
        41
    boat and chair forms, 40
    in diamond, 43
Cycloserine, 367
Cysteine, 346
Cytidine triphosphate, 378
Cytochrome C, 361
Cytosine, 374

2,4-D, 258
Dacron, 114, 278

DBCP, 175
DDT, 129, 172
Decongestants, 335
Dehydration, 64
    of alcohols, 194
Dehydrohalogenation, 63,
    169–71
Delrin, 228
Delta units, 402
Demerol, 339
Denaturation of proteins, 360
Denatured alcohol, 187
2-Deoxy-D-ribose, 145, 374
Deoxyribonucleic acid, 374–76
    replication, 383
    3-D structure, 380–82
DES, 309
Detergents, 114, 299–303
Developers, photographic, 190
Dexedrine, 334
Dextrans, 250
Dextrins, 250
Dextromethorphan, 338
Dextrorotatory, 139
Dextrose, 240
Diabetes, 253, 368–69
Diabinese, 253
Diabis, 253
Diacetyl, 297
Diamond, 43
Diastereomers, 146–47
Diazinon, 364
Diazonium salts, 323–28
    aliphatic, 323
    coupling reactions, 327
    preparation, 324
    replacement reactions, 325
Diborane, 92
p-Dichlorobenzene, 133
2,4-Dichlorophenoxyacetic acid,
    258
Diedrin, 128, 172–73
Dienes, 49
Diesel fuel, 67
Diethylstilbesterol, 309
Diet pills, 335
Dihydroxyaluminum sodium
    carbonate, 284
Diisopropylfluorophosphate,
    364
Dimethylamine, 314
2,4-Dinitrofluorobenzene, 349
2,4-Dinitrophenylhydrazine, 222
Dioctylphthalate, 34
Dioxin, 258
Dipolar ion, 347
Direct dyes, 332
Disaccharides, 243–48
Disinfectants, 185, 188, 302
Disperse dyes, 332
Disulfide, 199
Disulfide bridges, 353

DNA, 374–76
    replication, 383
    3-D structure, 380–82
    transcription, 384–87
DNFB, 349, 351
2,4 DNP, 222
Domagk, Gerhard, 322
Double bond, 10–11, 17, 33–34
Double helix of DNA, 380–82
Drake, Edwin, 66
Dry cleaning agents, 161–62
Drying oils, 298
D-Series, 144, 236
Dyes and dyeing, 328–34
    classification
        application, 330, 332–34
        structure, 330–32
    color, 328–30
Dymelor, 253
Dynamite, 188

E_1 and E_2 mechanisms, 169–70
Eastman, George, 230
E. coli bacteria, 390
Edman degradation, 351
Electromagnetic radiation, 328,
    393–94
Electron, 1
    orbitals, 2–3
    spin, 3
Electron configuration
    atomic orbitals, 2–3
    of carbon, 12
    outer shells, 7
    periodic table, 4–6
    stable octet, 6
Electron dot formulas, 9–11
Electronegativity, 6
Electrophiles, 76, 116–18
Electrophilic addition, 74–80
Electrophilic aromatic substitu-
    tion
    in Bakelite production, 229
    bromination of aniline, 319
    Friedel-Crafts, alkylation, 115,
        117
    halogenation, 115, 117
    nitration, 115, 118
    sulfonation, 115, 118
    using diazonium salts, 327–28
Electrophoresis, 348
Elements, 1
Eleostearic acid, 294–95
Elimination reactions
    of alkyl halides, 169–70
    E_1, E_2 mechanisms, 169–70
    general equation, 58
    in preparation of alkenes and
        alkynes, 63–65
Empirical formula, 26
Emulsion, 243

Enantiomers, 142–43, 147
   resolution, 153–54
Enediols, 241–42
Energy spectrum, 328, 393–94
Enolate ion, 224
Enols, 76, 224
Enzymatic methods, 352
Enzymes, 361–64
   classification, 362
Ephedrine, 334–35
Epinephrine, 154, 334, 369
Epoxides, 197–98
Epoxy Resins, 198
Equatorial hydrogens, 41
Ergosterol, 311
Ergot alkaloids, 339
Erythrose, 237
Esterification, 272–73
Esters
   in carbohydrates, 243
   condensation reactions, 272–
      73
   nomenclature, 263
   structure, 261
Estradiol, 309
Estrogens, 309
Ethanethiol, 199
Ethanol, 185–87
Ethanolamine, 198, 304
Ether, 164, 188
Ether, diethyl, 164, 188
Ethers, 32
   nomenclature, 180–81
   physical properties, 182–84
   preparations, 182
   reaction with hydrogen ha-
      lides, 194–96
   structure, 178
   sulfur analogues, 198–99
   Williamson synthesis, 164,
      171
Ethyl acetate, 275
Ethyl acetoacetate, 291
Ethyl alcohol, 181, 185–87
Ethylene, 127
Ethylene dichloride, 175
Ethylene glycol, 185, 187, 278
Ethylene oxide, 197–98
   Grignard addition, 220
Ethyl ether, 164, 188
Eucalyptus oils, 98
Eugenol, 104
Explosives, 187–89

FAD, 378–79
Familial hypercholesterolemia,
   308
Faraday, Michael, 104
Farnesol, 97

Fats and oils, 293–99
   fatty acid composition, 295
   physical properties, 295
   reactions, 296–99
Fatty acids, 293–95
FDC red 2, 331
Fehling's test, 211, 241
Fermentation, 186
   gasohol production, 127
Fertilizers, 22–23, 127
Fiberglass, 198
Fibrous protein, 358
Fischer projections, 143–44
   of carbohydrates, 238
Fischer-Tropsch process, 68
Flavinadenine dinucleotide,
   378–79
Flavin mononucleotide, 378–79
Fleming, Alexander, 366
FMN, 378–79
Food preservatives, 190, 266
Formaldehyde, 210
Formaldehyde polymers, 228–30
Formic acid, 257, 260
Formulas
   condensed, 27–29
   empirical, 26
   molecular, 26
   structural, 26
Fossil fuels, 70
Fragmentation ions, 408
Fragment peaks, 408
Free-radical polymerization, 82
Free radicals, 59
   in addition polymerization, 82
   in chain reactions, 62, 82, 89,
      163
   in halogenation, 62
   in rubber formation, 87–90
Freons, 162–64
Frequency, 393
Friedel-Crafts reaction, 115, 117,
   210, 229
Friedelin, 99
Fructose, cyclic form, 240
Functional isomerism, 31–33
Furan, 240
Furanose rings, 239–40

Galactose, 237
   in galactosemia, 254
Galactosemia, 254, 388
Gallstones, 308
Gangliosides, 304, 307
Garlic odor, 199
Gasohol, 127, 186
Gasoline, 67, 124–28
   octane boosters, 102
   production, 127–28
Genes, 383
Genetic code, 383–91

Geometric isomerism, 36–38
   cyclic compounds, 42
   double bonds, 36–38
   vision, 37–38
   Z,E configuration, 152–53
Geometry of molecules
   linear, sp, 17–19
   summary, 19–20
   tetrahedral, sp³, 14–15
   triangular, sp², 16–17
Geraniol, 95–96, 200
Gin, 186
Ginger, 96–97
Globular protein, 358
Glucagon, 255, 368
Glucose, 144, 237
   biosynthesis, 227
   blood levels, 311, 368
   cyclic form, 220–21, 237–39
   in starch and cellulose, 248–49
Glutamic acid, 346
Glyceraldehyde, 235–37
Glycerides, 293
Glycerol, 187, 293
Glycine, 346
Glycocholic acid, 308
Glycogen, 250
Glycoside, 242
Goodyear, Charles, 90–91
Grain alcohol, 185–87
Grapes, 257
Graphite, 43
Grignard reagent, 182
   carbonation, 265
Grignard synthesis of alcohols,
   182
Grignard, Victor, 217
GRS, 91
GTP, 378
Guanine, 374
Guanosine triphosphate, 378
Gulose, 237
Gum acacia, 250
Gum tragacanth, 250
Guncotton, 189, 251

Haber process, 127
Hair, 357
   permanents, 199, 353
Haloform reaction, 224–25
Halogenated hydrocarbons, 53
   elimination reactions, 63–65,
      169–71
   nucleophilic substitution,
      164–71
   physical properties, 160
   preparations, 60–63, 161
   uses, 161–64, 171–73
Halogenation
   of alcohols, 194–95
   of alkanes, 60–63

Halogenation (*con't.*)
 of alkenes, 75
 of alkynes, 76
 of aromatic compounds, 115,
  117
Hard water, 301
Haworth formulas, 239
Heating oil, 67
Heme, 359, 361, 364
Hemiacetals, 220–21
 in carbohydrates, 236–37, 242
Hemlock, 336
Hemoglobin, 360–61, 364–66
Heptachlor, 129, 172
Heroin, 338
Hertz, 393
Hesperetin, 253
Heterocyclic aromatic com-
  pounds, 109
Hexamethylene diamine, 276
Hexylresorcinol, 189
Hinsberg test, 322
Histamine, 110, 335–36
Histidine, 346
Histones, 377
Hofmann, Felix, 283
Homologous series, 57
Hormones
 insect juvenile, 174
 protein hormones, 368–69
 sex, 309
 steroid, 308–10
Horns, 357
Hückel 4*n* + 2 rule, 109
Human growth hormone, 391
Humectants, 187
Hybridization
 sp, 18
 sp², 16
 sp³, 15
 summary, 20
Hydration
 of alkenes, 75, 181
 of alkynes, 76, 210
Hydrazine, 352
Hydroboration, 92–93, 181
Hydrocarbons, 28
 classes of, 49
 halogenated, 53
 physical properties, 55–58
 saturated, 48
 unsaturated, 48
Hydrogenation
 of aldehydes and ketones, 182,
  214
 of alkenes, 75
 of alkynes, 76
 of oils, 296–97
Hydrogen-bonding
 in alcohols, 182–85
 in amines, 315–16
 in carboxylic acids, 264

in cellulose, starch, 250
in DNA, 380–82
in dyeing, 330
in nylon, 277
in protein structure, 356–60
Hydrogen cyanide, 13
Hydrogen sulfide, 23, 199
Hydronium ion, 22
Hydrophilic, 300
Hydrophobic, 300
 in proteins, 358
Hydroquinone, 190
Hydroxylamine, 222
Hygrine, 336
Hypercholesterolemia, 308
Hypoglycemia, 253–54

Idose, 237
Imidazole, 110
Imines, 221
Immunization, 390
Indigo, 114, 332, 334
Indole, 110, 314
Indole alkaloids, 339–40
Infrared spectroscopy, 395–400
 absorption assignments, 398
 sample spectra, 399–400
Ingold, C., 148
Ingrain dyes, 327, 332
Insecticides, 171–74
 carbamates, 173–74
 juvenile hormones, 174
 organochlorine, 172–73
 organophosphorus, 173
 pyrethrins, 174
 sex pheromones, 199–200
Insect juvenile hormones, 174
Insect pheromones, 199–200
Insect repellent, 6–12, 233
Instrumentation, 393–414
Insulin, 247, 253, 356, 368, 390
Insulin shock, 255
Integration (NMR), 406
Interferon, 391
Invert sugar, 246
Iodine number, 296
Iodine test for starch, 250
Iodoform reaction, 224–25
Ionic bonding, 6
Ionic valences, 8
Ions, 7
Isobutyl group, 52
Isocyanates, 279
Isoelectric point, 348
Isoleucine, 346
Isomerism
 conformational, 34–36
  cyclohexane, 40–42
 functional, 31–33
 geometric, 36–38, 42
  cyclic compounds, 42

double bonds, 36
 Z,E configurations, 152–53
optical, 137–56
 R,S configuration, 148–52
positional, 30–31
skeletal, 28–31
stereoisomerism, 137
structural, 137
Isomerization, 128
Isomers, 27
Isooctane, 125
Isoprene, 87, 95
Isopropyl alcohol, 181, 185
Isopropyl group, 51
Isoquinoline, 110, 337
Isotopes, 409
IUPAC nomenclature. *See* No-
  menclature, IUPAC

Jet fuel, 67
Juvenile hormones, 174

K_a's for carboxylic acids, 268
Kekule, August, 104
Keratin, 357
Kerosene, 67
Keta-enol tautomerism, 224
Ketones, 32
 addition reactions
  acetal formation, 220–21
  Grignard addition, 217–20
  hydrogenation, 214
  hydrogen cyanide, 213–14
  imine formation, 221–22
  LiAlH₄ reduction, 214–17
  mechanism, 213
  water, hydrogen halides, 212
 aldol condensation, 225–27
 α-hydrogens, 222–24
 haloform reaction, 224–25
 nomenclature, 207–10
 physical properties, 207
 preparations, 210
 reaction summary, 223
 reduction, 182
 structure, 206
Ketoses, 235
Knocking, 125
Kodak, 230
Kodel, 278

Lactase, 247
Lactic acid, 140–41, 213, 247,
  257
Lactobacillus acidophilus, 248
Lactones, 290
Lactose, 247
 and galactosemia, 254
Lactose intolerance, 247
Land, E. H., 137

Lard, 294-96, 199
Lassa fever, 390
Latex, 98
Lauric acid, 294-95
Leaded gasoline, 125
Lead poisoning, 199, 364
Le Bel, J. A., 140
Lecithin, 303
Lemon oil, 96
Leucine, 346
Levorotatory, 139
Levulose, 240
Lewis acids, 22
Lewis bases, 22, 317
Lewis electron dot formulas,
    9-11
Lidocaine, 285
Lignin, 252
Lignite, 66
Limonene, 96
Linoleic acid, 294-95
Linolenic acid, 294-95
Linoleum, 298
Linseed oil, 295
Lipids, 292-312
    fats and oils, 293-99
    phosphoglycerides, 303-4
    prostaglandins, 305
    saponifiable, 292
    sphingolipids, 304
    steroids, 304, 307-10
    vitamins, 310
    waxes, 292-93
Liquefied petroleum gas, 28
Lithium aluminum hydride,
    214-17
LNG, 67
Lobeline, 336-37
Local anesthetics, 188-89, 285
Lozenges, 188-89
LPG, 28, 67
LSD, 289, 339
L-Series, 144, 236
Lubricating oil, 67
Lucas test, 194
Lucite, 83, 86
Luminal, 284
Lycopene, 98
Lysergic acid, 339
Lysergic acid diethylamide, 339
Lysine, 346
Lysozyme, 358
Lyxose, 237

Magnesium trisilicate, 284
Ma Huang, 334
Malachite green, 331
Malaria, 365, 338
Malathion, 173, 364
Malonic acid, 284
Malonic ester synthesis, 279-80

Maltose, 244
Mannose, 237
    cyclic form, 239
Maple sap, 246
Margarine, 296-97
Marijuana, 191
Markovnikov's rule, 80
Mass spectrometry, 408-12
    fragmentation patterns,
        409-11
    molecular formulas, 409
Mechanical resolution of enan-
    tiomers, 153
Melmac, 83, 228
Membranes, 306-7
Menthol, 96
Menthone, 96
Meperidine, 339
Mercaptans, 198-99
Mercury poisoning, 199, 364
Merrifield protein synthesis,
    353-55
Merrifield, R. B., 355-59
Mescaline, 334-35
Meso compounds, 146-47
Messenger RNA, 385
Mestranol, 310
Metabolism
    aerobic, 216
    anaerobic, 216
Methadone, 339
Methane, 28
Methanol, 68, 185
Methapyrilene, 129, 336
Methionine, 346
Methyl acetyl carbinol, 297
2-Methyladenine, 377
Methylamine, 314
p-Methylaminophenol, 190
Methylation, 242
Methyl cellosolve, 198
Methylene blue, 332
Methylene chloride, 160
Methyl orange, 327
Methyl paraben, 190
Methyl salicylate, 104, 189, 262,
    282
Methyl tertiary butyl ether, 102,
    126
Micelles, 300
    in cell membranes, 307
Micrometer, 395
Micron, 395
Microwaves, 394
Milk of magnesia, 284
Mineral oil, 67
MMT, 126
Modified cellulose, 250-52
Molecular formulas
    definition, 26
    by mass spectrometry, 409
Molecular ion, 408

Molecular orbitals
    pi bonds, 12
    sigma bonds, 12
Molozonide, 93
Monomers, 81
Mononucleotide, 375
Monosaccharides, 235-43
    reactions, 241-43
Monosodium glutamate, 266
Mordant dyes, 332
Morphine, 338
Mothballs, 133
Motor octane number, MON,
    125
Movies, 3-D, 139
MTBE, 102, 126
Mueller, Paul, 172
Multiple bonds, 33
Mushroom toxin, 371
Mustard gas, 129, 199
Mutagens, 131
Mutarotation, 240-41
Mutations, 388
Mylar, 278
Myoglobin, 358-59, 361
Myrcene, 95
Myristic acid, 294-95

NAD^+, 216, 379
Nanometer, 397
Naphthalene, 111, 114
β-Naphthoxyacetic acid, 258
β-Naphthylamine, 313
Nasal decongestants, 335
Natural gas, 28, 48, 67
Nembutal, 284
Neoprene rubber, 92
Neosynephrine, 335
Nerve gases, 364
Neutron, 1
Newman projections, 34-35
Niacin, 217, 337, 379
Nicol prism, 137
Nicotinamide, 379
Nicotine, 109, 337
Nicotine acid, 216, 337, 379
Nicotine amide dinucleotide, 216
Night blindness, 38
Ninhydrin, 349
Nitration, 115, 118
Nitrile rubber, 101
Nitriles, 129, 320
    hydrolysis, 265
Nitro compounds, reduction, 320
Nitro dyes, 331
Nitro explosives, 188-89
Nitroglycerin, 187-88
Nitromethane, 189
Nitrosoamines, 389
Nitroso dyes, 331
Nitroso group, 329
Nitrous acid, 323

NMR, 401-8
Nobel, Alfred, 188
Nobel Prize, 188
Noble gas, 6
Noethindrone, 310
Nomenclature
  common, 49
    alcohols, 181
    aldehydes, 209-10
    alkenes, 55
    alkyl halides, 55, 160
    alkynes, 55
    amines, 315
    amino acids, 346
    carbohydrates, 235-36
    carboxylic acids, 260-61
    ethers, 181
    fats and oils, 294
    hydrocarbons, 55
    ketones, 209-10
    organic halogen compounds,
      160-61
  IUPAC, 49
    acid anhydrides, 262
    acid chlorides, 262
    alcohols, 178-80
    aldehydes, 207-9
    alkanes, 50-53
    alkenes, 54-55
    alkyl halides, 160
    alkynes, 54-55
    amides, 263-64
    amines, 314-15
    aromatic compounds, 111-
      13
    carboxylic acids, 258-60
    carboxylic acid salts, 263
    cyclic compounds, 50
    esters, 263
    ethers, 180
    halogenated hydrocarbons,
      53
    ketones, 207-9
    nonaromatic hydrocarbons,
      50-55
    organic halogen compounds,
      160
    phenols, 180-81
    summary, 415-18
    unsaturated hydrocarbons,
      54-55
Nonionic detergents, 302
Nonreducing sugars, 242
Norenthindrone, 310
Novocaine, 285
Nuclear magnetic resonance,
  401-8
  chemical shifts, 403
  integration, 406
  peak splitting, 406
  sample spectra, 404-5
  summary, 407

Nucleic acids, 373-92
  structures, 373-77
Nucleophiles, 164
Nucleophilic addition, 213
Nucleophilic substitution,
    164-69
  alcohols and ethers with hy-
      drogen halides, 194-96
  alkylation of amines, 320
  general reaction, 164
  S_N1 and S_N2 mechanisms,
      165-69
  Williamson synthesis of
      ethers, 182
Nucleoproteins, 377
Nucleosides, 374-75
Nucleotides, 374-76
Nucleotide vitamins, 378-79
Nucleus, 1
Nutmeg, 294
Nylon, 276-78
  nylon 6, 277
  nylon 66, 276
  nylon 6-10, 277

Ocimene, 95-96
Octane numbers, 124-26
Oils, 293-98
Oleic acid, 294-95
Olive oil, 294-95
Onion scent, 199
Opium, 338
Opsin, 37-38, 222
Optical isomerism, 137-56
  in biological world, 154-56
  in carbohydrates, 236
  without chiral carbons, 158
  in cyclic compounds, 148
  designation of configuration,
      148-52
  in proteins, 347
  summary of terms, 147
Optically active compounds,
    138-39, 147
Optical rotation, 139
Orange oil, 261
Orbitals
  atomic, 2-3
  molecular, 12
Organic chemistry, definition, xv
Organic compounds, classes,
    inside front cover
Organic halogen compounds,
    159-73
  applications, 161-64
  classes of, 159
  nomenclature, 160-61
  nucleophilic substitution,
      164-69
  physical properties, 160
  preparations, 161

Organochlorine insecticides,
    172-73
Organophosphorus insecticides,
    173
Orinase, 253
Orlon, 86
Ortho, meta, para, 112
Oxalic acid, 261
Oxidation
  of alcohols, 196-97, 210, 265
  of aldehydes, 210-12
  of alkenes, 93-94
  of alkylbenzenes, 123, 265
  drying oils, 298
  rancidification, 297-98
Oximes, 222
Oxiranes, 197-98
Oxytocin, 369
Ozone, 93
Ozone shield, 163
Ozonide, 93
Ozonolysis, 93-94, 210

PABA, 323
Pain relievers, 282-84
Paint, 298
Palmitic acid, 294-95
Palm oil, 295
Paper, 252
Paraffin wax, 67, 292
Para red, 343
Parathion, 173
Pasteur, Louis, 140
Pauling, Linus, 357
PBBs, 162
PCBs, 162
Peak splitting (NMR), 406
Peanut oil, 295
Peat, 66
Pellagra, 337, 379
Penicillamine, 364
Penicillin, 366
Penicillin G, 156
Pentachlorophenol, 135
Pentobarbital, 284
Pepper, 336
Peppermint oil, 96
Pepsin, 381
Peptides, 350
Periodic table, 4-6
  covalent bond formation, 9-10
  electron configuration, 4-6
  trends in electronegativity, 6
  trends in ionic bonding, 4-7
Perioxides, 82
Permanent hair wave, 199, 353
Pesticides, 171-74, 199
Petroleum, 48, 66-70
  fractionation, 67
  gasoline, 124-28
Peyote, 334

Phenacetin, 283
Phenanthrene, 111
Phenformin, 253
Pheniramine, 336
Phenobarbital, 284
Phenol, 111, 114, 188
Phenolic resins, 230
Phenolphthalein, 331
Phenols
  acidity, 191–93, 265
  electrophilic substitution, 192
  nomenclature, 180–81
  physical properties, 183
  structure, 178
  uses, 188–90
Phenylalanine, 144, 346
Phenylalkylamine alkaloids,
    334–36
Phenylephrine, 335
Phenyl group, 113
Phenylhydrazones, 222
Phenylketonuria, 388
o-Phenylphenol, 189
Phenylpropanolamine hydro-
    chloride, 335
Pheromones, 95–96, 199–200
Phosgene, 24
Phosphate esters, 243
Phosphatidic acid, 303
Phosphatidyl choline, 303–4
Phosphatidyl ethanolamine, 303
Phosphatidyl inositol, 303
Phosphatidyl serine, 303
Phosphoglycerides, 303–4
Phospholipids, 303
Photographic developers, 190
Physical properties
  alcohols, 182–85
  aldehydes, 207
  amines, 315–16
  aromatic compounds, 110
  carboxylic acids, 264
  ethers, 182–84
  fats and oils, 295
  hydrocarbons, 55–58
  hydrogen-bonding, 182–85
  ketones, 207
  organic halogen compounds,
      160
  phenols, 183
Pi bond, 12, 17, 36
Picric acid, 133
Pineapple oil, 261
α-Pinene, 96
Piperidine, 336
Plane-polarized light, 137–38
Plaque, 247
Plasticizers, 83, 162
Plastics, 83
β-Pleated sheet, 357–58
Plexiglas, 83, 86
Plunkett, Roy J., 87

Poisons, 364
Polar bonds, 19–20, 60
Polarimeter, 138–39
Polaroid sheet, 137
Polaroid sunglasses, 138
Polio, 390
Polyamides, 276–78
Polybutadiene rubber, 92
Polycarbonate plastics, 289
Polychlorinated biphenyls, 162
Polyesters, 278
Polyethylene, 84–85, 127
Polyhydric alcohols, 187–88
Polyisobutylene, 81
Polyisoprene rubber, 87–92
Polymerization, 127
Polymers
  addition, 81–92
    commercial examples, 84–87
    mechanism of formation,
        81–83
    rubber, 87–92
  condensation, 276–79
Polymethylmethacrylate, 86
Polypeptides, 350
Polypeptide synthesis, 353–55
Polypropylene, 85
Polysaccharides, 248–52
Polystyrene, 85
Polyurethanes, 278–79
Polyvinyl acetate, 101
Polyvinyl chloride (PVC), 83, 86
Polyvinylidene chloride, 87
Polyvinylidene fluoride, 101
p Orbitals, 3
Positional isomerism, 30–31
Potassium sorbate, 266, 297
Prelog, V., 148
Preparations
  alcohols, 181–82
  aldehydes, 210
  alkenes and alkynes, 63–65
  alkyl halides, 60–63, 161
  amines, 319
  carboxylic acids, 210
  ketones, 210
Primary structure of proteins,
    350–53, 356
Progesterone, 309
Prolactin, 369
Proline, 346
Proof alcohol, 186–87
Propane, 28
Propionaldehyde, 210
Propyl gallate, 190
Prostaglandins, 305
Prostanoic acid, 305
Prosthetic groups, 361
Proteins, 345–72
  biosynthesis, 384–87
  conjugated, 361
  definition, 350

denaturation, 360
enzymes, 361–64
primary structure, 350–53,
    356
quaternary structure, 360
secondary structure, 356–58
synthesis, 353–55
tertiary structure, 358–60
Proton, 1
Protoporphyrin group, 364
Purine, 110, 374
Purine alkaloids, 340
Putrescine, 313
PVC, 83, 86
Pyran, 240
Pyranose rings, 239–40
Pyrethrin insecticides, 174
Pyridine, 109, 336
Pyrimidine, 109, 374
Pyrolle, 110
Pyroxylin, 251
Pyrrolidine, 336

Quantum numbers, 2
Quaternary ammonium salt,
    302, 320–21
Quaternary structure of pro-
    teins, 360
Quinidine, 338
Quinine, 338
Quinoid group, 329
Quinoline, 110, 337

Racemic mixture, 142, 147
Radiofrequency, 394, 401
Radiowaves, 394
Rancidification, 297–98
Rayon, 251–52
RDX, 189
Reaction intermediates, 59
Reaction mechanisms, 59
Reaction types, 58
Reactive dyes, 332
Recombinant DNA, 390–91
Reducing and nonreducing
    sugars, 241–42
Reductions
  of aldehydes and ketones, 182
  of alkenes and alkynes, 75–76
  hydrogenation of carbonyls,
      214
  with lithium aluminum hy-
      dride, 182, 214–27
  of oils, 296–97
  to prepare amines, 319–20
Reforming, 128
Refrigerants, 163
Rendered fat, 299
Replication of DNA, 383
Research octane number (RON),
    125
Reserpine, 339–340

Resolution of enantiomers, 153–54
Resolution through diastereomers, 154
Resonance, 88–90
    allylic free radical, 88–90
    aromatic amines, 318
    benzene, 106–8
    carboxylate ion, 267
    energy, 105
    enolate ion, 223–24
    forms, 88–90
    hybrid, 88–90
    phenoxide ion, 102
Retinal, 37–38, 97, 222
Retinoic acid, 97
Rhodopsin, 37–38, 222
Riboflavin, 378–79
Ribonuclease, 358
Ribonucleic acid, 376–77
    messenger, 385
    ribosomal, 386
    3-D structure, 382
    transfer, 386
Ribose, 237, 377
    cyclic form, 240
Ribosomal RNA, 386
RNA, 376–77
    3-D structure, 382
    mRNA, 385
    rRNA, 386
    tRNA, 386
Roses, oil of, 104, 233
Rosin, 99
R,S designation of chiral carbons, 148–52
Rubber, 87–92, 98
Rubbing alcohol, 102, 185
Rum, 186
Rum flavoring, 262

Sabin-Salk vaccine, 390
Saccharide, 235
Saccharin, 129, 131, 253–254
Salicylamide, 282–83
Salicylic acid, 189, 282–83
Salol, 283
Sanger reagent, 349, 351
Saponification, 299
Saponification number, 299
Saran, 87
Saturated hydrocarbons, 28, 48, 50
Sawhorse diagrams, 34–35
SBR, 91–92
Schöenbein, Christian, 251
Scopolamine, 337
Seconal, 284
Secondary butyl, 52
Secondary structure of proteins, 356–58

Selinene, 97
Serine, 346
Sex attractants, 200
Sex hormones, 309–10
Sex pheromones, 200
Shale oil, 69
Shampoos, 302
Shapes of molecules, 13
    linear carbon, 17
    summary, 20
    tetrahedral carbon, 14–15
    trigonal carbon, 15
Shark liver oil, 97
Shellac, 228
Shortening, 296
Sialic acid, 304
Sickle cell anemia, 365, 388
Sigma bond, 12, 15
Silk, 357
Silver mirror test, 211, 241
Single bond, 10, 15
Skatole, 314
Skeletal isomerism, 28–31
Skunk scent, 199
Sleeping pills, 285, 335
Smallpox, 390
S$_N$1 and S$_N$2 mechanisms
    with alcohols and ethers, 194–96
    with alkyl halides, 164–69
    reaction rates, 166
    stereochemistry, 167
Soaps, 266
    production of, 299
    structure and action, 299–301
Soaps and detergents, 299–303
Socrates, 336
Sodium benzoate, 114, 266
Sodium bicarbonate, 266, 284
Sodium borohydride, 214
Sodium chloride, crystalline structure, 8
Sodium nitrite, 323–24, 389
Sodium pentothal, 284
Sodium salicylate, 282–83
Solubility, 184
Somatostatin, 255, 368
s Orbitals, 2
Soybean oil, 295
Spearmint, 98
Specific rotation, 139
Spectrophotometers, 395
Spectroscopy, 393–414
    infrared, 395–400
    mass spectrometry, 408–12
    nuclear magnetic resonance, 400–8
    ultraviolet, 397
    visible, 397
Spermaceti wax, 292–93
Sphingolipids, 304
Sphingomyelin, 304

sp Hybridization, 18
sp^2 Hybridization, 16
sp^3 Hybridization, 15
Squalene, 97
Stable octet, 6–7
Starch, 155, 248–50
Stearic acid, 294–95
Stereoisomerism, 137
    conformational, 34–36, 40–42
    geometric, 36–38, 42
    optical, 137–56
Steroids, 304, 307–10
Streptococcus mutans, 247
Strong and weak acids, 268
Structural formula, 26
Structural isomerism, 137
Strychnine, 339–40
Styrene, 101
Styrene-butadiene rubber, 91–92
Styrofoam, 85
Substitution reactions
    electrophilic aromatic, 114–23
    general equation, 58
    halogenation of alkanes, 60
    nucleophilic, 164–69, 196
Substrate, 362
Sucrose, 245–47
    fermentation, 186
Sugar, 245–47
Sugar beets, 246
Sugar cane, 246
Sugars (carbohydrates), 235–56
Sulfadiazine, 323
Sulfa drugs, 322–23
Sulfamerzine, 323
Sulfameter, 323
Sulfamethizole, 323
Sulfanilamide, 323
Sulfides, 198–99
Sulfonamides, 322–23
Sulfonation, 115, 118
Sulfonic acids, 111
Sunglasses, Polaroid, 138
Sunscreens, 323
Sweet acidophilus, 248
Syndets, 302
Synthetic detergents, 302
Synthetic fuels, 68
Synthetic gas, 68
Synthetic rubber, 91
Systox, 364

2,4,5-T, 258
Table sugar, 245–47
Tachycardia, 338
Tallow, 294–95, 299
Talose, 237
Tar, 67
Tar sands, 69
Tartaric acid, 146, 257
Taurocholic acid, 308